Introduction to Robotics in CIM Systems

Fifth Edition

James A. Rehg
The Pennsylvania State University
Altoona, Pennsylvania

Prentice
Hall

Upper Saddle River, New Jersey
Columbus, Ohio

Library of Congress Cataloging-in-Publication Data

Rehg, James A.
    Introduction to robotics in CIM systems/James A. Rehg.—5th ed.
       p. cm.
    Includes bibliographical references and index.
    ISBN 0-13-060243-4
    1. Robotics. 2. Computer integrated manufacturing systems.   I. Title.
TJ211.R422 2003
670.42'72—dc21   2001052086

Editor in Chief: Stephen Helba
Assistant Vice President and Publisher: Charles E. Stewart, Jr.
Assistant Editor: Delia K. Uherec
Production Editor: Tricia L. Rawnsley
Design Coordinator: Diane Ernsberger
Cover Designer: Robin Chukes
Cover art: Photoresearchers
Production Manager: Matthew Ottenweller
Electronic Text Management: Karen L. Bretz

This book was set in Zapf Calligraphic 802 and Swiss 721 by TechBooks. It was printed and bound by R.R Donnelley & Sons Company. The cover was printed by Phoenix Color Corp.

Pearson Education Ltd., *London*
Pearson Education Australia Pty. Limited, *Sydney*
Pearson Education Singapore Pte. Ltd.
Pearson Education North Asia Ltd., *Hong Kong*
Pearson Education Canada, Ltd., *Toronto*
Pearson Educación de Mexico, S.A. de C.V.
Pearson Education—Japan, *Tokyo*
Pearson Education Malaysia Pte. Ltd.
Pearson Education, *Upper Saddle River, New Jersey*

10 9

ISBN: 0-13-060243-4

To the three women who most influenced my life . . .
Marci, Rose, and Bettie

Preface

INTRODUCTION

An industrial robot is just another industrial machine. This statement has been used frequently by those in education and industry to calm fears about mass worker displacement or to encourage corporate management to adopt flexible automation. Like the numerical control turning center or the lathe that preceded it, the industrial robot is a machine designed to increase productivity, improve quality, and reduce direct labor cost, but the robot is *not* just another industrial machine.

Robots are versatile: They can be used in every industry that provides goods and services; they can be adapted to numerous job functions; they can change job functions easily; and they work with uncanny skill and unmatched endurance. Robots are different from any industrial machine in the history of automated production. The potential of the robot as an agent of change in manufacturing and in our daily lives has not been fully realized.

Employment in the robotics area has changed significantly from the early days of robots. Engineers and technicians working in the robotics field do not concentrate on just robot technology as they did when robots were first introduced. Today the engineer and technician must be capable of working on the entire production system, which often includes robot technology. The job might be in an automation system design company that designs, builds, and installs robot-based automation for other companies, or in an automation design department in a large manufacturing company that designs automation systems for its plants. The automotive industry is a good example of companies in the latter category. Engineers and technicians with robot and automation systems skills are also needed by companies using the technology to produce products. In these companies, they work either individually or in teams to make the automation system perform to specifications. This includes work on robots, programmable logic controllers, and computer numerical machines; system controller programming; troubleshooting system and production problems; performing system upgrades; and training operators on the proper use of the technology. Positions for robot technicians are not plentiful; however, the need for engineers and technicians who can design, develop, implement, and support automated production systems with robots is significant and expanding.

Although robots are unique, they share one element with other automated production equipment; namely, to be effective, they must be integrated into the total solution. Education in robotics must reflect this emphasis on the total system as well. The first edition of this text focused on the robot as part of an integrated production cell. The interfaces between the robot and cell devices were emphasized. In the second through fourth editions, the integration of the robot-automated cell with the computer-integrated enterprise was introduced. Now this edition takes the integration of the robot with the automated cell and system to another level. While robots remain the primary focus of the text, additional emphasis is placed on the hardware, software, and programming that support the implementation of automated work cells and manufacturing systems.

CHANGES TO THE FIFTH EDITION

Major additions and changes to the fifth edition include the addition of chapter goals and objectives at the beginning of every chapter. A *Career Spotlight* section was added at the start of every chapter as well. The Career Spotlight focuses on the career opportunities for the technology areas covered in the chapter. A major change in Chapter 1 was the addition of a robot safety section at the end. Safety is an important issue in robotics and in all manufacturing. While safety is addressed in great detail in Chapter 10, I was encouraged by users of the text to stress safety early. Another important skill for both engineers and technicians is the ability to troubleshoot technical systems. To help in learning these important skills, an introduction to troubleshooting was added to Chapter 3. That was followed with a discussion on troubleshooting sensor system at the end of Chapter 5. In addition, the sensor information in the chapter was revised and updated. Major updates and additions were made to the programmable logic controller and robot programming topics in Chapter 8. A full section was added covering the program commands for the Yaskawa robot. The chapter presentation gives students detailed information about the most frequently used commands and shows them how to translate a task point graph into a robot program using Yaskawa code. In addition, a number of new figures were added to help illustrate important concepts, and the text was changed at numerous places to make it easier to read and understand. I hope you find these changes useful and helpful.

FOR THE STUDENTS

More graduates of engineering and engineering technology programs are working in manufacturing automation because production systems have become increasingly complex and highly automated. As a result, students need to understand the theory and operation of robotics and automation as they apply to production systems in industry. The primary goal for this text was to create a clear and comprehensive text for students in two- and four-year engineering and engineering technology programs to learn industrial robotics and automation systems. Every effort was made to present the material in a logical order, to express the concepts in a fashion that a first-time reader could understand, and to keep the needs of the student foremost in every part of the text development. Authors often use technical terms to describe a new concept that have not been previously defined or that are not common knowledge for the students. A special effort was made in this text to not use any terms or technical language that were not introduced or defined earlier in the text. The text can be used in semester and quarter length courses.

I hope this text helps you build a comprehensive understanding of the concepts that embody industrial robotics and automation systems. I have a friend and former student who once said, "There are things that I know and things that I know I know." This is an interesting observation on learning, and I asked him what he meant by "know I know." He said that certain lessons were presented in such a way that he could remember some of the concepts only as they were being presented. The lesson material was never internalized, though; he only knew it as it was presented. However, after working example problems and spending time thinking about a concept, he understood that idea in a new way. He had internalized the concept, theorem, or algorithm so that he could use it to solve problems that were different from the one in which it was originally presented. His depth of understanding meant that he would never forget the material. The concept just made sense; it was like remembering his name. To reach this level of understanding of a subject takes effort and time, and it often takes revisiting the concept numerous times. I hope you will reach that level of *knowing you know* some of the concepts presented in this text. If the presentations of some material help you get by previous learning barriers, I would like to know. Please e-mail me at jamesa@rehg.org and share the learning experience. On the other hand, if you think that some area could be presented more clearly, I would like to know about that as well. I hope you enjoy the text and find it useful; it was written for you.

CHAPTER CONTENT

Industrial robots and the concept of a work-cell system are introduced in **Chapter 1**. A brief history of robots is included along with a rationale for the renewed interest in robot applications in the 1990s and in the new century. The definition of an industrial robot, a description of a basic robot system, and the new terms used to describe its operation are included, along with an introduction to the manufacturing systems that are most appropriate for robot automation. A safety section was added to emphasize the importance of robotics and manufacturing safety. In **Chapter 2,** the different types of robot systems are classified by arm geometry, power sources, control techniques, and path control. The advantages and disadvantages of each type of system are discussed. In addition, a section on drive systems is included, along with a section on design guidelines for automated cells. Numerous quantitative problems are included throughout and at the end of the chapter. Additional automated cell design problems are included at the end of the chapter.

Chapter 3, Automated Work Cells and CIM Systems, is a chapter focusing on the basic process used in the implementation of a single work cell or an entire CIM system. In addition, a detailed discussion of system troubleshooting has been added. Flexible and fixed automation cells and systems are described and the use of robotics in each is discussed. Two industry case studies are presented as examples of automated system design, and a number of numerical problems are provided at the end of the chapter. End-of-arm tooling is covered in **Chapter 4.** The many different types of grippers are classified into several categories. The wrist interface section is integrated into the tooling section, and coverage of robot tool changers has been expanded. In addition, the section on active and passive compliance includes force/torque sensing in the tool interface. Sections cover gripping force calculations, as well as collision systems. The use of numerical problem examples and problems at the end of the chapter has been expanded.

Chapter 5, Automation Sensors, covers limit switches and proximity and photoelectric sensors. Sections cover the design of trip dogs for limit switch applications, the operation of proximity and capacitive sensors, and sensors that support the field-bus standard. All sections have been revised and updated with new figures and descriptions. Numerous example problems have been added, along with problems at the end of the chapter. In **Chapter 6,** Work-Cell Support Systems, the devices and systems traditionally used in automation cells to support the robot are explained. Topics include vision, material handling, automatic storage and retrieval systems, part feeding, inspection, and automatic tracking. The vision section encompasses back lighting and front and structured lighting concepts. Additional case studies are available at the end of the chapter.

Chapter 7, Robot and System Integration, focuses on the control architecture required in the automated enterprise. The major topics are the enterprise network, numerical control, servo and nonservo robot controllers, and simple and complex sensor interfaces. Case study problems are available at the end of the chapter. **Chapter 8** covers the programming issues associated with an automated system. Topics include cell control software, programming the programmable logic controller, robot reference frames, and servo and nonservo robot programming using on- and off-line programming software. The chapter was significantly expanded by the addition of a section on programming using the Yaskawa command language.

Chapter 9 covers work-cell justification and applications. Justification is explained in terms of payback, return on investment, and cash flow methods with an introduction to future value of money and discounting techniques. A spreadsheet demonstrating a discounted cash flow process is included on the CD-ROM supplied with the Instructor's Manual. Robot applications are described by means of industrial application case studies.

Chapter 10 includes safety standards, a thorough description of proximity sensing devices, light curtains, calculation of safe distances, sensing mats, interlock devices, risk assessment and estimation, hazard reduction, personal protection equipment, design guidelines, and safety justification. Numerous examples and end-of-chapter problems and cases are also included. **Chapter 11** deals with the problems created by automation with the human interface. In addition, the chapter describes self-directed work teams and how they function.

Chapter 12 contains the West-Electric (W-E) case study. Parts of the West-Electric case study can be used as an activity at the end of each chapter (except Chapter 1), or the case can be used as a capstone project at the end of the course. In either implementation, students follow the progress of an industrial design team as they automate one part of the manual production system. Students are also invited to join in the design process and create an automated cell for an adjacent manual production process. In their solution, students have an opportunity to use skills learned in each chapter of the text, plus skills from computer-aided design and programmable logic controllers courses. This case study uses a videotape to show all the manual manufacturing processes as a starting point for the introduction of automation. The tape, *Forging*, is number 150 and is available from Genium Publishing Corporation in Schenectady, New York, 1-800-243-6486. Numerous cell design problems are included. The robot work-cell design component of the text is integrated into the W-E case study at two levels. The first level is a study of the design process used by the W-E automation team on the automation of the upset forging work cell. The second level is the design of an automated cell to support the extrusion process. Some of the detail is intentionally missing from the W-E team design so that class discussion of the case can take place. The production information for the entire turbine blade line is provided so that other work-cell projects are possible from the same case.

Appendixes A, B, and C include robot data sheets, vendor Internet sites, and justification spreadsheet documentation, respectively.

CD-ROM SUPPORT

The Instructor's Manual for the text comes with a CD-ROM containing items useful in developing lecture material and using the case studies. The CD-ROM includes PowerPoint transparencies for every chapter.

ACKNOWLEDGMENTS

Many people have contributed to the preparation of this manuscript. I would like to thank the many companies that have contributed material and have been supportive and helpful. A very special thanks to Bettie Steffan for suggestions and editing on the initial project. I would like to thank the reviewers for this edition: Alfred A. Gates (Central Connecticut State University), Charles A. Haven (Ivy Tech State College), and Thomas Sanford (Mission College). Thanks also to the students and instructors in industrial and engineering technology programs across the country who have provided many suggestions for the fifth edition.

James Rehg

About the Author

James A. Rehg, CMfgE, is an associate professor of engineering at Penn State–Altoona. He earned a BS and MS in electrical engineering from St. Louis University and has completed additional graduate work at Wentworth Institute, University of Missouri, South Dakota School of Mines and Technology, and Clemson University. Before moving to Penn State, he was director of the Computer Integrated Manufacturing project and department head of CAD/CAM and Machine Tool Technology at Tri-County Technical College, and previous to that he was director of Academic Computing and the Manufacturing Productivity Center at Trident Technical College. Professor Rehg also served as director of the Robotics Resource Center at Piedmont Technical College and department head of Electronic Engineering Technology at Forest Park Community College. His industrial experience includes work in instrumentation at McDonnell Douglas Corporation and consulting in the areas of computer-aided design, robotics, computer-integrated manufacturing, and programmable logic controllers.

Professor Rehg has written five texts on robotics and automation and many articles on subjects related to training in automation and robotics. His most recent text is *Computer-Integrated Manufacturing,* 2nd ed., with coauthor Henry Kraebber of Purdue University, published by Prentice Hall in 2000. Professor Rehg has received numerous state awards for excellence in teaching, including the outstanding instructor in the nation by the Association of Community College Trustees and the Penn State Engineering Society Outstanding Teaching Award in 1998.

The fact is, that civilization requires slaves. The Greeks were quite right there. Unless there are slaves to do the ugly, horrible, uninteresting work, culture and contemplation become almost impossible. Human slavery is wrong, insecure, and demoralizing. On mechanical slavery of the machine, the future of the world depends.

Oscar Wilde

Contents

3   Automated Work Cells and CIM Systems   119

5   Automation Sensors   202

Introduction to Industrial Robots

CHAPTER GOALS AND OBJECTIVES

The goal of the first chapter is to provide a broad overview on the subject of industrial robotics. In addition to the hardware-specific topics, the chapter also describes how robotics offers a solution to productivity, profitability, and market share issues every manufacturer faces. At the completion of this chapter you should be able to:

■ Describe how the robot industry was born and the two major reasons why robots became a factor in manufacturing automation.

■ Name the external and internal challenges present for every manufacturer and describe how order-winning and order-qualifying criteria help to identify the best production solution.

■ Define *robotics* and *computer integrated manufacturing*.

■ Classify a production operation into one of five categories.

■ Name and describe all the elements of a robot system.

■ Define eleven prerequisite terms used most frequently in robotic literature.

■ Explain safe robot practices.

1-1 INTRODUCTION

The introduction of new technology in industrial nations causes changes throughout the social structure of every country in the world. The nature and degree of those changes are proportional to the effects a new technology has on the production of goods and services. Technology in agriculture, for example, caused farming

CAREER SPOTLIGHT

The broad nature of the topics covered in this chapter correspond to equally broad career opportunities in manufacturing automation in general and robotics in particular. Careers span the range from system designer to system maintenance and support. The subject of the work can be as specific as a particular type of robot arm or as broad as an entire manufacturing system. Employers include the designers and suppliers of automation components, the vendors that sell and maintain the manufacturing hardware and software, and the manufacturers who are the end users of the automation. In the chapters that follow, robot and automation issues are addressed in sufficient detail to prepare you for a career in many segments of this broad work area. If you master and enjoy the concepts brought to light in this text then there is high likelihood that you will find a rewarding career in automation and robotics.

to change from an employer of 80 percent of the population in 1890 to an employer of 2 percent of the population in 2000. Although there was a reduction in the number of farmers, an increase occurred in the number of other jobs required to support the mechanized farm industry.

Today, the application of computer technology in automation is expanding at a rate that is exceeded only by the growth of the Internet. Networks of industrial computers connect individual manufacturing cells into flexible manufacturing systems (FMSs), and FMSs are linked to form computer-integrated manufacturing (CIM) systems spanning entire enterprises. Although changes in the robotics industry and the introduction of new robot products have slowed from the rapid development pace of the early 1980s, robots continue to have greater change potential than any industrial machine in the past. A study of robot systems and the integration of robots into FMS and CIM systems will be addressed in the following chapters. The first step is to determine what industrial robots are, what they can do, and what they are currently unable to do.

1-2 HISTORY OF THE INDUSTRY

A brief review of robot development is important because it puts the current machines and interest in them into a historical perspective. The following list of dates highlights the growth of automated machines that led to the development of the industrial robots now available.

1801 Joseph Jacquard invents a textile machine that is operated by punch cards. The machine is called a programmable loom and goes into mass production.

1892 In the United States, Seward Babbitt designs a motorized crane with gripper to remove ingots from a furnace.

1921 The first reference to the word *robot* appears in a play opening in London. The play, written by Czechoslovakian Karel Capek, introduces the word *robot* from the Czech *robota,* which means a serf or one in subservient labor. From this beginning the concept of a robot takes hold.

1938 Americans Willard Pollard and Harold Roselund design a programmable paint-spraying mechanism for the DeVilbiss Company.

1939 Isaac Asimov's science fiction writing introduces robots designed to help humanity and work safely. Three years later he would formulate his *Three Laws of Robotics,* and the view of robots as gothic menaces, presented in earlier works, starts to change.

1946 George Devol patents a general-purpose playback device for controlling machines. The device uses a magnetic process recorder. In the same year the computer emerges. American scientists J. Presper Eckert and John Mauchly build the first large electronic computer, called the ENIAC, at the University of Pennsylvania. A second computer, the first general-purpose digital computer, dubbed Whirlwind, solves its first problem at the Massachusetts Institute of Technology (MIT).

1948 Norbert Wiener, a professor at MIT, publishes *Cybernetics,* a book that describes the concept of communications and control in electronic, mechanical, and biological systems.

1951 A teleoperator-equipped articulated arm is designed by Raymond Goertz for the Atomic Energy Commission.

1954 The first programmable robot is designed by George C. Devol, who coins the term *Universal Automation.* The patent application, labeled *Programmed Article Transfer,* was issued as U.S. Patent 2,988,237 in 1961. Devol is joined by Joseph F. Engelberger in 1956; they shorten the name to *Unimation* and form the first successful robot manufacturing company.

1959 Planet Corporation markets the first commercially available robot.

1960 Unimation is purchased by Condec Corporation, and development of Unimate Robot Systems begins. American Machine and Foundry, later known as AMF Corporation, markets a robot, called the Versatran, designed by Harry Johnson and Veljko Milenkovic.

1961 The first Unimate robot is installed to unload a die casting machine.

1962 General Motors (GM) installs the first industrial robot on a production line. The robot selected is a Unimate.

1968 Stanford Research Institute builds and tests a mobile robot with vision capability and names it Shakey. Kawasaki Heavy Industries receives a license from Unimation to manufacture the robot in Japan for sale under its brand name.

1970 At Stanford University a robot arm is developed that becomes a standard for research projects. The arm is electrically powered and becomes known as the Stanford Arm.

1971 Japanese Industrial Robot Association (JIRA) is started to promote the use of robots in Japanese industries.

1973   The first commercially available, minicomputer-controlled industrial robot is developed by Richard Hohn for Cincinnati Milacron Corporation. The robot is called the T3, The Tomorrow Tool.

1974   Professor Scheinman, the developer of the Stanford Arm, forms Vicarm Inc. to market a version of the arm for industrial applications. The new arm is controlled by a minicomputer.

1975   Robot Institute of America is formed to help US industries effectively implement robots in factory automation.

1977   ASEA Brown Boveri Robotics Inc., a European robot company, offers two sizes of electric-powered industrial robots with a microcomputer controller.

1978   With support from GM, Unimation develops the Programmable Universal Machine for Assembly (PUMA) robot using technology from Vicarm Inc.

1980   The robot industry starts a period of rapid growth, with a new robot or company entering the market every month.

1981   GM brings the Japanese robot manufacturer Fanuc Robotics into the US market through a joint venture to develop, manufacture, and market industrial robots in the United States. The new company is called GM-Fanuc Robotics.

1983   The robot industry enters a maturing period as industry recognizes that robots and the other automation hardware must be integrated into a unified system. The number of major robot manufacturers falls to twenty-five or fewer.

1984   Direct-drive robot arms are introduced by Adept Corporation with electric-drive motors connected directly to the arms, eliminating the need for intermediate gear or chain drives.

1986   Robot applications and installations continue to grow but with increased emphasis on the integration of the robot into work-cell, FMS, and CIM systems.

1990   ASEA Brown Boveri Robotics Inc. purchases the robotics division of Cincinnati Milacron, and all future robots will be ASEA machines.

1991   The emergence of a global economy, the emphasis on producing competitive products, and use of new microelectronics and display technology in robotic systems spark renewed interest in the robot in integrated automation systems.

1992   After helping to make Fanuc Robotics the sales leader in the United States (annual sales of $200 million in a $700 million market), GM sells its interest because of financial troubles.

1994   Robot sales in the United States for the first 6 months set new records in number of units sold (4,355) and in total value ($383.5 million). The two largest companies were Fanuc Robotics Corporation and ASEA Brown Boveri (ABB), with annual sales of about $200 million and $120 million, respectively.

2002 The use of industrial robots continues to increase and the consolidation in the robot industry appears to have ended. The number of companies integrating robots and automation hardware into production cells and systems has tripled in the last two years.

In the beginning, the domestic robot industry was dominated by US manufactures (Unimation, Cincinnati Milacron, Westinghouse, IBM, General Electric, Prab, Brinks, and others). Today, the major manufacturers of robots used by US industries are international companies and include Fanuc Robotics Corporation, ABB, Motoman, Seiko, Panasonic Factory Automation, Kawasaki Robotics, Reis Robotics, Sony Component Products, and Nachi Robotic Systems. All of these companies are Japanese except ABB, which is a Swedish industrial conglomerate, and Reis, which has its headquarters in Germany. Adept Technology, Inc. is the largest US-based robot manufacturer. Japanese dominance results from the strong adoption of robot technology by their industries. Automotive products companies are the largest users of robots, but the demand for robot automation in other industries is growing. The primary driving forces behind the renewed interest in robots are increased productivity, lower unit cost, and greater reliability; for example, today's prices for robots are half what they were 10 years ago for comparable machines, and reliability is five times better.

1-3 FIFTY-YEAR-OLD INDUSTRY

As our review of history showed, robots are not new; in fact, mechanical components of robot-like devices date back to the early part of the 20th century. Robot technology, available in 1960, was not adopted on a large scale in the United States until the late 1970s. There are two reasons for the delay in the application of robots in US industry.

The first reason relates to *economics* and the second to *hardware*. In the late 1960s and early 1970s, economics was the primary barrier preventing the rapid deployment of robots into industrial applications. The return-on-investment (ROI) policy in most US industries required that new machines pay for themselves in 1 to 2 years. So the ROI policy required that the yearly saving from the robot installation over two years had to be equal to or greater than the cost of the robot project. Early robot applications focused on replacement of human labor; as a result, direct and indirect labor costs were the primary component used to calculate the ROI. In the 1960s and early 1970s, the hourly cost of the robot equaled or exceeded that of a human operator (Figure 1–1); therefore, the robot could not be justified economically. In 1960 the cost of operating a robot was more than $9 per hour, whereas the overall cost of a human operator was less than $5 per hour. Figure 1–1 shows that the hourly cost advantage switched to the robot in the early 1970s when the United States experienced several years of double-digit inflation. In addition, the cost of human labor increased at that time because of enhanced benefit packages and health plans. This dramatic increase in the cost of human labor in the mid-1970s started the trend to use robots in manufacturing and established the industrial robot industry in the United States. Roger Smith, chairman of

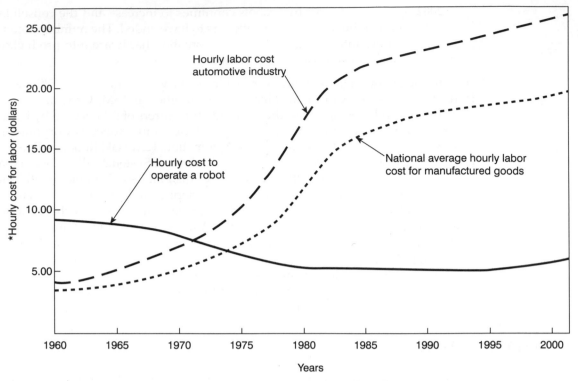

*Labor cost includes fringe benefits for human operators and initial cost plus support cost for robots

Figure 1–1   Robot and Human Labor Cost.

the board of GM at that time, remarked, "Every time the cost of labor goes up a dollar an hour, a thousand more robots become economical." The demand for robots in the 1980s would have continued to rise if it were not for a significant change in the labor picture midway through the decade. In the mid-1980, yearly increases in labor cost were small across most of the manufacturing sectors and dropped in some areas. The reasons for this leveling in the hourly rate included a reduction in the demand for labor as a result of increased use of automation, implementation of more efficient production systems, downsizing of the manufacturing and service workforces, and little growth in the price of basic resources and raw materials due to noninflationary fiscal policies in the major manufacturing countries.

The second factor that influenced the emergence of the robot industry in the late 1970s was the development of the *microprocessor,* or computer on a chip. The robots available in 1970 had remarkable capability, but their versatility pales before that of machines currently being offered by robot manufacturers. The primary difference is not mechanical but electronic sophistication. The microprocessor, developed by Intel Corporation in 1971, provided robot manufacturers with a source of very inexpensive intelligence to drive their machines. The first computer-controlled industrial robot, the Cincinnati Milacron T3, used a single minicomputer to direct all the robot motion and machine control within the work cell. The machine had tremendous capability but was limited by the speed of its single computer, which had to do all the calculations. New robot controllers use a network of powerful microprocessors to solve simultaneously the many concurrent problems associated with robot motion and control. The addition of microprocessors and improved software provided intelligence and operating speed not found on early robots.

Industrial robots became a significant production tool in the late 1970s and early 1980s due to increased labor cost and the development of the microprocessor. However, the introduction of robotics in industry slowed in the mid-1980s as a result of the leveling of labor cost and a greater emphasis on the development of integrated production systems. Robots proved that they could satisfy unique manufacturing problems, but robots could not be stand-alone solutions because industry was moving toward integrated manufacturing cells and production systems. This move to integrated solutions was driven by pressure on companies to meet a host of external and internal challenges.

1-4 INTEGRATED SYSTEMS—MEETING THE EXTERNAL AND INTERNAL CHALLENGES

To remain competitive in global markets, US manufacturers must recognize and meet two challenges, one *external* and the other *internal*. It is important to understand the nature of these challenges facing manufacturing, because the challenges drive the use of integrated manufacturing systems and the selection of specific robotic solutions.

External Challenges

External challenges result from various forces and conditions outside the enterprise. The challenges illustrated in Figure 1–2—*new/niche market entrants, traditional competition, suppliers, global economy, costs of money, Internet,* and *customers*—are common for most global manufacturing enterprises. Let's examine each of the items in more detail.

Companies have always had product lines called their cash cows: high-revenue-generating products that carry the rest of the product line. The success of the cash cow is a result of market dominance, technological superiority, patent protection, or the absence of competition. IBM's mainframe computer line in the late 1960s and early 1970s and the Windows® operating system from Microsoft are examples of this phenomenon. The production and design technology available today, however, makes it possible for small companies to develop a product in a *market niche* and compete with an established producer. In the IBM example, a former employee recognized the profit potential of the mainframe computer market niche and set up a company, Amdhol, to compete with IBM in this area. The Amdhol computers could run IBM software and work with IBM peripheral devices, so only high-profit computers were sold without the need to develop application software or supporting equipment.

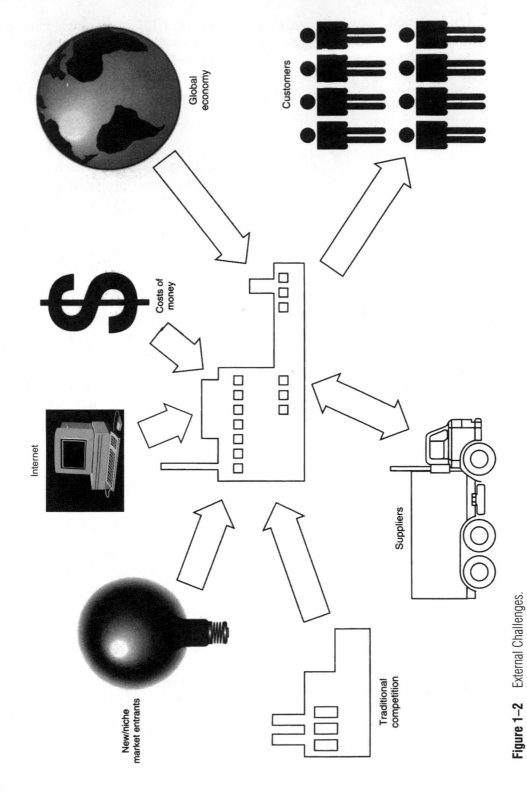

Figure 1–2 External Challenges.

(*Source:* COMPUTER INTEGRATED MANUFACTURING, 2nd ed. by Rehg and Kraebber, p. 6. Reprinted by permission of Pearson Education, Inc., Upper Saddle River, NJ.)

Traditional competition, the second challenge, continues to be a problem for manufacturing. Manufacturers often counted on the lack of technology and inefficiency of competitors as a buffer for their own shortcomings. Those days are gone. The technology to increase market share is affordable and available to everyone, so the challenge to stay ahead of the traditional competition is stronger today than ever.

Another external challenge is present in the *purchaser-supplier relationship.* In the 1980s many companies purchased parts and subassemblies from vendors who offered the lowest bid. Price continues to be a factor; for example, a study by a midwest injection molding die manufacturer indicates that a purchase price for parts from a vendor that is 9 percent higher than what the competition is paying can result in a 22 percent decrease in profits over the product's life. In addition to offering a competitive price, however, suppliers must meet minimum defect levels, provide predictable and reliable delivery, and shorten product design time. The challenge for companies is to establish fewer but more collaborative purchaser-supplier relationships in every aspect of the business.

In the 1990s manufacturers must market to a true *global economy;* no one country is capable of controlling the marketplace. Most manufacturers recognize the good and bad news associated with a global marketplace. Global customers offer an expanded base for marketing products; however, competition is increasing from every corner of the world. The challenges for US companies are to provide products that meet world-class standards and to market the goods and services to the global economy just as effectively as international competition.

Another external challenge presented to manufacturers is the *costs of money.* With interest rates that change with the economy, the company that uses capital resources effectively will be the long-term winner. The cost to introduce and sustain a product in the marketplace has reached staggering proportions. Manufacturing methods that reduce the cost of doing business, such as inventory reduction, become the tools for survival.

Another recent challenge is the business change created by the *Internet.* The development of the World Wide Web (WWW) is changing the way companies do business. The Web is providing companies with a direct access to customer households. Through either the proliferation of the home computer or through the development of the *set-top box* for the cable TV market, the Web has entered many homes around the globe. The challenge for companies is to determine the best strategy in this quickly changing market area. A good example is the electronic commerce (e-commerce) area. Before Amazon.com initiated the sale of books over the Web, the future for traditional bookstores appeared to be unchallenged. Now all of the major booksellers have a Web presence because the percentage of books purchased over the Web continues to increase. Companies that embrace the Internet technology intelligently will gain an advantage over their competitors who ignore the power of the Web or misjudge their correct relationship with this new technology.

The last and most difficult external challenge for manufacturing is provided by the *customers.* Today's sophisticated shopper buys on the basis of quality, service, cost, performance, and individual preference. If a manufacturer cannot meet all the needs of the customer, then another supplier is usually ready to step in for the sale.

Companies must respond to the external challenges (new/niche market entrants, traditional competition, suppliers, global economy, costs of money, Internet, and customers) in two steps:

1. Recognize that these challenges exist and admit that the problems they create must be solved, because they will never go away.
2. Develop a manufacturing strategy to minimize the negative impact of the external challenges on the success of the business.

Internal Challenge

The internal challenge is to develop a manufacturing strategy that satisfies the following definition:

A manufacturing strategy is a plan or process that forces agreement between the market-driven corporate goals and the production capability of the company.

The goal of the manufacturing strategy is to increase product market share. However, to win orders and increase market share, the enterprise must be competitive in price, quality, design lead time, delivery reliability, flexibility, customer service, innovation, and any other factor that customers use to make a purchasing decision. These competitive qualities were identified and called *order-winning criteria* by Terry Hill, a British manufacturing engineer. Hill defined order-winning criteria as *the minimum level of operational capability required to get an order.* For example, if quality is an order-winning criterion, then the manufacturing unit must provide that feature for a product to sell. When design innovation is the criterion, then the design group is responsible. Order-winning criteria such as after-sales service, delivery reliability, and flexible financial policies demand action from areas outside manufacturing. Order-winning criteria affect every unit of the enterprise because they give every unit or group in the enterprise its marching orders for success.

Hill also defined a second term, *order-qualifying criteria*. It is important to make the distinction between the criteria that win product orders and those that qualify a product to enter the race. An example best describes this difference. In the 1970s the order-winning criterion for the color television market in the United Kingdom (UK) was price. The Japanese entered the market with products that were competitive in price but far superior to the UK products from the standpoints of quality and reliability. Demand for imported televisions grew because of higher quality and reliability; quality became the new criterion on which purchases were made. UK manufacturers lowered the price on domestic sets to regain market share; however, the price could not drop low enough for the UK companies to sell sets, because customers were now buying because of quality, not price. Quality and reliability were the new order-winning criteria for selling televisions in the UK market. By the 1980s, UK companies matched the quality and reliability standards set by the Japanese; as a result, the order-winning criterion reverted back to *price*, and *quality* became the order-qualifying criterion.

The enterprise must meet the order-qualifying criteria to get into the market or to maintain its present market share. After the qualifying criteria are satisfied, the enterprise turns its attention to the criteria on which orders are won. Market share is increased when the order-winning criteria are understood and executed better than the competition. Therefore, in the manufacturing model developed by Hill, the enterprise must take these actions:

- Analyze every product and agree on the order-qualifying and order-winning criteria for the current market conditions for every product.
- For every product, project the order-winning criteria in the market in the future.
- Determine the *fit* between the criteria necessary to succeed in the marketplace and the current capability in manufacturing and across the enterprise.
- Change or modify either the marketing goals or the manufacturing process choices and enterprise infrastructure to force consistency between the market demands and the marketing goals, on the one hand, and the enterprise's capability to meet the demands and goals, on the other.

Meeting the Internal Challenge

A review of the order-winning criteria indicates that manufacturing has the responsibility to satisfy a part of the criteria most of the time. For manufacturing to step up to this responsibility, automated systems using robotics are required to improve shop floor standards. A table of manufacturing standards compiled by the consulting company Coopers and Lybrand is given in Figure 1–3, where the US average is compared with the world-class standard. How well a company compares with the best in the world is an indication of how ready that company is to meet the criteria necessary to win orders. The values used for the world-class standards and US averages are open for debate, because they change with industry group and geographic area studied. Nevertheless, it is important for every company to determine how well it is performing compared with other industries in its market sector.

The following sections describe each standard and indicate some of the order-winning criteria that are affected.

Setup Time.   Setup time is the length of time required to prepare a machine for production. If it takes a long time to get a machine ready for production, then it is not economical to produce a small number of parts. As a result, long setup time causes large production lot sizes (the size of the minimum number of parts produced) and results in higher part and product costs due to large inventories. Also, with more time allocated to setting up the machine, less time is spent in production on the machine, so more equipment and facilities are required and capital expenditures increase.

| | Manufacturing standards | |
Attribute	World-class standards	US average
Setup time		
System	< 30 minutes	24 hours
Cell	< 1 minute	
Quality		
Captured	1500 ppm	3-5%
Warranty	300 ppm	2-5%
Cost of quality	3-5%	15-25%
Manufacturing/total space	> 50%	25-35%
Inventory		
Product velocity	> 100 turns	2-4 turns
Material residence time	3 days	3 months
Flexibility	270 parts	25 parts
Distance	300 feet	> 1 mile
Uptime	95%	65-75%

Figure 1–3   Internal Challenges.

(*Source:* COMPUTER INTEGRATED MANUFACTURING, 2nd ed. by Rehg and Kraebber, p. 14. Reprinted by permission of Pearson Education, Inc., Upper Saddle River, NJ.)

Every machine on the factory floor has an hourly cost to the manufacturer called the *burden rate*. An understanding of machine burden rate is required to see the interactions among setup time, lot size, part cost, and inventories. The burden rate is a cost rollup that includes:

- Recovery of the original cost of the process machine and associated support hardware, such as robots and material handling equipment
- Maintenance costs (fault and preventive)
- Labor costs (direct and indirect)
- Programming for computer numerical control (CNC) type machines and robots
- Any other costs associated with having the machine in the production system

The burden rate is determined by dividing the capital cost by the useful life of the equipment, and then adding on the recurring hourly cost. A typical range of burden cost for an automated production system might be $100 to $500 per hour, depending on the level of automation present. Every hour of the day these costs accumulate. One part of the cost of manufactured items produced on the machine is set by the burden rate of the production machine. The costs associated with setup time are also calculated using these burden costs. The following example describes the impact that setup has on production.

Example 1–1

A production system has a burden rate of $200 per hour. The setup time for a machined casting is 2.6 hours, and the system can produce a finished part every 6 minutes.

a. For a lot size of 500 parts, how much of the part cost is associated with setup cost?
b. A lot size of 100 parts is required to reduce inventory levels. What new setup time is required to maintain the same setup cost on the parts?
c. The finished part cost is the sum of part production cost, setup cost, and a 30 percent margin. What percentage of the final cost results from the setup time for the 500 and 100 lot size production?

Solution

a)
$$\text{Setup time cost} = \text{setup time} \times \text{burden rate}$$
$$= 2.6 \text{ hrs} \times \$200/\text{hr} = \$520$$

$$\text{Setup cost per lot size part} = \frac{\text{setup time cost}}{\text{lot size}}$$

$$= \frac{\$520}{500} = \$1.04$$

b) Determine the total setup time cost for a lot size of 100 at a per part setup cost of $1.04.

$$\text{Total setup time cost} = \text{lot size} \times \text{setup cost per part}$$
$$= 100 \times \$1.04 = \$104$$

$$\text{Setup time (new)} = \frac{\text{setup time cost}}{\text{burden rate}}$$

$$= \frac{\$104}{\$200/\text{hr}} \times \frac{60 \text{ min}}{\text{hr}} = 31.2 \text{ min}$$

c) Determine parts produced per hour, part production cost, and the total part cost.

$$\text{Part production per hr} = \frac{60 \text{ min}}{\text{part production time (min)}}$$

$$= \frac{60 \text{ min/hr}}{6 \text{ min/part}} = 10 \text{ parts/hr}$$

$$\text{Part production cost} = \frac{\text{burden rate}}{\text{parts production per hr}}$$

$$= \frac{\$200/\text{hr}}{10 \text{ parts/hr}} = \$20/\text{part}$$

$$\text{Finished part cost} = (\text{part production cost} + \text{setup cost}) \times 1.3$$
$$= (\$20 + \$1.04)1.3 = \$27.35$$

$$\text{Setup percentage} = \frac{\text{setup cost}}{\text{finished part cost}} \times 100$$

$$= \frac{\$1.04}{\$27.35} \times 100 = 4\%$$

The flexibility provided by robots reduces the setup time in many flexible manufacturing cells. The order-winning criteria affected include *price, delivery speed,* and *flexibility.*

Quality   Quality is expressed as a percentage in Figure 1–3 and includes captured, warranty, and cost of quality. *Captured quality* is a measure of the number of defects that are found before the product is shipped to the customer. *Warranty quality* represents defective parts discovered after the product is shipped to the customer. *Cost of quality* represents the total quality cost to the enterprise in percent of sales. Robots offer many quality advantages; for example, GM uses robots to put down the front windshield sealant and install the glass (Figure 1–4). The high

Figure 1–4   Robot Installation of Windshield After Sealer Is Put down by Robot Automation.
(*Source:* COMPUTER INTEGRATED MANUFACTURING, 2nd ed. by Rehg and Kraebber, p. 398. Reprinted by permission of Pearson Education, Inc., Upper Saddle River, NJ.)

degree of placement accuracy and repeatable performance of the robot ensure that no water leaks will occur. In this case, the robot was selected over a human operator because of the quality needed.

Example 1–2

Quality is often represented either as a percentage of bad parts or is measured with units of part per million (ppm), as illustrated in Figure 1–3. Defective parts are converted to ppm units by first finding the ratio of bad parts to total parts and then multiplying that ratio by 1,000,000. Determine the number of bad parts for the following lot sizes using the US average (use the small percent values in the range) and world-class standard values for captured quality listed in Figure 1–3.

a) 10,000 parts
b) 5000 parts

Solution

a) Determine bad parts for quality represented as a percentage of parts and in ppm notation.

$$\text{Bad parts} = \text{total parts} \times \frac{\% \text{ bad parts}}{100}$$

$$= 10{,}000 \text{ parts} \times 0.03 = 300 \text{ parts (US)}$$

$$\text{Bad parts} = \text{total parts} \times \frac{\text{ppm bad parts}}{10^6}$$

$$= 10{,}000 \text{ parts} \times \frac{1500}{1{,}000{,}000}$$

$$= 15 \text{ (world-class standard)}$$

b) Bad parts $= 5000 \text{ parts} \times 0.03 = 150 \text{ parts (US)}$

Bad parts $= 5000 \times 0.0015 = 7.5 \text{ parts (world-class standard)}$

The order-winning criteria affected by this standard are *quality* and *price*.

Manufacturing Space Ratio.   Manufacturing space ratio is a measure of how efficiently manufacturing space is utilized. The total footprint, or floor area, of the process machines plus the area of the assembly workstations is divided by the total area occupied by manufacturing.

Example 1–3

An automated work cell used by a manufacturer covers 300 square feet of factory floor space. The CNC process machine in the cell has a footprint that measures 94 inches by 62 inches. The remaining space includes pallets for raw materials and finished parts, an operator's desk, walking space around the machine, and an inspector's bench.

a) Calculate the manufacturing space ratio for the work cell.

b) Calculate the total work cell area if the cell conforms to the world-class standard listed in Figure 1–3.

Solution

a) Convert machine footprint to square feet, then find the ratio.

$$\text{Machine footprint (sq ft)} = \frac{94 \text{ in.}}{12 \text{ in.}} \times \frac{62 \text{ in.}}{12 \text{ in.}}$$

$$= 40.48 \text{ ft}^2$$

$$\text{Assembly area} = 0$$

$$\text{Manufacturing space ratio} = \frac{\text{process machine footprint} + \text{assembly area}}{\text{total work cell area}}$$

$$= \frac{40.48 \text{ ft}^2}{300 \text{ ft}^2} \times 100 = 13.5\%$$

b) $$\text{Total work cell area} = \frac{\text{process machine footprint}}{\text{manufacturing space ratio (world-class)}}$$

$$= \frac{40.48 \text{ ft}^2}{50\%}$$

$$= 80.96 \text{ ft}^2$$

The maximum value for the ratio is 1 or, when expressed as a percentage, 100%. Therefore, as the ratio approaches the maximum, the production operation is more efficient. The order-winning criterion affected is *price*.

Inventory. Manufacturing inventory can be grouped into two broad categories: *movement* and *organization*. Movement inventories exist because time is required to manufacture products or to transport goods from one location to another. Inventories for transport are sometimes called *in-transit* or *pipeline*. Movement inventories during manufacturing are called *work-in-process* (WIP) inventories. The type of manufacturing system used often dictates the level of movement inventory.

Organization inventories decouple successive stages in the production and distribution systems to minimize disturbance to the system. This decoupling prevents production problems at one stage in a production sequence from causing problems at later stages in production. Organization inventories are classified into the following three groups: cycle stock, safety stock, and anticipation stock.

Cycle stock inventories are created whenever products are produced or raw material is purchased in larger quantities than are currently required. The inventory costs associated with the operation are less than the costs to manufacturing of producing or buying smaller quantities. For example, larger than ideal lot size production is justified because the inventory costs are less than the costs associated with reducing machine setup time.

Safety stock provides protection against irregularities and uncertainties in the demand or supply stream. The most common example is the stocking of extra items of raw materials due to the uncertainty attached to production quality. In many manufacturing systems it is difficult to predict accurately how many production parts will be scrapped due to quality issues, so excess inventory is ordered, called safety stock.

Anticipation stock is associated with products whose markets exhibit seasonal patterns of demand and whose supply stream is relatively constant. It is often not economically feasible to produce a sufficient number of products on a weekly or monthly basis to supply a seasonal demand. Examples are Christmas toys, calendars for the following year, and lawn mowers for the spring.

Inventory is not necessarily evil; after all, some inventory is necessary to reduce disruptions in manufacturing and in fulfilling customer orders. The goal is to reduce the inventory level continuously while maintaining continuous and smooth production plus on-time and complete order shipments to the customer.

Inventory levels are measured in terms of *inventory turns*, or *velocity*, and in *residence time*. The number of inventory turns for a product is equal to the annual cost of goods sold divided by the average inventory value. The residence time is the average number of days that a part or raw material item spends in production. The goal is reduced inventory; therefore, in terms of the measurement standard, inventory turns need to rise while residence time must fall. Inventory turns are determined from a relationship between the annual cost of the goods sold and the average annual inventory investment expressed as follows.

$$\text{Inventory turns} = \frac{\text{annual cost of goods sold}}{\text{average annual inventory investment}}$$

Determination of the number of inventory turns is illustrated in Example 1-4.

Example 1–4

A manufacturer reports that it cost $2,000,000 annually to produce a product and the average annual inventory investment for the product is $100,000. Find the number of inventory turns for the product.

Solution

$$\text{Inventory turns} = \frac{\text{annual cost of goods sold}}{\text{average annual inventory investment}}$$
$$= \frac{\$2,000,000}{\$100,000} = 20$$

An inventory turns value of 20 implies that new inventory to produce the product is delivered 20 times in a year. The number of inventory turns indicates the level of investment in raw materials and parts that a product imposes on the manufacturer. The higher the number of turns, the less the investment in raw material and parts. If this standard is performed well, parts and raw materials spend less time in the plant. As a result, the order-winning criteria that benefit are usually *price, quality, delivery, speed,* and *delivery reliability.*

Flexibility. *Flexibility* is a measure of the number of different parts that can be produced on the same machine. Improved efficiency in this standard results from good part design, innovative fixture design, and selection of the optimum production machines. The robot supports this standard well, because the reprogrammable robot is one of the most flexible machines in manufacturing. Excellence in this standard primarily aids the *price* order-winning criterion.

Distance. *Distance* measures the total linear feet that a part travels through the plant from raw material in receiving to finished products in shipping. A high value for this standard indicates more cost for the product and decreased quality as a result of handling. Keeping the value of this standard low helps the *price* and *quality* order-winning criteria.

Uptime. *Uptime* is the percentage of time a machine is producing to specifications compared with the total time production can be scheduled. An increase in uptime means that less equipment is required for the same level of production. The total hours of operation without machine failures has increased fourfold over the past 10 years. Improvement in this standard helps the *price* order-winning criterion.

1-5 THE PROBLEM AND A SOLUTION

We have seen that the problem facing the enterprise is twofold: (1) the external challenges (niche market entrants, traditional competition, suppliers, global economy, cost of money, Internet, and customers) are present and will not go away; (2) customer order-winning and order-qualifying criteria drive the market. The enterprise must develop a manufacturing strategy to win orders based on the criteria present in the marketplace. In other words, the enterprise must change.

One way for manufacturers to change is to improve their performance on the following six worldwide standards:

1. Design and manufacturing lead time by product
2. Inventory turns by product
3. Setup times on production equipment
4. Output/productivity by product per employee
5. Total quality and level of rework
6. Number of suggestions by product for improvements per day per employee

The move to robot automation in the 1970s was driven by labor cost; companies had to become more productive. The emergence of global markets in the 1980s has moved the focus beyond productivity to satisfying the order-winning criteria of customers who can shop from around the globe. This new class of customers requires a *quality product, wide product selection, frequent product improvements,* and *new models* on a regular basis. Satisfying the six enterprise standards is necessary to meet the demands of customers.

Manufacturers are using a process called *CIM* (pronounced *sim*), computer-integrated manufacturing, to change their operational philosophy. The CIM process—first defined by Joseph Harrington, Jr., in 1973—guides manufacturers through the major changes required in the organization and helps identify the automation hardware and software necessary to meet the order-winning criteria for their products. CIM systems, supported by robotics, can produce a competitive product consistent with customer expectations.

1-6 DEFINITION OF ROBOTICS AND COMPUTER-INTEGRATED MANUFACTURING

Many single-purpose machines, often called *hard automation,* have some features that make them look like robots. Without some definition, it would be difficult to set apart robots from the millions of other automated machines so that they can be studied. The International Standards Organization (ISO) defines an industrial robot, in standard ISO/TR/8373-2.3, as follows:

> *A robot is an automatically controlled, reprogrammable, multipurpose, manipulative machine with several reprogrammable axes, which may be either fixed in place or mobile for use in industrial automation applications.*

The key words are *reprogrammable* and *multipurpose,* because most single-purpose automation machines do not meet these two requirements. Reprogrammable implies two elements: (1) the robot's motion is controlled by a written program, and (2) the program can be modified to change significantly the motion of the robot arm in real time. This real-time programming flexibility is demonstrated, for example, when the pickup point for a randomly placed part is located by a vision system's camera, and the robot's program is changed while its arm is moving to the part.

Multipurpose means a robot is able to perform many different functions, depending on the program in memory and tooling at the end of the arm. For example, in one company a robot could be tooled and programmed to do welding, and in a second company the same type of robot could be used to stack boxes on pallets.

The Computer and Automation Systems Association (CASA) of the Society of Manufacturing Engineers (SME) defines CIM as follows:

> *The integration of the total manufacturing enterprise through the use of integrated systems and data communications coupled with new managerial philosophies that improve organizational and personnel efficiency.*

CIM describes a new approach to manufacturing, management, and corporate operation. Although CIM systems can include many advanced manufacturing technologies such as robotics, computer numerical control (CNC), computer-aided design (CAD), computer-aided manufacturing (CAM), and just-in-time production (JIT), it goes beyond these technologies. CIM is a new way to do business that includes a commitment to total enterprise quality, continuous improvement, customer satisfaction, use of a common database for all product information with

Figure 1–5   CASA/SME Wheel.

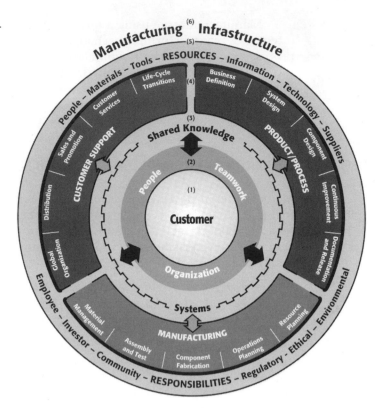

every department participating, removal of communication barriers among all departments, and the integration of enterprise resources. The CIM concept model (Figure 1–5) by CASA/SME was updated and the new CIM wheel has the following functions:

1. The hub of the wheel, titled *customer,* is the primary target for all marketing, design, manufacturing, and support efforts in the enterprise. Only with a clear understanding of the marketplace and the customer can the enterprise be successful.

2. The next layer on the wheel focuses on the means of organizing, hiring, training, motivating, measuring, and communicating to ensure teamwork and cooperation in the enterprise. The techniques used to achieve this goal include: self-directed teams, teams of teams, organizational learning, leadership, metrics, rewards, quality circles, and corporate culture.

3. This section focuses on the shared corporate knowledge, systems, and common data that is used to support people and processes. The resources used include manual and computer tools to aid research, analysis, innovation, documentation, decision making, and control of every process in the enterprise.

4. The three main categories of product/process definition, manufacturing, and customer support make up this section of the wheel, and they include 15 key processes that form the product life cycle.

5. The enterprise has resources (people, materials, tools, information, technology, and suppliers) and responsibilities (employee, investor, the community, as well as regulatory, ethical, and environmental obligation).

6. The final part of the wheel is the manufacturing infrastructure. This infrastructure includes: customers and their needs, suppliers, competitors, prospective workers, distributors, natural resources, financial markets, communities, governments, and educational and research institutions

Many companies use robots in the manufacturing sections of the CIM wheel and have integrated them into the CIM system. As a result, the study of robotics must address the techniques used to interface robots to the CIM system.

1-7 MANUFACTURING SYSTEM CLASSIFICATION

In many early implementations, the robot was simply dropped into the production area to replace the human operator. Little effort or planning was done to integrate the robot into the manufacturing system, because the robot was viewed as just a human replacement. Now industry understands that the robot is just one part of the total manufacturing system, and the robot must be integrated into the system along with the other automation hardware, software, and process machines. As a result, an understanding of the manufacturing system is critical for selection of the correct robot model and supporting hardware. The study of manufacturing systems starts with a classification of manufacturing based on how products are produced.

The classification process divides all production operations into the following five groups: *project, job shop, repetitive, line,* and *continuous.* Overlap cannot be avoided between some of the categories, and most manufacturers use two or more of the manufacturing systems in the production of an entire product line. Classification of companies into these groups requires a detailed analysis and evaluation of the production operations. As a result of the classification process, the activity in the factory automation area of the CIM wheel (Figure 1–5) can be understood better. If the manufacturing system requirements are well defined, then success in the selection and integration of the robot hardware and software is assured. Distinguishing characteristics of each classification category are described next.

Project

The most distinctive characteristic of this category is that products are complex, and production quantities are often just one unit. Products coming from a project-type manufacturing system include oil refineries, large office buildings, cruise ships, and large aircraft. In each case individual products may be similar

Figure 1–6 Types of Plant Layout:
(a) Fixed-Position Layout; (b) Process
Layout; (c) Product-Flow Layout.
(*Source:* COMPUTER INTEGRATED MANUFAC-
TURING, 2nd ed. by Rehg and Kraebber, p. 47.
Reprinted by permission of Pearson Education,
Inc., Upper Saddle River, NJ.)

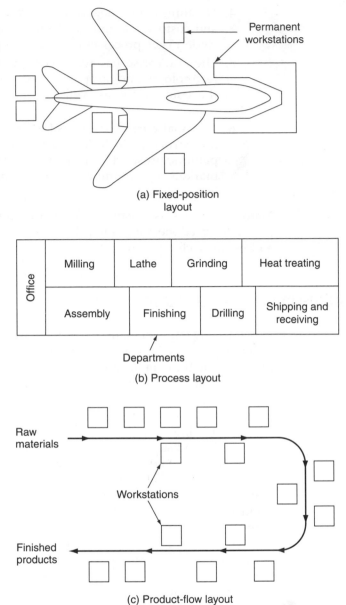

(a) Fixed-position
layout

(b) Process layout

(c) Product-flow layout

but they usually are not identical. The layout of the production area for this manufacturing system is called *fixed position* (Figure 1–6a). Because of their size and weight, products like ships and refineries remain in one location and the production equipment and parts are moved to them. Few robot applications are found in project-type manufacturing because the production and assembly equipment must be mobile.

Job Shop

Job shop production quantities, called *lot sizes,* are small, as are the size and weight of the parts produced. As a result, the parts are moved or routed between fixed production work cells for manufacturing processing. The classic machine shop with lathes, mills, grinders, and drill presses is the example most often used for the job shop classification. The production equipment layout for the job shop (Figure 1–6b) is frequently called *job shop layout* or *process layout.* Other distinguishing features include less than 20 percent repeat production on the same part, noncomplex products, and intensive movement of the products between machines and departments. Opportunities for robot applications in the job shop are present but limited by the high variation in parts and products.

Repetitive

The repetitive-type manufacturing system has the following unique characteristics: orders for repeat business approaching 100 percent, contracts with customers for multiple years, moderately high product volume with production quantities varying over a large range, and little variation in the routings of parts between production machines. The plant layout could be either the process layout in Figure 1–6b or more likely the product-flow layout in Figure 1–6c. For example, a supplier to Ford uses the repetitive-type manufacturing system to make three different water pump models with a combined volume of 10,000 pumps per week. A contract for 3 years on a product like this is very common. The production machines in this type of manufacturing system are usually special-purpose, automated systems with robots frequently integrated into the process.

Line

The line-type manufacturing system has several distinguishing characteristics: the delivery time required by the customer is often shorter than the total time it takes to build all the individual parts of the product, the product has many different options or models, and an inventory of subassemblies is normally present. Car and truck assembly is an example of this category. Manufacturing facilities in this group usually use the product-flow (Figure 1–6c) type of plant layout. Robots are frequently used to perform assembly tasks as the product moves through the line (see Figure 1–4).

Continuous

Continuous manufacturing systems have the following characteristics: the time required for manufacturing the product is longer than the time the customer is willing to wait for delivery of the product, product demand is predictable, product inventory is always present, high production volume is present, and products have few options. This type of production system always uses the product-flow type of plant layout (Figure 1–6c) with the production line limited to one or just a few different products. The production of nylon carpet

Figure 1–7 High-Speed, In-line, Fixed Automation System for Small Parts.
(Courtesy of METO-FER Corp., Pittsburgh, Penn.)

yarn is an example of this type of industry; chemical compounds enter on one end of the system and a finished product, nylon filament thread, flows out of the other end. Production of electrical components (for example, switches and fuel pumps for the automotive industry) also involves this type of manufacturing system. The linear flow manufacturing system in Figure 1–7 illustrates high-speed line assembly. Robots used in these applications are capable of high-speed and high-volume operation.

The chart in Figure 1–8 offers comparative data on the characteristics of the manufacturing systems just described. Take some time and study the chart. We will consult it later to see how the characteristics of the manufacturing system dictate the type of robot automation most frequently used.

The number of manufacturing classification categories is not critical; however, the classification system is important in understanding what makes automotive production different from nylon carpet yarn production. A specific robot model may satisfy the automation requirements of just one manufacturing classification category. So a successful robot implementation requires complete knowledge of the manufacturing system to be automated, and classification is the first step.

	Project	Job shop	Repetitive	Line	Continuous
Process speed	Varies	Slow	Moderate	Fast	Very fast
Labor content	High	High	Medium	Low	Very low
Labor skill level	High	High	Moderate	Low	Varies
Order quantity	Very small	Low	Varies	High	Very high
Unit quantity cost	Very large	Large	Moderate	Low	Very low
Routing variations	Very high	High	None	Low	Very low
Product options	Low.	Low	None	Very high	Very low
Design component	Very large	Large	Very small	Moderate	Small

Figure 1–8   Manufacturing System Characteristics.

1-8 ROBOT SYSTEMS

Integrating robots into the repetitive, line, and continuous manufacturing systems requires knowledge of both the manufacturing and robot systems. Therefore, the definition of the robot system is the logical starting point. The Robotic Industries Association defines a robot system as follows:

An industrial robot system includes the robot(s) (hardware and software) consisting of the manipulator, power supply, and controller; the end-effector(s); any equipment, devices, and sensors with which the robot is directly interfacing, any equipment, devices, and sensors required for the robot to perform its task; and any communications interface that is operating and monitoring the robot, equipment, and sensors.

As the definition indicates, a robot system is more than the hardware; it includes any devices interfaced to the robot for control of the work cell. In the chapters that follow, the larger robot system detailed in the definition will be studied. At this time, however, let us consider the *robot hardware*.

Robot Hardware

A basic robot system is illustrated in the block diagram in Figure 1–9. The hardware includes a mechanical arm to which the end-of-arm tooling is mounted, a computer controller with attached teaching device, work-cell interface, and program storage device. In addition, a source for pneumatic or hydraulic power is part of the basic system. The work-cell controller and other external devices that are part of the manufacturing system connect to the robot through the robot work-cell interface.

Mechanical Arm

The arm is a mechanical device driven by electric-drive motors, pneumatic devices, or hydraulic actuators. The basic drive elements will be either linear or rotary

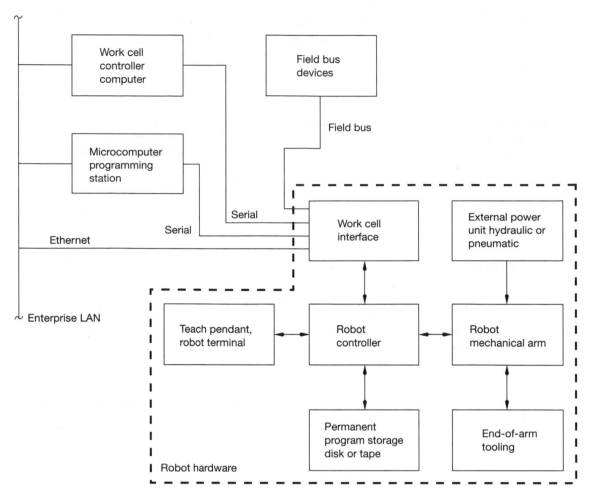

Figure 1–9   Robot Hardware and Data Interfaces.

actuators. The combination of motions included in the arm determine the type of arm geometry present. The basic geometries include rectangular, cylindrical, spherical, and jointed-spherical. Figure 1–10 shows a jointed-spherical mechanical arm with all the motions indicated. Note that all six arm motions are rotational, and that five of the six are produced by rotational actuators. The elbow extension is the only rotary motion produced by a linear actuator. Study the figure until you can see how the linear actuator produces rotary motion in axis 3.

The six arm motions are divided into two groups. The first group includes the arm sweep, shoulder swivel, and elbow extension: these are called the *position motions*. With a combination of these three motions, the arm can move to any required position within the work area. The second group of motions, associated with the wrist at the end of the arm, includes the pitch, yaw, and roll. A combination of

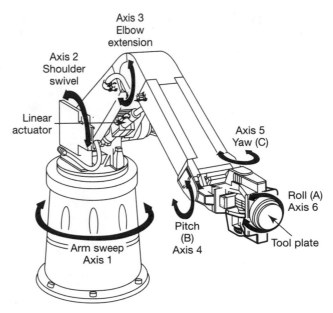

Figure 1–10 Jointed-Spherical Mechanical Arm.
(Courtesy of Cincinnati Milacron Corp.)

these three motions, called *orientation*, permits the wrist to orient the tool plate and tool with respect to the work.

The Fanuc arm in Figure 1–11 shows all of the axes. A complete description of the arm geometries, drive systems, and controller operation is presented in Chapter 2.

Production Tooling

The robot arm alone has no production capability, but the robot arm interfaced to production tooling becomes an effective production system. The tooling to perform the work task is attached to the tool plate at the end of the arm. The tool plate, which is usually a part of the wrist, is identified in Figure 1–10.

The tooling is frequently identified by several names. The terms used to describe tooling in general are *end-of-arm tooling* and *end effector*. If the tooling is an open-and-close mechanism to grasp parts, it is referred to as a *gripper* (Figure 1–12). In this text the terms *gripper* and *end-of-arm tooling* will both be used to refer to the tool on the robot arm. A complete discussion of robot end-of-arm tooling is presented in Chapter 4.

External Power Source

The external power required to operate a robot system includes sources to drive the arm motion (*electrical, hydraulic,* or *pneumatic*) and electricity for the electronic controller. Because most grippers are activated by compressed air, a source of compressed air is required for most systems. Large robot arms using

Figure 1–11 Fanuc ARC Mate Robot.

(Courtesy of Fanuc Robotics North American, Inc.)

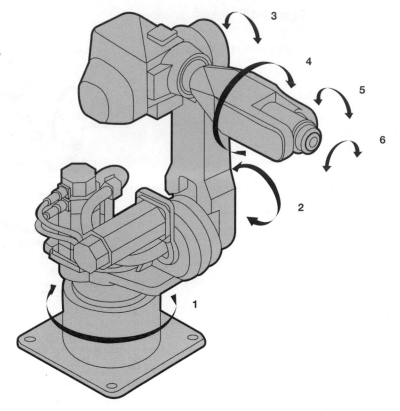

hydraulic actuators for motion require a hydraulic power source, and in some cases a compressed air source will also be necessary for the gripper. All electric-drive arms require only electrical power for motion, but many need compressed air for tooling.

Figure 1–12 Fibro Manta Gripper.

Robot Controller

Of all the hardware building blocks of a robot system, the controller is the most complex. It is also the unit with the greatest degree of variation from one manufacturer to the next. Figure 1–13 is a block diagram of a typical controller used on electric robots.

The controller, basically a special-purpose computer, has all the elements commonly found in computers, such as a central processing unit (CPU), memory, and input and output devices. Most controllers have a network of CPUs, usually standard microprocessors, each having a different responsibility within the system. The distributed microcomputer network in the controller has the primary responsibility for controlling the robot arm and controlling and/or communicating with the work cell in which it is operating. The controller receives feedback from the arm on current joint position and velocity and responds with outputs to the servo drives to change the current position or velocity based on the program stored in memory. The controller can also communicate with external devices

Figure 1–13 Robot Controller Block Diagram.

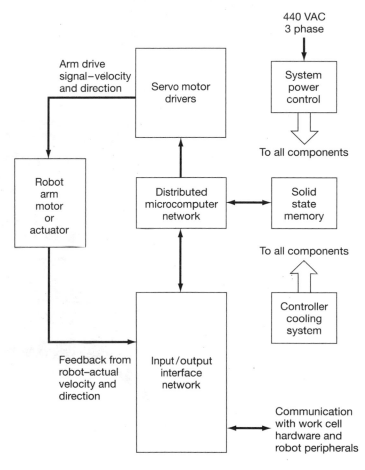

through the *input/output interface* (Figure 1–13). The operation of controllers is discussed in detail in Chapters 2 and 7.

Teach Stations

Teach stations on robots may consist of teach pendants, programming units, controller front panels, and DOS- or Windows-based software running on microcomputers. The teach pendant in Figure 1–14 is an example of a handheld terminal with a flat screen, multiline, touch-sensitive display. Some robots permit a combination of programming devices to be used in programming the robot system, whereas others provide only one system programming unit. In addition, some robots have a UNIX or Microsoft operating system in a microcontroller that is used for program creation. Some robot vendors have programming units that are teach pendants with enhanced program creation and editing capability. All of the vendor teach stations support three activities: (1) robot *power-up* and preparation for programming, (2) *entry* and *editing* of programs, and (3) *execution* of programs in the work cell. The development of a program includes keying in or selecting menu commands in the robot controller programming language, moving the robot arm to the desired position in the work cell, and recording the position and command in memory. Chapter 8 includes a detailed analysis of the programming process.

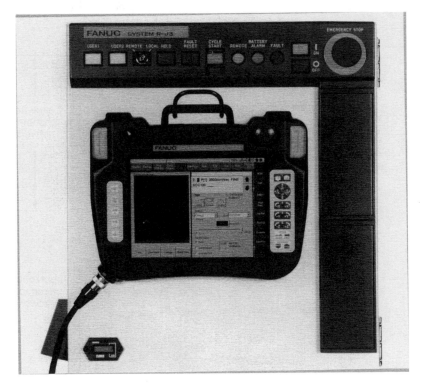

Figure 1–14   Teach Pendant and Controller Controls for a Fanuc Robot.
(Courtesy of Fanuc Robotics North American, Inc.)

ABB Controller Interface

Analysis of the interface for an ABB Maestro System 4 robot controller (Figure 1–15) indicates the computing power and interface options available in current controllers. The interface to the robot production program has three partitions: one for the work cell operator, a second for the application programmer, and a third for the individual responsible for the process performance. Each has access to information on a need-to-know basis. *Operators* can perform some program editing functions using the programming unit. In addition, they can use ABB's Quick Teach software system to simulate the operation of the programming unit for training.

The *application programmer* has all the capability of the work cell operator with the ability to fully edit any program and to work in the off-line programming

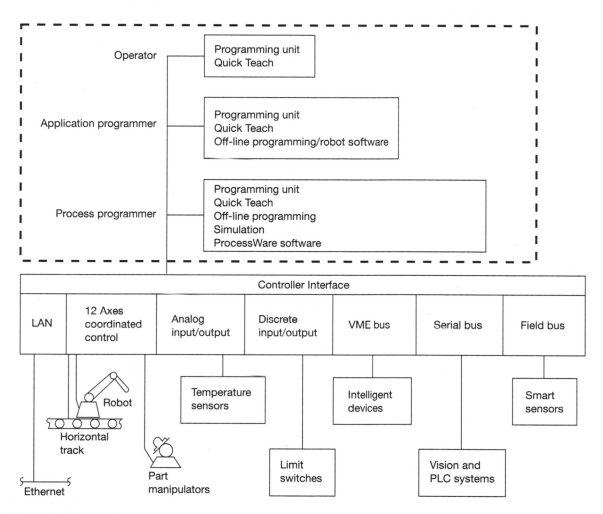

Figure 1–15 ABB S-4 Control Interface.

mode. In the off-line mode, the logic and command flow for a complete program for the robot application can be developed without any requirement for interaction with the robot controller.

The *process programmer* has all the capabilities of the previous two levels plus the ability to use ABB ProcessWare software to develop software rapidly for specific robot applications like welding. In addition, there is the ability to run simulations of the application to verify desired arm moves and production results.

The data interface (Figure 1–15) is equally robust. The controller provides for 12 axes of coordinated control that includes six on the robot, one for a linear track to move the robot horizontally in the work cell, and five for other work piece or robot manipulators. The controller supports both analog and digital data input and output so the controller can interface directly with process sensors like analog temperature probes and discrete devices like limit switches. A VME-bus interface supports the large inventory of VME instrumentation and control cards available from third-party vendors for intelligent control of external devices. There are four serial links (each can be configured as either an RS232 or 485) for serial data exchange between the controller and other work cell devices like vision systems, work cell and host computers, programmable logic controllers (PLCs), smart gauges, and bar code concentrators. In addition, the controller can be set up as a node on an Ethernet LAN or can communicate directly with PLCs and field bus devices using any of the following field bus protocols: Flex I/O, Interbus-S, Profibus, and Controller Area Network (CAN).

The control present in this robot controller makes it an ideal partner in the control hierarchy found in many of the automated work cell and flexible manufacturing systems used by global manufacturers.

1-9 SOME BASIC TERMS

With the introduction of every new technology there follows a new vocabulary that must be learned for the user to be literate in that field. Robotics is no exception. Most of the robotics vocabulary can be learned as you read the text, but a few terms are prerequisite. Specifications for the AdeptOne robot from Adept Technology, Inc. in Figure 2–17 are given in Figures 1–16 and 1–17. For each of the definitions that follow, find the corresponding value for the AdeptOne robot.

Work Envelope.   The work envelope (Figure 1–17) is the space in which the robot gripper can move with no limitations in travel other than those imposed by the joints. Figure 1–18 shows a work envelope or work volume for a jointed-spherical geometry robot in contrast to that of a human worker.

Axis Numbering.   The following axis numbering convention is used to describe the axes of motion of a robot (see Figures 1–10 and 1–11).

AdeptOne Robot

Product Specifications

Load
Payload:	·20 lbs (9.09 kg)
Downward force:	40 lbs (18.2 kg)

Joint 4 inertia
standard:	96 lb-in^2 (280 kg-cm^2)
maximum:	1000 lb-in^2 (2900 kg-cm^2)

Positioning (X,Y)
Resolution:	0.0006"
Repeatability:	0.001"
Accuracy††:	0.003"

End-of-Arm User Connections
User solenoid valves:	2
Additional 12VDC solenoid drivers:	2
User signal lines:	10

Calibration Method
Uses local calibration indices

Cycle Time
Maximum velocity (no load): 30 ft/sec

The robot tool performs a continuous-path motion from location "a" to location "b" and back to "a". The path consists of all straight-line segments as shown:

Payload	Cycle Time
1 lb	0.9 sec
13 lbs	1.3 sec
20 lbs	1.7 sec

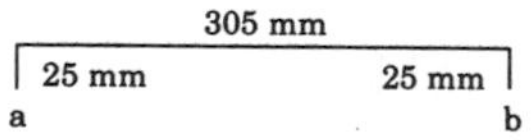

305 mm

25 mm | 25 mm
a | b

Note: These cycle times apply to a standard AdeptOne. With the optional Adept HyperDrive performance enhancements, tool-tip velocity and cycle times improve. Refer to the Adept HyperDrive datasheet for details.

† maximum possible using "J4 Tuner" software and running at reduced speed.
†† over a 17" X 17" area using Adept's High-accuracy Positioning System (HPS).

Joint Motion
Joint 1:	300°
Joint 2:	294°

Joint 3
standard:	7.7"
optional:	11.6"
Joint 4:	554°

Robot Brakes
Joints 1, 2, and 3 use fail-safe mechanical brakes

Robot-to-Controller Cable   15'

Design Life   42,000 hours

Robot Weight   400 lbs

Options
- Adept HyperDrive
- High-accuracy Positioning System (HPS)
- Force Sensing Module
- Clean room package (Class 10)
- Fifth axis (servo pitch axis)
- 11.6" vertical stroke
- Robot-mounted camera calibration
- 25' robot-to-controller cable set
- Conveyor-tracking hardware

System Requirements

Controller
Compatible with Adept A-series or S-series MC and CC controllers.

Power
The Adept controller supplies all necessary power and control.

Air
80 psi minimum, 120 psi maximum, 1 CFM, clean dry air (excluding end effector requirements).

Environment
41 – 122°F (5 – 50°C), 5 – 90% relative humidity (noncondensing).

Product specifications are subject to change without notice
Adept, AdeptOne, Adept MC, Adept CC, and HyperDrive are
trademarks of Adept Technology, Inc.
Revision date 9/92.

Figure 1–16   AdeptOne Robot Specifications.
(Courtesy of Adept Technology, Inc.)

1. Start at the robot mounting plate (base).
2. Label the first axis of motion encountered as axis 1.
3. Progress from the base to the end-effector, numbering each axis encountered successively.

Position Axes.   The position axes for the robot in Figure 1–10 are labeled 1, 2, and 3. An ancillary device often used to provide an additional axis, *linear movement* of the

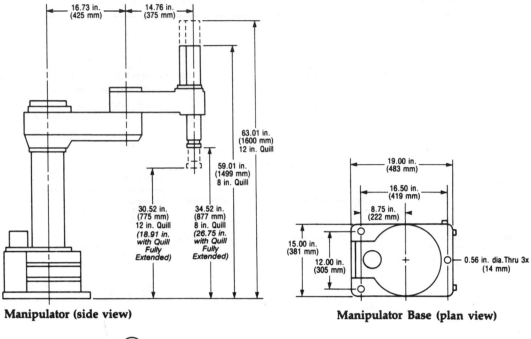

Manipulator (side view)

Manipulator Base (plan view)

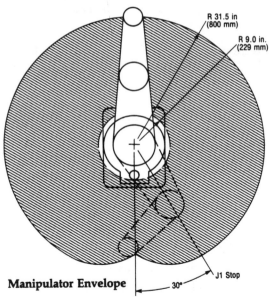

Manipulator Envelope

Figure 1–17 AdeptOne Robot Work Envelope and Arm Dimensions.

(Courtesy of Adept Technology, Inc.)

Figure 1–18   Human and Robot
Work Envelope.
(Courtesy of TPC Training Systems)

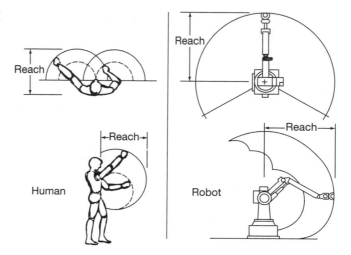

entire robot along the floor, is shown in Figure 2–8. The position axes, defined for each robot geometry, move the tooling to the point where work must be performed.

Orientation Axes.   The orientation axes, labeled 4, 5, and 6 in Figure 1–10, orient the tooling so that it is aligned with work. *Pitch, yaw,* and *roll* are illustrated in Figure 1–19 and compared with those of the human wrist (Figure 1–19a). Three techniques are used in robotics to duplicate the human movement. The first configuration (Figure 1–19b) was described earlier.

The second (Figure 1–19c) uses a *roll 1, pitch/yaw,* and *roll 2* configuration. Note that the center axis produces yaw in the vertical position and pitch in the horizontal position. The three-roll wrist, a third type of wrist actuator, is illustrated in Figure 1–20 and shown in Figure 1–4. A front view of the wrist is presented in Figure 1–20 with the shaft to connect the wrist to the arm on the top and the shaft to the tool plate on the left. Three concentric shafts enter the wrist from the robot arm through the shaft at the top of the figure and interface with complex gearing inside the ball to provide

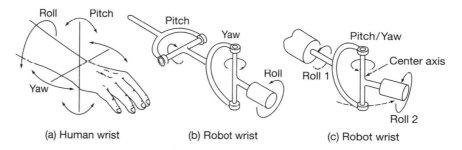

Figure 1–19   Human and Robot Wrist with Orientation Axes.
(Courtesy of TPC Training Systems)

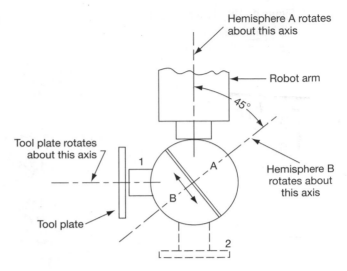

Figure 1–20 Operation of a Three-roll Wrist.

Hemisphere A rotates about this axis

Robot arm

45°

Tool plate rotates about this axis

Hemisphere B rotates about this axis

Tool plate

the three wrist orientation motions. The tool plate (left side of the ball in Figure 1–20) rotates to provide a roll motion. The line or separation through the sphere shaped object indicates that it is assembled from two hemispheres that can rotate with respect to one another. The entire sphere can rotate at the end of the arm where the top hemisphere (labeled A) connects to the arm shaft. The lower half of the sphere (labeled B) can rotate with respect to the upper half of the sphere to change the orientation of the tool plate. Note in Figure 1–20 that a 180° rotation of the lower hemisphere (B) would move the tool plate from position 1 to position 2. As a result, pitch and yaw motions can be achieved by a combination of rotations by the upper hemisphere (A) at the arm and the motion of the lower hemisphere (B) with respect to the upper hemisphere. All three orientation motions are achieved with a small wrist geometry.

Degree of Freedom. Every joint or movable axis on the arm is a degree of freedom. A machine with six movable joints, such as the one in Figure 1–10, is a robot with 6 *degrees of freedom* or *axes*. Orientation of the tool by the wrist involves a maximum of 3 degrees of freedom, and up to 4 degrees of freedom are used for positioning within the work envelope. A range of 4 to 7 degrees of freedom is typical for industrial robots.

Coordinate Systems. All points programmed in the work cell are identified by a *base coordinate system* that consists of three translation coordinates—X, Y, and Z— and three rotational coordinates—A, B, and C. The number of work-cell coordinates required to define a programmed point is determined by the number of degrees of freedom present on the robot. Figure 1–21 shows the coordinate system frequently used by robots with 6 degrees of freedom.

Accuracy. Accuracy is best explained by an example from target shooting with a rifle. The targets in Figure 1–22 indicate the results when two different rifles are fired at the center of the targets. The rifle producing the results in Figure 1–22a is

Figure 1–21 Robot Coordinate
System.

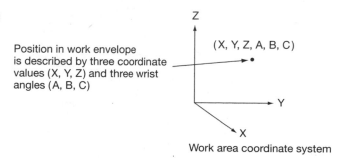

Position in work envelope
is described by three coordinate
values (X, Y, Z) and three wrist
angles (A, B, C)

(X, Y, Z, A, B, C)

Work area coordinate system

not accurate because all shots missed the center; however, the rifle is repeatable because all shots hit the target in the same area. The second rifle was both accurate
and repeatable, because it placed all shots (Figure 1–22b) in a close group in the
center of the target.

In robotics, *accuracy* is the degree to which a robot arm can move to a specific
translation or *position* point in the work cell when the point coordinates are: (1)
entered from an off-line programming station, (2) calculated inside the program,
(3) received from a vision system, or (4) generated in a work-cell simulator. For example, a vision system could specify that the robot tooling should move to a point
described as X = 50.00 inches, Y = 45.30 inches, Z = 10.01 inches (tooling position), A = 0.00 degrees, B = 90.00 degrees, and C = 100.00 degrees (tooling orientation). This move implies that the tool center point moves from the robot's 0, 0,
0 point to a point 50.00 inches along the X axis, 45.30 inches along the Y axis, and
10.01 inches along the Z axis. The wrist then changes by the degrees specified. The
accuracy specification describes how close the tool center point will be to the point
described by the vision system. In general, robot accuracy is the difference

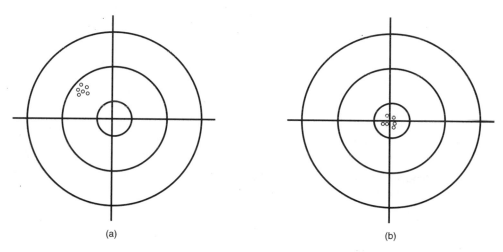

(a)   (b)

Figure 1–22 Rifle Targets: (a) Repeatable and (b) Accurate and Repeatable.

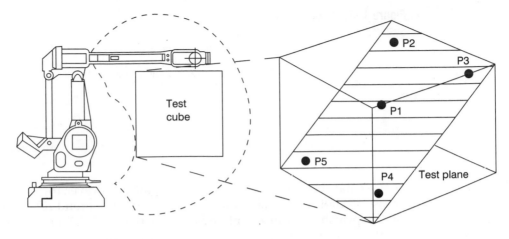

Figure 1–23   ISO Test Cube.

between actual location of the robot tool and the location specified by the translation point entered by one of the four methods specified earlier. Robot accuracy is
usually an order of ten worse than the arm's repeatability.

Industry standards are in place to assist robot manufacturers in determining
the accuracy of robot systems. The ISO standard uses a test cube with an inclined
test plane containing five programmed points (Figure 1–23). The test results for
three ABB robots are shown in Figure 1–24.

ROBOT TYPE	IRB 2400/10	IRB 4400/60	IRB 6400/2.4-120
Rated load	10 kg/22 lbs	60 kg/132 lbs	120 kg/264 lbs
Rated speed	1 m/s/39"/s	1 m/s/39"/s	1 m/s/39"/s
Repeatability	0.05 mm/0.002"	0.07 mm/0.003"	0.1 mm/0.004"
Linear path accuracy	0.46 mm/0.02"	0.83 mm/0.033"	2.1 mm/0.083"
Linear path repeatability	0.14 mm/0.006"	0.29 mm/0.011"	0.53 mm/0.021"
Path velocity fluctuation	1.6%	2.2%	2.4%
Minimum positioning time	to 0.2 mm/0.008"	to 0.4 mm/0.016"	to 0.4 mm/0.016"
On 35 mm linear path	0.21 sec	0.2 sec	0.23 sec
On 350 mm linear path	0.55 sec	0.6 sec	0.59 sec
Average power consumed on ISO test paths	330 W	700 W	1000 W

"Rated speed" designates test speed, according to the ISO standards.

Figure 1–24   ISO Test Results at Rated Load and Speed.

Observing the robots in the chapter figures indicates that robots have a base on which the arm is built. The geometry requires that each axis or degree of freedom be built on the previous axis, with axis 1 attached to the base, axis 2 attached to axis 1, and so on, until the tool plate is attached to the end of the last axis. The degree of inaccuracy is a result of the buildup of the mechanical tolerances on the arm elements as they are assembled. For example, the AdeptOne robot in Figure 1–16 has an accuracy of 0.003 in., which implies that the arm elements are manufactured and assembled with less than 0.003 in. variation from the base to the tool plate on every robot produced.

Repeatability.   Repeatability is the degree to which a robot system can return to a specific programmed position point in the work cell. Frequently, a robot is taught the required gripper location by moving the arm to the location with the teach pendant and then pushing the *program point* button. The repeatability specification indicates how well the robot arm can return to the taught point on each subsequent cycle of the program execution. The best repeatability on assembly robots is ±0.0005 inches.

A robot's repeatability is not affected by the tolerance buildup that affects accuracy because the tool-plate reference point is aligned with a desired work envelope point either visually during programming or with mechanical devices. Any manufacturing or assembly variation in the arm linkages is eliminated by the visual alignment of the tooling with the desired work envelope location. The errors associated with repeatability are produced by robot axes that do not have tight joint movements or when looseness or sloppiness is introduced in the mechanical linkages due to wear.

Tool Center Point.   The *point of action* for the tool mounted to the robot tool plate is called the *offset* or *tool center point* (TCP), see Figure 1–25. With an offset of 0, 0, 0 for L, A, and B, respectively, the TCP is located at point TCP1 in Figure 1–25. With L = 10.00 inches, A = 5.00 inches, and B = 4.00 inches, the TCP is located at TCP2. When a tool is mounted to the tool plate, the distance from TCP1 to the

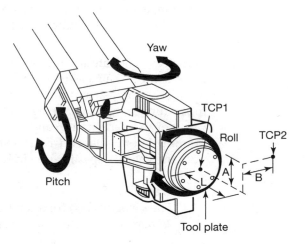

Figure 1–25   Tool Center Point.
(Courtesy of Cincinnati Milacron Corp.)

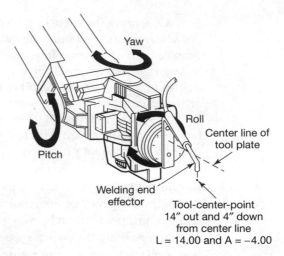

Figure 1–26 Robot with Welding Tooling.

Yaw

Roll

Center line of tool plate

Pitch

Welding end effector

Tool-center-point
14" out and 4" down
from center line
L = 14.00 and A = –4.00

action point on the tool is included in the robot program. For example, the welding torch in Figure 1–26 has a TCP of L = 14.00 inches, A = –4.00 inches, and B = 0.00 inches. In the welding example, the robot will control the motion at the welding electrode as the arm moves through the programmed points. In some robot systems the offset is part of the programming language. In the Seiko DARL language, for example, the define offset command is **DEF TL** $< 1 - 9 >$ X_1, Y_1, X_2, Y_2, Z_1. Up to nine define offset tool statements can be used to specify the X, Y, and Z coordinates for the TCP.

Precise identification of the tool center point (TCP) or offset is critical to system accuracy and repeatability. If the robot tooling collides with part of the work cell during production, the offset of the TCP can be changed. Calibration of the new offset causes lost production time. Reis robot systems overcome this problem with a special function called *TCP measurement*, which automatically determines the hand length, tool length, and tool angle of unknown tools. In the case of tool collision, a simple move to a reference point from four different directions establishes the new TCP (Figure 1–27).

Velocity.   The rate at which the robot can move each axis and the TCP under program control is a measure of the machine's velocity. Velocity is expressed in linear or angular English and metric units. Arm velocity between programmed points is much higher than the average velocity over a number of programmed points because of the deceleration and acceleration of the TCP as it passes through a programmed point. Direct-drive robots like the AdeptOne have excellent tool velocity; note the velocity specification in Figure 1–16.

Example 1–5
The velocity (V_{NL}) for a robot is rated at 20 feet per second no load with a 25 percent reduction at maximum payload. The robot moves through

Figure 1–27 Reis TCP
Measurement.
(Courtesy of Reis Machines Inc.)

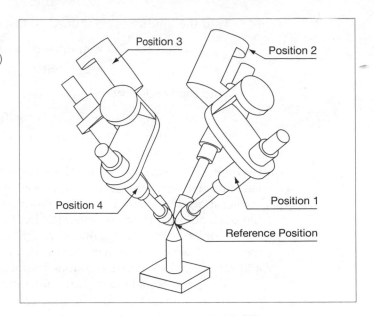

200 inches with close to maximum payload and 125 inches with no part in the gripper. There are 23 programmed points and 6 seconds of programmed delays. Test data indicates that each programmed point takes an average of 0.1 seconds and the velocity (V_{GL}) with the gripper empty is 18 feet per second. Find the fastest possible time for one cycle.

Solution

Find velocity (V_{ML}) for full payload.

$$V_{ML} = V_{NL} \times 0.75$$
$$= 20 \text{ ft/sec} \times 0.75 = 15 \text{ ft/sec}$$

Find the time for maximum load moves.

$$T_{ML} = \frac{\text{distance}}{V_{ML}}$$

$$= \frac{200 \text{ in.} \times \dfrac{1 \text{ ft}}{12 \text{ in.}}}{15 \text{ ft/sec}} = 1.11 \text{ sec}$$

Find the time for gripper empty moves.

$$T_{GL} = \frac{\text{distance}}{V_{GL}}$$

$$= \frac{125 \text{ in.} \times \dfrac{1 \text{ ft}}{12 \text{ in.}}}{18 \text{ ft/sec}} = 0.58 \text{ sec}$$

Find the time for programmed points.

$$T_{PP} = \text{number of points} \times 0.1 \text{ sec/point}$$
$$= 23 \text{ points} \times 0.1 \text{ sec/point} = 2.3 \text{ sec}$$

Find total cycle time.

$$T_T = T_{ML} + T_{GL} + T_{PP} + T_{\text{Delay}}$$
$$= 1.11 \text{ sec} + 0.58 \text{ sec} + 2.3 \text{ sec} + 6 \text{ sec} = 9.99 \text{ sec}$$

Payload.   The *rated* payload is the weight that the robot is designed to manipulate under the manufacturer's specified performance conditions of speed and acceleration-deceleration over the entire work envelope. The center of gravity of the payload must be within offsets or locations specified by the manufacturer. The *maximum* payload is the maximum weight that the robot can manipulate at a specified speed, acceleration-deceleration, center of gravity location (offset or location), and repeatability under continuous operation over a specified work envelope.

The payload or load capacity of a robot (Figure 1–16) often determines if the machine is suitable for a specific task. The payload is the total combined weight of the gripper or end-of-arm tooling and the part to be moved (payload = tooling weight + part weight).

Example 1–6
A robot must move a 14-pound load in a material handling application. If the maximum payload (P_M) for the robot is 25 pounds, what is maximum allowed weight for the gripper (W_G) if a 25 percent safety factor is required?

Solution
Find the maximum payload for safety factor.

$$P_S = P_M \times 0.75$$
$$= 25 \text{ lbs} \times 0.75 = 18.75 \text{ lbs}$$
$$W_G = P_S - \text{material weight}$$
$$= 18.75 \text{ lbs} - 14 \text{ lbs} = 4.75 \text{ lbs}$$

Additional terms associated with robot hardware and software are introduced as needed throughout the text.

1-10 ROBOT SAFETY GUIDELINES

The safety of personnel and equipment is an important consideration in any manufacturing area and is especially critical when machines like robots are present. Safety results from a detailed plan to operate in a safe manner; as a result, safety considerations must be part of every machine and work-cell design. Safety design considerations are covered in great detail in Chapter 10; however, safety guidelines are included here because it is important to introduce safe robot operating procedures at the start of any robot study.

Work Cell Safety Design Requirements

Much can be said about the design requirements for a safe work cell; most of the issues are addressed in Chapter 10. Some fundamental issues, however, should be known from the start.

- The maximum reach or work envelope of the robot should be clearly marked on the floor with high contrast safety tape or paint.
- A drive-active warning device should be clearly visible from any point around the total perimeter of the robot work cell. The warning device should be a flashing light or strobe that is automatically triggered anytime the drive motors, valves, or actuators that produce robot motion are powered. Most robot controllers provide a set of contacts that can be used to control a drive-active warning device.
- No hardware should be in the robot work envelope except that portion of the process machines that must be accessed by the robot tooling.
- Personnel should be protected from inadvertent exposure to the robot arm by barriers placed around the work cell. The barriers can be work-cell equipment or standard safety fence. In situations where production requirements prevent hard barriers from being erected, safety curtains using sensors should be used to warn of an intrusion and power down the robot to a safe condition.
- Emergency stop switches should be located at strategic points around the work cell to power down the robot to a safe condition when a problem situation is detected.
- Signs should be used to clearly indicate that a programmable device is present and movement of the device could occur at anytime.
- Provision should be made to allow maintenance workers to lock off (sometimes using lock-out procedures) master control power panels and device controllers when they are making changes to the system.

Guidelines for Safe Robot Use

There are a number of ways that a robot can cause injury to personnel, but the two most common are injury from a blow to some part of the body by the robot arm or tooling, and having some part of the body trapped between the robot arm and a fixed object. While both types of accidents are serious, the second usually causes the most serious injury and can be avoided.

- The general rule to avoid serious injury is to *avoid pinch points*.

This means that you must be constantly alert to avoid any position in the work cell that puts any part of your body between the robot arm and some fixed or unmovable object. Most impact accidents from a robot happen because personnel in the work envelope make the following incorrect assumptions:

- *If the arm is not moving it is assumed that the program has been halted.* In many robot applications the robot program halts robot motion until a signal from some external machine indicates that the robot should proceed with the next motion. For example, a robot program to move parts to a conveyor could have a pause programmed in the routine if the previous part was still in the drop position. If the previous part gets jammed on the conveyor, the robot would halt at a position some distance from the conveyor. The situation may not occur for months, then one day the cell seems to be shut down due to a jam. An operator who moves between the robot and conveyor to remove the jam could be injured when the robot proceeded after the jammed part was removed. The operator made two mistakes: first, assuming that the robot was off because it was not moving, and, second, walking into a pinch point.

- *If the arm is repeating one pattern of motion, then it is assumed that the same pattern will continue.* Robot programs often have subroutines that allow the robot motion to be changed by an external signal. For example, a robot may have a program that moves castings from a die cast machine to a trim press. What may not be apparent is that a sensor determines if the casting is good before the move to the trim press is executed. If the casting is bad, a subroutine is used to move the part to a scrap bin. Most parts are good, so the repeated motion to the trim press could be incorrectly assumed until that one bad part out of a hundred is generated.

- *If the arm is moving slowly, it is assumed that slow movement will continue.* Programmers have a wide range of speeds that they can use in a robot program. The slow speed for one move is no guarantee that the same speed will be used for the next move.

- *If you program the arm to move, you assume it will move the way you wanted.* Robot programs are developed using two steps: (1) the points or the desired locations in the work cell are taught and given names to identify the locations, (2) the program commands are organized into the program structure to move the robot to the programmed points at the correct speed. It is not uncommon to get taught points interchanged or incorrectly identified. When you think the robot is going to one location, it actually moves to an entirely different one.

Robot accidents can occur during three different activities: *regular operation, programming*, and *maintenance*. A study, Figure 1–28, conducted by the Japanese indicates the frequency of accidents in eight different situations that cover these three areas. Some of the causes involve failure or incorrect action of the robot system or some peripheral machine or device. System failure cannot be completely eliminated; however, a good preventive maintenance program can eliminate some accidents attributed to machine or component failure. A better-educated work force would reduce the injury rate in many of the groups where incorrect

Figure 1–28 Causes of Accidents in Robot Work Cells.

Cause	Percent
Incorrect action of the robot during normal operation because of system fault	5.6
Incorrect action or operation of peripheral equipment in the robot work cell during normal operation	5.6
Careless approach of robot by operator, programmer, maintenance personnel, or unauthorized employee	11.2
Human error in teaching or testing	16.6
Incorrect action of the robot during manual operation because of system fault	16.6
Incorrect action of the robot during testing and repairing	16.6
Incorrect action or operation of peripheral equipment in the robot work cell during testing and repairing	16.6
Other causes	11.2

action or operation was due to the operator or maintenance personnel not following correct procedures.

There are few reasons for an operator to be in the work envelope of a robot. Production systems and procedures can be developed that keep the operator out of harm's way. Most maintenance procedures do not require the maintenance personnel to be in the work envelope during testing, so with good maintenance procedures and education the maintenance-related accidents can be greatly reduced. The only activity that requires a person to be close to the robot arm and tooling is programming taught points.

During the programming process it is often necessary for the programmer to be visually close to the tooling to assure that alignment with the part or fixture is present before the point is programmed. During this time the programmer is moving the tooling with motion buttons on the teach pendant. When manually taught points is the programming technique required for the robot cell, the programmer must be close to the robot with the drive system active. To minimize the risk of injury the following guidelines should be followed.

- Always have an operator or other programmer present with a hand on the emergency stop when working close to the robot tooling.
- Never stand in a pinch point when performing any programming activity.
- Always use a low-arm-move speed when programming requires close proximity to the tooling.

- Never change electrical cables or input/output signal wires with the controller power on.
- Never stand in the work envelope when you send the tooling to a taught point to verify location accuracy. Signal the move to the programmed point from outside the work envelope and check the point accuracy after all motion has stopped.
- Never stand in the work envelope when testing a new or edited robot program.
- If possible, keep one hand on the robot arm as you move the robot with the teach pendant. If a failure in the system causes a sudden motion of the robot in your direction, your hand and arm will cause you to be pushed away minimizing direct contact with the moving robot arm.

While accidents can never be completely eliminated, the following four guidelines will reduce them significantly.

- Use effective perimeter warning devices, barriers, and interlocks around work cells and robot-type devices
- Provide general training for all shop floor employees concerning the dangers inherent in robot systems.
- Provide machine-specific safety training for operators who work in robot work cells.
- Establish a comprehensive preventive maintenance program for the robot and work cell implemented by maintenance personnel trained in the hardware and software used in the cell.

In addition to safety guidelines for the robot, a good industrial safety program must also consider the hazards associated with injury from electrical shock and the dangers from broken or leaky high pressure air and hydraulic lines.

Injury from a robot can often be avoided by remembering these guidelines and the **three Rs** of robotic safety. **R**obots **R**equire **R**espect.

1-11 ROBOT STANDARDS

In 1984 the Robotic Industries Association (RIA) established the R15 Executive Committee on Robotic Standards and the Automated Imaging Association (AIA). One year later the A15 Standards Committee in the AIA was formed to deal with vision-related standards. The robot standards follow the procedures of the American National Standards Institute (ANSI).

R15 Standards

The following standards are addressed by the R15 subcommittees:

- **R15.01 ELECTRICAL INTERFACE.** ANSI/RIA R15.01: *American National Standard for Industrial Robots and Robot Systems—Common Identification Methods for Signal and Power Carrying Conductors.*

- **R15.02 HUMAN INTERFACE.** ANSI/RIA R15.02: *American National Standard for Industrial Robots and Robot Systems—Human Engineering Design Criteria for Robot Control Pendants.*

- **R15.03 MECHANICAL INTERFACE.** Two standards: (1) ANSI/RIA R15.03/1: *Circular Mechanical Flange Interface;* (2) ANSI/RIA R15.03/2: *Shaft Mechanical Interface.*

- **R15.04 COMMUNICATION/INFORMATION.** Companion standard to the ISO 9506 standard on Manufacturing Message Specification.

- **R15.05 PERFORMANCE.** ANSI/RIA R15.05: *American National Standard for Industrial Robots and Robot Systems—Evaluation of Point-to-Point and Static Performance Characteristics.* Additional standards on dynamic characteristics and reliability are also available.

- **R15.06 SAFETY.** ANSI/RIA R15.06: *American National Standard for Industrial Robots and Robot Systems—Safety Requirements.*

- **R15.07 SIMULATION/OFFLINE PROGRAMMING.** ANSI/RIA R15.07: *American National Standard for Industrial Robots and Robot Systems—Simulation/Offline Programming—Terms and Notations for Characterizing Industrial Robots.*

A15 Standards

The following standards are addressed by the A15 subcommittees:

- **A15.01 SYSTEM COMMUNICATION.** Communication standards and protocols for exchange of information between machine vision systems and other devices.

- **A15.05 PERFORMANCE.** ANSI/AIA A15.05/1—1989: *Automated Vision Systems—Performance Test—Measurement of Relative Position of Target Features in Two-Dimensional Space.*

- **A15.07 TERMINOLOGY.** Terminology standards to facilitate machine vision communication including person to person, data communication, and user interfaces.

- **A15.08 SENSOR INTERFACES.** These are a series of machine vision camera interface standards.

- **A15.09 MARKING AND LABELING.** Marking and labeling standards for machine vision systems.

The terminology and definitions used throughout the text are consistent with the RIA and AIA standards listed above.

1-12 SUMMARY

Automation has changed the workplace, and computer-integrated manufacturing (CIM) will bring even greater change. A study of the history of robotics indicates that the technology is maturing and has a role in the CIM factory of the new century. While economics and technology caused the initial move to robots in the 1970s, the drive to meet customer demands fuels the move to current robot applications. Customer demand, niche market entrants, traditional competition, suppliers, the global economy, and the costs of money are external challenges that all manufacturers must address to be successful. These external challenges are best addressed with a manufacturing strategy that works to enhance the order-winning and order-qualifying criteria present in the marketplace. These criteria are established by the customer and dictate the product values that must be present in the marketplace to win orders for products or services. Several world-class standards are available to establish how ready a manufacturer is to compete globally. These standards include setup time, quality, manufacturing space ratio, inventory control, manufacturing flexibility, production distance, and machine uptime. These seven world-class standards, along with product and design lead time and employee productivity, are critical yardsticks for measuring success in manufacturing. All producers must know how their operations compare with the competition in each of these categories so that corrections can be made in areas of weakness.

The definition of *robot hardware* includes two very important terms: *reprogrammable* and *multifunctional*. To qualify as a robot, the machine's hardware must be capable of changing its functional characteristics by changing the program that drives it through its motions. The basic robot system includes a mechanical arm, special tooling attached to the arm at the tool plate, one or more computers in a controller, a teach station, a program storage device, and a source of pneumatic or hydraulic power. Robots are no longer stand-alone production devices. Integration of robots into effective work cells is now the key to achieving production efficiencies.

A broader definition of enterprise integration is covered by computer-integrated manufacturing (CIM) concepts. Commitment to CIM requires an enterprise commitment to total quality, continuous improvement, customer satisfaction, use of a single production database, removal of communications barriers, and integration of all enterprise resources. CIM concepts can be applied to all five production operation groups (project, job shop, repetitive, line, and continuous) to make the systems more competitive.

A study of robot automation starts with some basic terms used to describe system operation and performance. These include *work envelope, axis numbering, position axes, orientation axes, degrees of freedom, coordinate system, accuracy, repeatability, tool center point, velocity,* and *payload.* An understanding of these terms unlocks useful information found on robot system data sheets.

Robot safety is an important topic because the machines are capable of harm to both humans and other machines and it is necessary to be in the work envelope of a robot during programming and maintenance. Basic requirements for safe operation include: a clearly marked work envelope, "robot on" warning devices, protection of humans and equipment through barriers and careful machine placement, ample emergency stop switches, clearly stated warning signs, and safety conscious maintenance practices. The most common cause of human injury from a robot is the failure to avoid pinch points. Most impact accidents occur because personnel make incorrect assumptions with regard to the next movement of the robot arm. Impact accidents occur during regular operation, programming, and maintenance. Seven guidelines are provided in the chapter to minimize the risk of injury during programming, and four guidelines are provided to reduce the risk of accidents in general. Keeping the three Rs in mind, Robots Require Respect, will go a long way toward ensuring a safe robot environment.

QUESTIONS

1. Identify an early robot-like device that demonstrates the mechanical operation found in later industrial robots.
2. Who was awarded the first patent for an industrial robot, and in what year was the patent approved?
3. What were the names of the first three robot manufacturers?
4. What company was the first to drive a robot with a powerful minicomputer?
5. Describe the two major events that caused the emergence of robots in the 1970s.
6. Why didn't implementation of robots in the 1980s follow earlier predictions?
7. Describe the six major external factors that affect manufacturing, and identify the factor with the greatest influence on the products offered.
8. What steps are necessary to overcome the effects of the external challenges?
9. What are order-winning and order-qualifying criteria, and how do they affect decisions on adding manufacturing automation and robots to the shop floor?
10. Describe the seven shop-floor standards that drive efficient manufacturing, and identify those that are supported most effectively by robot technology.
11. Why has the emergence of the global marketplace forced companies to move to integrated solutions and adopt CIM-type managerial philosophies?
12. What is the CASA/SME definition of CIM?
13. What is the ISO definition of a robot?
14. What operational feature makes robots different from other hard-automation machinery?
15. What is the relationship between robotics and CIM?
16. How are current robot work-cell implementations different from the cells constructed in the 1970s?
17. How do the characteristics of the five basic manufacturing systems affect the adoption of robot automation?
18. What are the major elements of a robot system?
19. What is the difference between position and orientation arm motions?
20. What are the three names used to identify robot tooling?
21. What function does the external power source serve in the basic robot system?
22. Describe the basic elements found in robot controllers.
23. What are the three basic operations performed from robot teach stations?

24. Define *repeatability* and *accuracy* with respect to robot operation and explain how they are different.

25. Define the following robot terms: *work envelope, degrees of freedom, orientation axes, position axes, tool center point, maximum operational payload, velocity,* and *work-cell coordinates.*

26. List seven design guidelines that would improve work-cell safety when robots are present.

27. What is a pinch point, and how can pinch points be avoided?

28. Describe the four incorrect assumptions concerning robot arm movement that are most frequently responsible for impact accidents.

29. What are the three activities during which robot accidents occur?

PROBLEMS

1. A production system has a burden rate of $350 per hour and a setup time of 3.4 hours. What is the setup cost per part for a lot size of 200 parts?

2. The system described in Problem 1 must move to a 75-part lot size. How much must the setup time be reduced to keep the setup cost per part constant?

3. A casting costs $10, the burden cost on the machine that finishes the casting is $300 per hour, time for the part on the machining center is 15 minutes, and the setup time is 2.8 hours. If the castings are machined in lot sizes of 50, what percentage of the part cost is due to setup time?

4. A study to replace the operator in Problem 3 with a robot indicates that the burden rate for the cell would increase to $345 per hour, the part machining time would drop to 12 minutes, and the setup time would drop by 0.4 hours. Would this additional automation decrease part cost? What is the part cost for the ideal lot size of 25 on this new cell?

5. Repeat the calculations in Example 1–2 using the warranty quality standard values from Figure 1–3.

6. Compare the results of the defective part calculations from Example 1–2 and Problem 5. What is the significance in manufacturing when the projected defect rate in a lot size production is less than one part?

7. A work cell includes two 48-inch-square pallets for raw casting and finished castings, an operator's work area used to insert bushings into the machined casting that is 30 inches by 48 inches, a tool rack that covers 12 square feet, wide walkways and machine clearances that cover 45 square feet, and the production machine that measures 58 inches by 97 inches. Determine the manufacturing space ratio for the work cell.

8. A company has ten work cells like the one described in Problem 7. Due to increased demand, another cell must be added but there is no factory floor space. What manufacturing space ratio would be needed in each cell to put eleven work cells in the space that now supports ten? What two areas in the work cell could most likely be reduced to allow for an eleventh work cell?

9. A company has an annual cost for goods sold of $12 million and an average inventory value of $500,000. What are the inventory turns for this company? What average inventory value would produce an inventory turns value of 300?

10. What is the significance to product cost or manufacturing profit of the two different inventory turns values in Problem 9 when the cost of money is considered?

11. A manufacturer plans to use an AdeptOne robot (Figures 1–16 and 1–17) to assemble electric motors. If the weight of the motor armature is 4 pounds, what is the maximum allowed weight for the tooling? What is the maximum tooling weight for a 30-percent safety factor?

12. Use the robot specifications in Figures 1–16 and 1–17 and calculate the time required to move at maximum velocity over a 25-inch straight path with no load.

13. The program for a robot moves the TCP with a 10-pound payload through 425 inches of motion with the part in the tooling and 350 inches with the gripper empty. The total program has 45 programmed points. Calculate the time for the robot to complete one cycle moving at 30 percent of maximum velocity when the part is in the gripper and 60 percent when the gripper is empty. Assume time to pass through a program point is 0.08 seconds and the maximum velocity is the same as the AdeptOne.

14. Repeat Problem 11 for an ABB model IRB 2400L and a Fanuc model LR Mate 100*i* robot using the robot specifications in Appendix A. Compare the results from the three robots.

15. Repeat Problem 12 for an ABB model IRB 2400L and a Fanuc model LR Mate 100*i* robot using the robot specifications in Appendix A. Compare the results from the three robots.

16. Repeat Problem 13 for an ABB model IRB 2400L and a Fanuc model LR Mate 100*i* robot using the robot specifications in Appendix A. Compare the results from the three robots.

PROJECTS

1. Create a two-dimensional or three-dimensional wire frame or solid model of the AdeptOne robot arm (Figure 1–17) or a robot in Appendix A using Auto-CAD or any other available CAD software.

2. Use primary and/or secondary research to gather data on five manufacturers of different products in your city, region, or state that use automated manufacturing systems with robots integrated into the production process. The data should include the company name, address, and phone number; products manufactured; and type of manufacturing system used.

3. Using available resources, determine the current order-winning and order-qualifying criteria for products identified in Project 2.

4. Prepare a report on the concept of order-winning and order-qualifying criteria described in the book, *Manufacturing Strategies,* by Hill.

5. Prepare a report on the definition of CIM as described by Harrington in his book *Computer Integrated Manufacturing*. Include a comparison with the SME definition provided in this chapter.

6. Determine the AdeptOne specification for each term defined in Section 1-9.

7. Develop a safety poster that could be mounted in a work cell to indicate safe practices that should be used during programming.

8. Develop a safety-tips sheet that could be given to operators working in a robot work cell to help them recognize safe and unsafe work habits.

9. Develop a safety-tips sheet that could be given to maintenance personnel working in a robot work cell to help them recognize safe and unsafe work habits.

Robot Classification

CHAPTER GOALS AND OBJECTIVES

The primary goal of this chapter is to provide a classification framework with a place for every robot used in manufacturing. The classification process serves several purposes. First, it provides a framework for the study of robots that ensures that all-important concepts and technologies are covered. Second, it gives students a structure into which they can categorize the hundreds of different robot models available to manufacturing. At the completion of this chapter you should be able to:

- Classify a robot based on arm geometry, power sources, applications, control techniques, and path control.
- Recognize and describe the motion of the Cartesian, cylindrical, spherical, and articulated robot arm configurations.
- Describe the operation of the hydraulic, pneumatic, and electromotive force sources used to power robot systems.
- Calculate the torque delivered to the robot arm through the application of the following drive systems: belt and pulley, chain, gears, ball-screw, and harmonic.
- Recognize and describe the two major application areas of assembly and nonassembly.
- Describe the operation of servo (closed-loop) and nonservo (open-loop) robot systems.
- Calculate the transfer function and other operational parameters for the three positional feedback devices: potentiometer, optical encoder, and resolver.

The career spotlight for this chapter focuses on careers with the primary robot manufacturers. Robot manufacturers can be classified in a manner similar to the way we classify their machines, namely, by power source, application, and control techniques. The measure that divides all the robot manufacturers into roughly two groups is control techniques: servo and nonservo type robots. There are many suppliers in each group, with ABB, Adept, Fanuc, Kawasaki, Motoman, Nachi, Panasonic, Reis, Sankyo Seiki, and Seiko the primary manufacturers of servo type machines. An even larger number of companies provide nonservo robots or components used to build open-loop systems. Companies like Festo, Mack, METRO-FER, and Tol-O-Matic provide nonservo robot systems. The web address for many of the manufacturers is listed in Appendix B. A visit to a company site often shows a link that is provided to a page listing current career opportunities. Career opportunities exist for the two- and four-year graduate, with the job title and responsibility adjusted for the number of years of education and related job experience.

Career areas include: designer (design of new robot arms and controllers), application development (development of customer solutions using robots and other automation components), system test (testing robots and application solutions), field service (assembly of systems and troubleshooting problems at the user's site), industry technical support (support for industry groups using many robots [e.g., automotive]), and technical sales (selling the technology to the end user). The field service and technical sales areas usually require a lot of travel in the United States and often overseas. In some cases for the field service and sales positions, you can work out of the city of your choice as long as it is close to a major airport. Most of the major robot manufacturers have their corporate headquarters in overseas locations. However, in most cases they have a major facility in the United States because of the many robot applications in our domestic companies.

A career with a robot manufacturer implies that the job would focus on one brand of robot and would normally include additional education by the company on their specific product. However, the company would expect you as their new employee to be well versed in electrical and mechanical systems and in general robotics concepts, thus enabling you to quickly learn the specifics of their system.

- Describe the operation of internal positioning devices: fixed hard stops, adjustable hard stops, limit switches, and stepper motors.
- Describe the operation of four types of path control: stop-to-stop, continuous, point-to-point, and controlled path.
- Apply the robot selection criteria and robot survey in the design of an automated work cell.

2-1 INTRODUCTION

Work today tends to put everything into categories, groups, and classifications. The robot industry is no exception. In this chapter the five machine classification groups (*arm geometry, power sources, applications, control techniques,* and *path control*) used by the robot industry are explained so that we can develop a working knowledge of current robots. In addition, the classification technique used by the

International Standards Organization is described. This chapter also provides an opportunity to compare and contrast the primary system components that make each vendor's hardware unique.

2-2 ROBOT ARM GEOMETRY

In general, the fundamental mechanical configurations of robot manipulators are categorized as *Cartesian, cylindrical, spherical,* and *articulated.* The Cartesian is divided into traverse axes and gantry, and the articulated is divided into horizontal and vertical. A description of these configurations follows.

Cartesian Geometry

A robot with a *Cartesian* geometry can move its gripper to any position within the cube or rectangle defined as its work envelope. Two configurations form this geometry, *gantry-* and *traverse-*type machines. Figure 2–1 is an example of the rectangular work volume that a robot must use to load a conveyor from supply bins. A Cartesian gantry robot, like that illustrated in Figure 2–2 and pictured in Figure 2–3, could work in this application. In Figure 2–2, the 3 degrees of freedom used to position the arm in the X, Y, and Z directions are indicated by

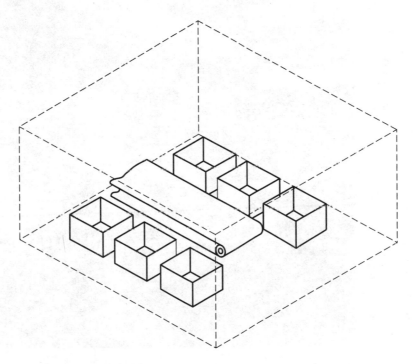

Figure 2–1 Rectangular Work Volume.

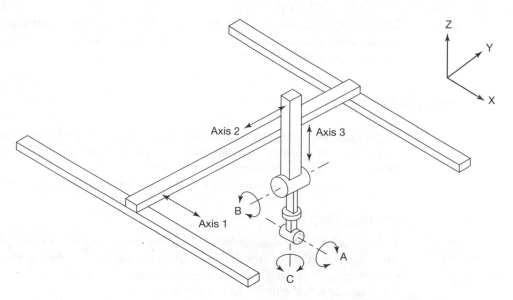

Figure 2–2   Rectilinear Geometry Robot.

Figure 2–3   Large Gantry Robot to Load and Unload Process Machines.

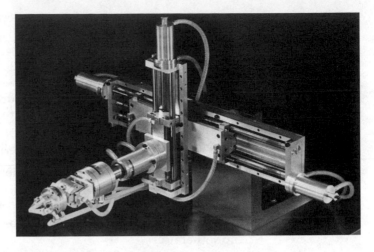

arrows on axes 1, 2, and 3; the 3 degrees of freedom for orientations A, B, and C of the tool mounting plate on the wrist are also shown by arrows. Note that the gantry robot in Figure 2–3 is used to load and unload a metal turning center from the top. The positioning movement in the X, Y, and Z direction is illustrated. Cartesian traverse-type robots are pictured in Figures 2–4 and 2–5. A common characteristic of the traverse style is the mounting of positional axes on top of each other. For example in Figure 2–4, the horizontal axis that moves the gripper out and in is attached to the vertical Z axis, and that combination is

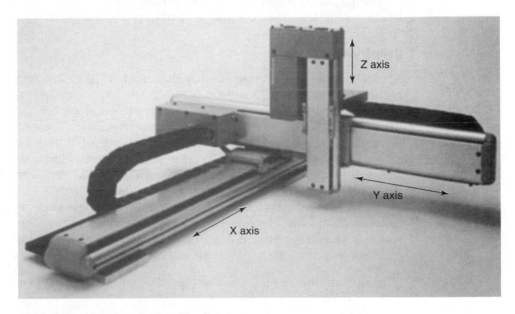

Figure 2–5 Cartesian Configuration Robot.
(Courtesy of Sankyo Seiki [America] Inc.)

connected to a horizontal axis that moves the arm left and right in the cell. Verify this construction in Figure 2–4, and determine how the axes are stacked in Figure 2–5.

The power for movement in the X, Y, and Z directions is most often provided by pneumatic linear actuators or by ball-screw drives. However, rotary actuators are used to provide the pitch, roll, and yaw turning motion at the tool mounting plate; the positioning and orientation actuators can be powered by hydraulic, pneumatic, or electric sources.

Cartesian coordinate geometry has the following advantages:

- Very large work envelopes are possible because travel along the X axis (Figures 2–3 and 2–5) can be increased easily. Systems have been developed with work envelopes more than 80 feet long.
- Overhead mounting leaves large areas of manufacturing floor space free for other uses.
- The position control system is often less complex.

The disadvantages of this type of robot geometry include the following factors:

- Access to the work envelope by overhead crane or other material-handling equipment may be impaired by the robot-supporting structure.
- On some models the location of drive mechanisms and electrical control equipment overhead makes maintenance more difficult.

The primary applications for Cartesian coordinate systems are in material handling, part handling related to machine loading and unloading, assembly of small products, and in electronic printed circuit board assembly.

Cylindrical Geometry

A *cylindrical* geometry robot, like that illustrated in Figure 2–6, can move its gripper within a volume described by a cylinder. The cylindrical geometry arm is positioned in the work area by two linear movements along the Z axis and in the R direction, and one angular base rotation about the Z axis. The motions of the Z and R axes are illustrated in Figure 2–7a and b. Movement of the robot in the cylindrical geometry work cell requires a controller that can coordinate the motion of all of the axes during a move. As a result, the controller must be more sophisticated than the controller used for Cartesian geometry systems.

The axes on cylindrical geometry robots are driven pneumatically, hydraulically, or electrically. The axis 1 rotation for the cylindrical geometry robot in Figure 2–6 is less than 360 degrees because of mechanical design limitations.

Some of the advantages of cylindrical geometry are as follows:

- Deep horizontal reach into production machines is possible with a long R axis.
- The vertical structure of the machine conserves floor space.
- A very rigid structure is possible for large payloads and good repeatability.

Figure 2–6 Cylindrical Geometry Robot.

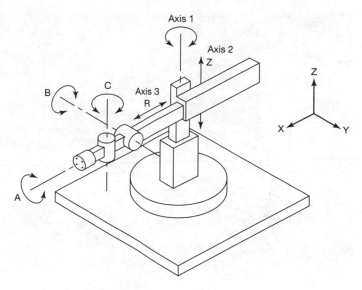

The singular disadvantage is the limited reach to left and right because of the mechanical constraints. This is often overcome by mounting the robot on a movable platform that can be positioned anywhere along the Y coordinate. Figure 2–8 illustrates this type of linear base.

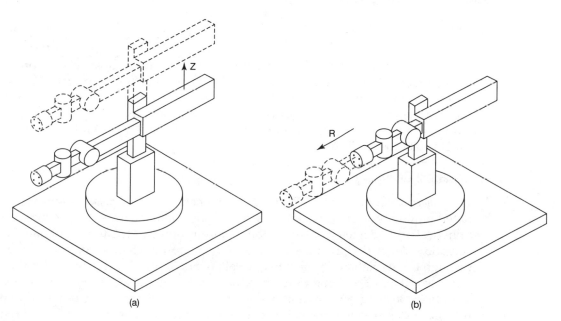

(a)

(b)

Figure 2–7 Z Axis Movement (a) and R Axis Movement (b).

Figure 2–8   Jointed-spherical Robot Mounted to a Long Linear Axis.

The cylindrical type of arm geometry can be used for most applications, but it is especially desirable when deep horizontal reach in front of the robot is necessary or when the manufacturing layout consists of machines to be serviced by the robot in a circle with a small radius. The die-cast application in Figure 2–9 is an example of this type of production layout. A small pneumatic cylindrical geometry robot is pictured in Figure 2–10.

Spherical Geometry

The *spherical* geometry arm (Figure 2–11), sometimes called *polar*, requires coordinated motion in every positioning axis for movement in the X, Y, or Z directions. Spherical arm geometry positions the robot through two rotations and one linear actuation. As in the previous cases, the orientation of the tool plate is achieved through three rotations in the wrist (A, roll; B, pitch; and C, yaw). In theory, the rotation about axis 2 could be 180 degrees or greater, and the waist rotation about axis 1 could be 360 degrees. Then if axis 3, robot reach, went from the retracted to the fully extended position, the volume of operating space defined would be two concentric half spheres. The actual work envelope for one of the most frequently used

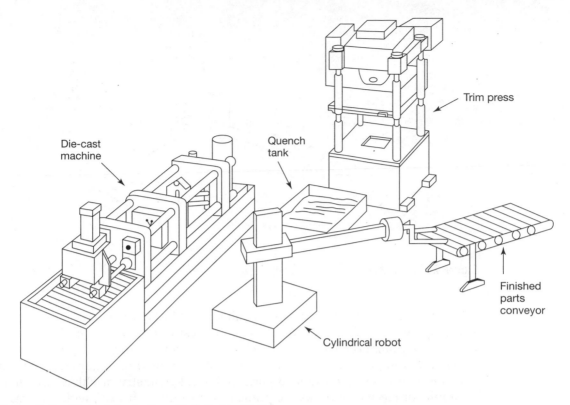

Figure 2–9   Die-cast Application with Cylindrical Robot.

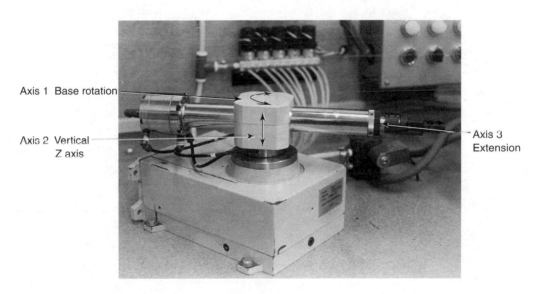

Figure 2–10   Seiko Model 700 Robot.

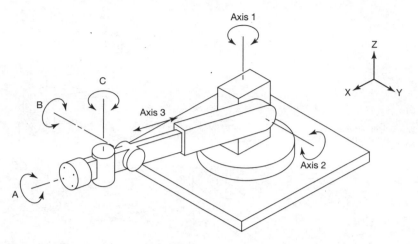

Figure 2–11 Spherical Geometry Robot.

early spherical geometry machines is illustrated in Figure 2–12. Note that the working volume is much less than the theoretical volume of the machine in Figure 2–11. Again, this results from mechanical design constraints. Figure 2–13 is a small spherical geometry robot designed to be mounted on a numerical control machine tool to load and unload parts. Spherical geometry machines use either hydraulic or electric drives as the prime movers on the six axes, with pneumatic actuation used to open and close the gripper.

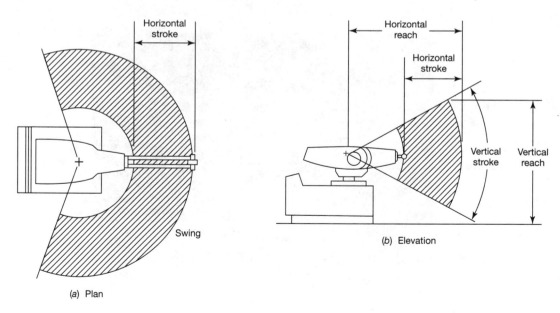

Figure 2–12 Westinghouse/Unimation 2000 Work Envelope.

Figure 2–13 GMF Machine-tending Robot.
(Courtesy of GMF Robotics)

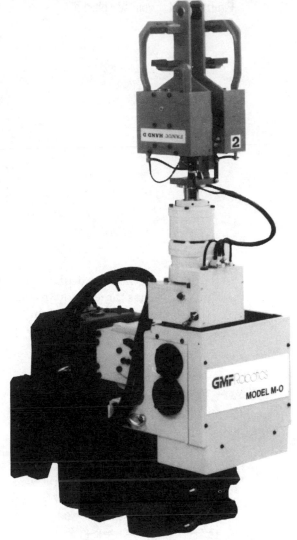

The advantages and disadvantages listed for cylindrical geometry can also be applied to spherical geometry, with the following exception: cylindrical geometry robots are more vertical in structure and spherical robots are low and long in size. This is especially true for spherical machines designed to provide long horizontal reach. Spherical geometry is not used on most current robots.

Articulated Geometry

Articulated industrial robots, often called *jointed arm, revolute,* or *anthropomorphic* machines, have an irregular work envelope. This type of robot has two main variants: *vertically* articulated and *horizontally* articulated.

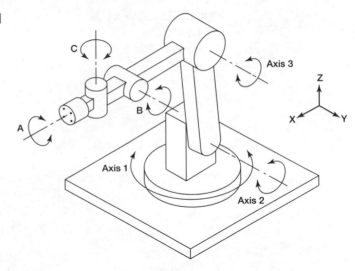

Figure 2–14 Jointed-spherical Geometry Robot.

The *vertically* articulated robot, sometimes called a *jointed-spherical arm* (Figure 2–14), has three major angular movements consisting of a base rotation (axis 1), shoulder (axis 2), and forearm (axis 3) joint. The irregular work envelope is illustrated in Figure 2–15. As in the previous arm designs, the orientation of the tool

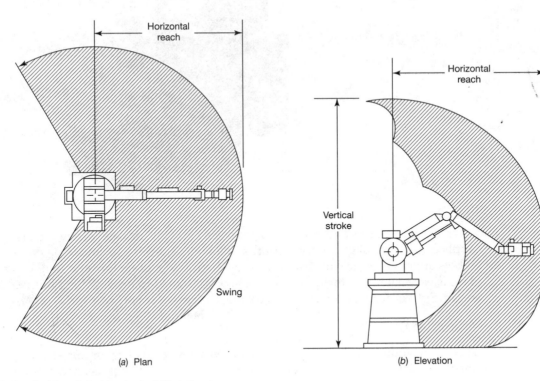

(a) Plan

(b) Elevation

Figure 2–15 Jointed-spherical Work Envelope.

plate is provided by the three rotations in the wrist. Straight-line motion along any of the three coordinates, X, Y, or Z, requires the coordinated movement of a minimum of three joints; therefore, sophisticated servo controllers are generally required for this type of arm geometry. Electric drives with feedback control systems are used on most machines. The human-like movements of the jointed-spherical arm create the following advantages for robotic applications:

- Although it occupies a minimum of floor space, the robot achieves deep horizontal reach.
- A good size-to-reach ratio is achieved, a result of the arm's ability to fold up when in the retracted position.
- High positioning mobility of the arm allows the robot to reach into enclosures and around obstructions.

This type of robot has one drawback: the construction and control requirements result in higher cost for the machine. An example of a jointed-spherical configuration is illustrated in Figure 2–16. Note the *three-roll wrist* that generates the motion for all three wrist axes in a small mechanical package and the *ball-screw* drive (covered by the flexible rubber sock) that lifts the arm.

There are two possible variations to the vertically articulated geometry just described: (1) an additional rotary motion axis (axis 4) in the forearm that rotates the forearm link; and (2) an additional linear movement axis (axis 4) in the forearm that extends and retracts the forearm link.

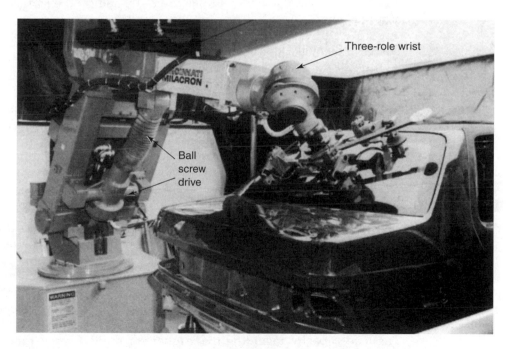

Figure 2–16 Jointed-spherical Robot Produced by Cincinnati Milacron.

Figure 2–17 AdeptOne SCARA Robot.

(Courtesy of Adept Technology, Inc.)

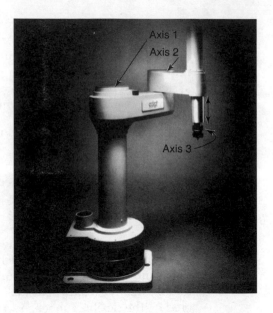

The three position movements for the *horizontally* articulated robot include: two angular movements consisting of an arm and forearm rotation, and one linear movement that is a vertical motion. Horizontally articulated arms are implemented with two mechanical configurations: (1) the *Selective Compliance Articulated Robot Arm (SCARA)* illustrated in Figure 2–17, and (2) the *horizontally base-jointed arm* in Figure 2–18.

Selective Compliance Articulated Robot Arm (SCARA)

The SCARA machine (Figure 2–17) has two horizontally jointed arm segments fixed to a rigid vertical member. Positions within the cylindrical work envelope are achieved through changes in axes 1 and 2. Vertical movement of the gripper plate results from the Z axis located at the end of the arm. SCARA machines usually have only one wrist axis orientation movement, rotation. This arm geometry is frequently used in electronic circuit board assembly applications because this geometry is particularly good at vertical part insertion.

Horizontally Base-jointed Arm

The horizontally base-jointed arm uses the same construction as the SCARA with one exception: the vertical Z axis is located between the rigid vertical base and the shoulder joint of the upper part of the arm (see Figure 2–18). In the Reis RH series machine pictured in Figure 2–18, the upper arm, forearm, and wrist all move vertically on the base. Compare this geometry with the cylindrical machine covered earlier. This geometry provides all the benefits of the cylindrical robot but with an arm that folds back to reduce the floor space required for the machine. The load capacity, including the gripper, for the RH series robots ranges from 16 kg to 300 kg.

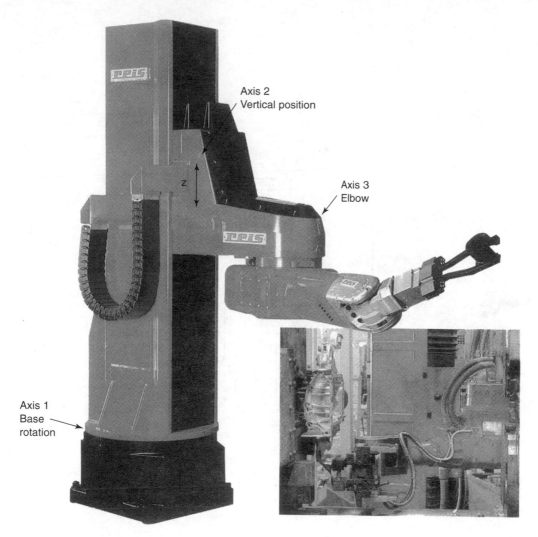

Figure 2–18 Reis RH Series Robot.
(Courtesy of Reis Machines Inc.)

2-3 POWER SOURCES

The three primary power sources used to drive manufacturing systems, namely, *hydraulics, pneumatics,* and *electromotive force,* are also used as prime movers in current robots. This section explains the classification of robotic arms by power source.

Hydraulic Power

A basic robot hydraulic power system is illustrated in Figure 2–19; study the diagram carefully. The pump and tank provide oil at high pressure for the system;

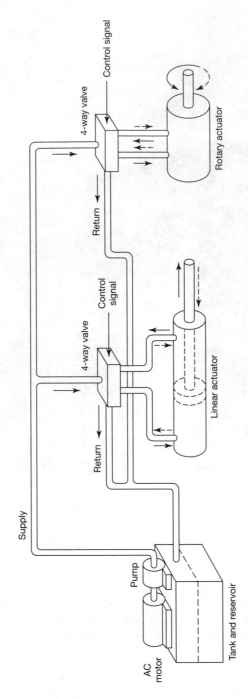

Figure 2–19 Robot Hydraulic System.

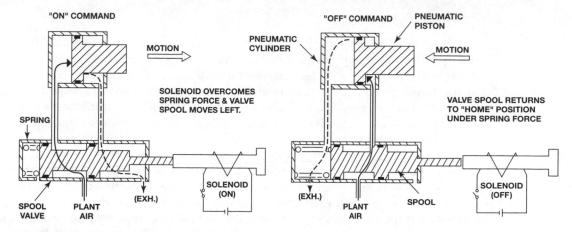

Figure 2–20 Pneumatic Actuators and Control Valves.
(Courtesy of Mack Corporation)

four-way control valves, called spool valves, switch the flow of high-pressure oil; and one or more actuators produce the desired motion. A cross section of a spool valve is provided in the pneumatic system in Figure 2–20. The two actuators commonly used are the linear type for straight-line motion and the rotary type where rotational torque is desired. In the case of the linear actuator, the high-pressure hydraulic oil is forced into one end of the hydraulic cylinder. When the chamber fills, the high-pressure oil causes the piston to move; this movement forces the oil on the other side of the piston to flow out of the cylinder. This oil is returned to the pump through the spool valve and return lines. In Figure 2–19 the solid arrows indicate the direction of oil flow and the resulting actuator movement. To reverse the actuator action, the fluid must flow in the direction of the dashed-line arrows. As indicated, the four-way valves, which control actuator movement, are driven by electrical signals from the robot control system.

Hydraulic actuators have one primary advantage: a very high power-to-size ratio that affords large load capability. That single advantage is offset by several disadvantages:

- Even the best hydraulic system will leak eventually.
- Hydraulic oil can become a fire hazard in arc welding applications.
- The additional equipment in the form of motor, pump, tank, and controls increases maintenance, energy, and robot costs.
- A higher noise level is associated with hydraulic systems.
- Hydraulic maintenance skills are also required for repairs.
- Regular testing of the hydraulic fluid is required to determine the wear on actuators.

The increased payload capability of new electric robots has significantly reduced the number of hydraulic robots used in manufacturing.

Pneumatic Power

The basic system components in pneumatics are the same as those identified for the hydraulic system (Figure 2–19). The primary difference is that the power is being transferred by a gas under pressure rather than by oil. Study the section views of the linear pneumatic actuators and four-way spool valves illustrated in Figure 2–20. The solenoid is an electrically actuated linear actuator that moves the round spool inside the valve against the spring. The two positions that the spool can assume permit air to be forced to either side of the piston in the robot actuator. The control valves for the hydraulic robot described in the previous section operate in a similar fashion. In most robotic applications, the pneumatic actuators can be classified as linear or rotary. In each case the actuator has two usable positions, one at each end of the travel. The positions for linear devices are retracted and extended, and rotary actuators come to rest at the full clockwise rotation and the full counterclockwise position. Feedback control systems are rarely used to stop a pneumatic actuator between hard stops, because the compressibility of the gas (air) that drives the actuators makes repeatability unpredictable. The repeatability is very good, however, if the actuator is driven against fixed stops at each extreme of travel. In some systems the fixed stops are adjustable slide blocks that stop the actuator as it extends or retracts. The gas serves to force the movable part of the actuator against the block; robot tool position is established by the block setting. A robot using this fixed-stop approach is called a *pick-and-place* device or *bang-bang* machine. The limited positioning capability of the pneumatic robot does not limit the use of this type of robot. In fact, more pneumatic robots are used in manufacturing worldwide than any other type.

The pneumatic power source has the following advantages:

■ Compressed air is available in most manufacturing areas.
■ Pneumatics is an inexpensive, well-developed technology.
■ System leaks do not contaminate the work area.
■ Fast operation and short cycle times are common characteristics.

The primary disadvantage has been the inability to drive the pneumatic system using feedback control to provide proportional operation and multiple stops. Figures 2–4 and 2–10 show pneumatic-powered robots for small applications.

Electric Power

The electric system includes a source of electric power and an electric motor. In most applications the motors are servomotors, but stepper motors are used on some robots where the payload is small. The servomotor can be dc or ac, with the latter becoming the more popular type. The servomotor is a precision electric motor that causes rotational motion in proportion to a supplied electrical command signal from a servo amplifier. DC servos are used primarily for higher power applications where larger torque is necessary. The dc device has a higher maintenance requirement with periodic replacement of brushes, and it generates radio frequency (rf) interference during operation.

AC servomotors tend to be more stable in their operation, lighter in weight, more rugged, and require less maintenance. However, they do not have the torque capability of their dc counterparts. The ac servo is a two-phase induction-type motor because the rotor is inductively coupled to the stator rather than through brushes as is the case in the dc servo. The motor is controlled by two ac voltages different in phase that are applied to two different windings. The electric drive is divided into two classifications: direct-drive and reduction-drive systems.

The electric motor with a *reduction-drive* system provides excellent rotational torque and angular positioning using the standard *belt* and *pulley, gear train,* or a *harmonic-drive* system described in Section 2-4. The motor and gears (A and B) in Figure 2–21 illustrate how the gear-type system would function in a servomotor-type robot arm. The gears usually provide torque amplification because the motor gear (B) has fewer teeth than the gear (A) to which the arm is attached. For example, in Figure 2–21 the servomotor gear must rotate through three revolutions (1080 degrees) while the arm moves through one-half of a revolution (180 degrees). A belt-drive system is illustrated in Figure 2–22. These techniques are frequently used on jointed-spherical arm geometries because all the axes require angular-drive motors. When linear motion is required from an electric motor, a *ball-screw* drive mechanism, described in Section 2-4, is used.

The *direct-drive* system, such as that used in the AdeptOne robot in Figure 2–17, has the motor driveshaft connected directly to the axis. Degrees of rotation of the motor shaft and degrees of angular change in the axis have a one-to-one relationship. Consequently, direct-drive systems can produce only angular motion.

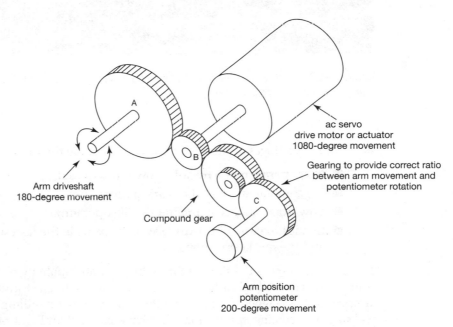

Figure 2–21   Position Feedback Potentiometer Gearing.

Figure 2–22 Belt- and Pulley-drive System on End of Forearm of Jointed-spherical Robot to Produce Orientation Motion in Wrist.

The advantages of the all-electric drive include the following:

■ No generation of hydraulic power is required.
■ No contamination of the work space occurs.
■ Low noise level is maintained while operating.
■ In addition, the direct-drive systems provide the fastest robot arm motion and the quickest response.

The disadvantage is the limited lifting or payload capability of the electric system compared with its hydraulic counterpart. As a result, electrically powered robots are primarily designed for machine tending, material handling, part assembly, or welding and coating applications that have payload and end-of-arm tooling typically less than 250 pounds.

2-4 DRIVE SYSTEMS

The primary function of drive systems is the transmission of power from one location to another. The most visible example of this occurs in power transmission lines used by electric utilities to transfer the electric energy generated at a power plant to the electric receptacle in your home. A second example, closer to robotics, is the transmission used in cars to transfer the rotational energy from the engine to the wheels. In robotics and factory automation, power must also be transferred. For example, the study of physics indicates that work is performed when a robot arm applies a force to lift a part over some distance. If the part is lifted in a specific period of time, then the work per unit of time provides the power consumed in the operation. The power to perform these operations in robots originates in hydraulic pumps for hydraulic robots, air compressors for pneumatic robots, and electric motors for electric robots. However, to make a robot axis move, the power must be transferred from the electric motor source to that axis by the drive system. Before drive systems are introduced, the concept of torque must be understood.

Torque

The mechanical system in Figure 2–23 has a lever arm with a *force* (F) applied at a *distance* (L) from a shaft that is the pivot point for the system. A *moment of force,* called *torque,* is created; it has English units of foot pounds and is described by the following equation.

$$\text{Torque} = L \,(\text{feet}) \times F \,(\text{pounds})$$

Example 2–1
Find the torque generated by a force of 40 pounds acting 1.4 feet from the shaft in Figure 2–23.

Solution

$$F - 40 \text{ lbs}$$
$$L = 1.4 \text{ ft.}$$
$$\text{Torque} = 1.4 \text{ ft} \times 40 \text{ lbs}$$
$$= 56 \text{ ft-lbs}$$

Figure 2–23   Moment of Force.

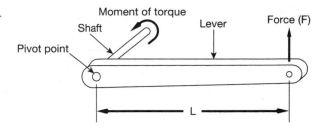

Example 2–2

A servomotor with a maximum stall torque of 15 ft-lbs is attached directly to the shaft in Figure 2–23. If L is 8 inches, what force (F) would have to be applied to keep the lever from rotating when the maximum motor torque is applied in the direction illustrated in Figure 2–23?

Solution

$$L = 8 \text{ in.}$$

$$= 8 \text{ in.} \times \frac{1 \text{ ft}}{12 \text{ in.}} = 0.67 \text{ ft}$$

$$\text{Torque} = 15 \text{ ft-lbs}$$

$$F = \frac{\text{torque}}{L}$$

$$= \frac{15 \text{ ft-lbs}}{0.67 \text{ ft}} = 22.39 \text{ lbs}$$

Note: In both examples, the lever weight is assumed to be zero.

In the previous section, direct-drive and reduction-drive robot systems were described. Assume the lever in Figure 2–23 is the forearm of a robot with a direct-drive axis motor attached to the shaft, and the gripper is attached to the end opposite the motor. Given the part and gripper weight, it would not be difficult to calculate the motor torque required to lift the part. However, if the motor torque is not sufficient to lift the forearm with the gripper and part present, then a reduction-drive system would be installed between the motor output shaft and the pivot point of the forearm. A reduction drive system reduces the rotational speed of the output but increases the torque at the output. The most commonly used reduction-drives in robots include: *belts, chains, gears, ball screws,* and *harmonic drives.*

Belts

Belt drives use a pulley that is matched to a *v-groove, synchronous,* or *flat* belt. A v-groove belt and pulley drive system is illustrated in Figure 2–24. The pulley on the drive spindle turns the belt, using only the friction between the belt and pulley to

Figure 2–24 V-Belt Drive and Pulley.

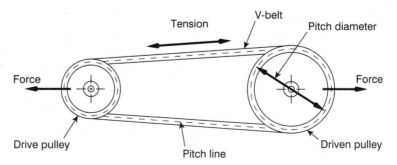

prevent slipping. The slanted sides of the belt are forced into the slanted sides of the pulley by the tension created when the drive and driven pulleys are forced apart. Pulley and v-groove belt systems are used for many electric motor driven devices, such as fans and pumps. However, they are not used as often in robot drive systems because of the possibility of slippage between the belt and pulley. Slippage is especially likely when the direction of rotation of the drive pulley changes frequently.

The relationship between the output and input of a system, like the v-groove belt drive in Figure 2–24, is called the *system transfer function*. Therefore, a transfer function for any system is defined as the change in system output divided by the change in system input. The units for the transfer function are dictated by the variables used in the numerator and denominator. For example, the pulley system in Figure 2–24 has an angular rotation input at the drive pulley and a resulting angular rotation at the output or driven pulley. The following example illustrates this concept.

Example 2–3

The drive pulley for the system in Figure 2–24 has a pitch diameter (the diameter at the center line of the belt) of 5 in. (D_I) and a pitch diameter of 8 in. (D_O) for the output pulley. Determine the transfer function (TR) for the pulley system and the output revolutions per minute (rpm) when the input is turning at 10 rpm.

Solution

a) Find the number of output pulley revolutions for one revolution of the input pulley.

$$1 \text{ rev. of input} = \pi \times D_I \text{ in. belt movement}$$
$$= 5\pi \text{ in. belt movement}$$
$$1 \text{ rev. of output} = \pi \times D_O$$
$$= 8\pi \text{ in. belt movement}$$

The output pulley turns $5\pi/8\pi$ of a revolution for each revolution of the input pulley.

$$TR = \frac{\text{Change in output rev.}}{\text{Change in input rev.}}$$

$$= \frac{\frac{5}{8} \text{ rev.}}{1 \text{ rev.}} = \frac{5}{8}$$

b) Find the output rpm for an input of 10 rpm.

$$TR = \frac{\Delta \text{ output rpm}}{\Delta \text{ input rpm}}$$

$$\text{Output rpm} = TR \times \text{input rpm}$$

$$= \frac{5}{8} \times 10 \text{ rpm} = 6.3 \text{ rpm}$$

Example 2-4

Determine the torque at the output for the pulley system in Example 2-3 when the input moment of torque is 15 in.-lb. How could the transfer function be used in this calculation?

Solution

a) Find the force on the belt at input and transfer the force to the output to calculate the equivalent torque.

$$F_I = \frac{torque}{L_I}$$

$$L_I = \frac{D_I}{2} \text{ in.} = \frac{5}{2} \text{ in.} = 2.5 \text{ in.}$$

$$F_I = \frac{15 \text{ in.-lb}}{2.5 \text{ in.}} = 6 \text{ lbs}$$

$$F_O = F_I$$

$$= 6 \text{ lb}$$

$$Torque = F_O \times L_O$$

$$L_O = \frac{D_O}{2} \text{ in.} = \frac{8}{2} \text{ in.} = 4 \text{ in.}$$

$$Torque = 6 \text{ lb} \times 4 \text{ in.}$$

$$= 24 \text{ in.-lbs}$$

b) We can show that the *TR* for torque is

$$TR = \frac{\Delta \text{ torque in}}{\Delta \text{ torque out}}$$

$$Torque \text{ out} = \frac{15 \text{ in.-lb.}}{\frac{8}{5}} = 24 \text{ in.-lbs}$$

The two previous examples illustrate the relationship between the rotational velocity and torque in every pulley system. If the output pulley has a lower rpm than the input pulley, then the output will have an increase in torque.

A second type of belt is the *flat belt*. It is like the v-groove except that the sides are normal to the belt surfaces and not slanted. The operation of a pulley system with a flat belt is identical to the v-groove; however, flat belts have a higher likelihood of slippage.

A third type of belt, called the *synchronous belt*, is illustrated in Figures 2-22 and 2-25. The belt has evenly spaced teeth on the interior surface that fit into grooves in the pulley. This type of belt is often called a *timing belt* because it is used in smaller automotive engines to synchronize the timing between valve opening and piston position. The motor pulleys, synchronous drive belts, and drive shaft pulleys in Figure 2-22 are used to transfer wrist motor rotation to the wrist axes.

Figure 2–25 Synchronous Belt Drive.

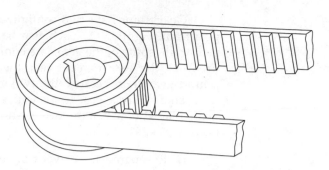

Synchronous belt drives are frequently used in robots because the positive contact and locking action of the teeth prevent belt slippage even under quick direction changes in the drive pulley.

Chains

Chain drive system operation is the same as the belt drives just described, but the belt and pulley are replaced with a chain and sprocket.

Gear Drives

Automation drive systems use several different types of gear geometries that include *spur gears, worm gears,* and *bevel gears.* Spur gears (Figure 2–26) get their name from the gear teeth that stick out like spurs from the gear surface. Spur gears are used in *gear trains* (Figures 2–21 and 2–26). Gear trains are a combination of two or more gears that produce the correct change of force and produce the correct direction of rotation. Every gear system has a gear called the *drive* or *driver* gear and an output gear called the *driven* gear. Some gear trains include an idler gear which only changes the direction of rotation at the output but has no affect on speed or Torque. In the gear train illustrated in Figure 2–21 the drive gear marked with a B and the driven gears are marked A and C. In the gear train illustrated in Figure 2–26, the driver, idler, and driven gears are marked.

Figure 2–26 Spur Gear Train.

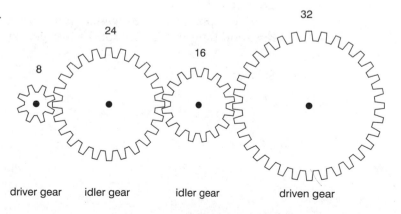

Gear trains have two configurations called *simple* and *compound*. In a simple gear train (Figure 2–26) each gear has its own shaft for rotation. In the compound gear train, some of the gears are connected to the same shaft. For example, in Figure 2–21, the gear train between the B drive gear and the C driven gear is a compound gear train because the two gears between B and C are connected to the same shaft. Simple gear trains with more than two gears (Figure 2–26) have idler gears placed between the driver and driven gears. Idler gears serve several functions:

■ To separate the driver from the driven gear
■ To reverse the direction between the input and output of the gear train
■ To provide power take-off points between the input and output

The most important parameter of a gear train is the *gear ratio* (GR). It is defined as follows for a simple (two-gear) gear train:

$$GR = \frac{N_O}{N_I}$$

where
N_O = number of teeth on the driven gear
N_I = number of teeth on the driver gear

The transfer function for a simple gear train is defined as follows:

$$TR = \frac{V_O}{V_I} = \frac{\theta_O}{\theta_I} = \frac{1}{GR}$$

where
V_O = rpm of driven gear
V_I = rpm of driver gear
θ_O = angular rotation of driven gear
θ_I = angular rotation of driver gear
GR = gear ratio

Example 2–5
A driven spur gear has 34 teeth and the driver has 6 teeth. Find the gear ratio and transfer function for the two-gear system.

Solution
a) $N_O = 34$, $N_I = 6$

$$GR = \frac{N_O}{N_I}$$

$$= \frac{34}{6} = 5.67$$

b) $TR = \dfrac{1}{GR}$

$\quad = \dfrac{1}{5.67} = 0.18$

Example 2–6

Find the gear ratio and transfer function for the gear train in Figure 2–26.

Solution

a) $N_1 = 8, N_2 = 24, N_3 = 16, N_4 = 32$

$$GR = \frac{\cancel{N_2}}{N_1} \times \frac{\cancel{N_3}}{\cancel{N_2}} \times \frac{N_4}{\cancel{N_3}} = \frac{N_4}{N_1}$$

$$= \frac{32}{8} = 4$$

b) $TR = \dfrac{1}{GR}$

$\quad = \dfrac{1}{4} = 0.25$

The previous example illustrates that the gear ratio for a simple gear system with more than two gears is the product of all of the gear ratios. Note that the idler gears cancel out of the final result, and the final gear ratio is determined by the driver and driven gears.

Example 2–7

Determine the gear ratio and transfer function for the gear train between gear B and gear C in Figure 2–21. Assume that gear B has 10 teeth (N_B), C has 24 (N_C), the large compound gear has 35 (N_{CL}), and the small compound gear has 8 (N_{CS}).

Solution

a) $N_B = 10, N_{CL} = 35, N_{CS} = 8, N_C = 24$

$$GR = \frac{N_{CL}}{N_B} \times \frac{N_C}{N_{CS}}$$

$$= \frac{35}{10} \times \frac{24}{8} = 10.5$$

b) $TR = \dfrac{1}{GR}$

$\quad = \dfrac{1}{10.5} = 0.10$

Figure 2-27 Worm Gear.

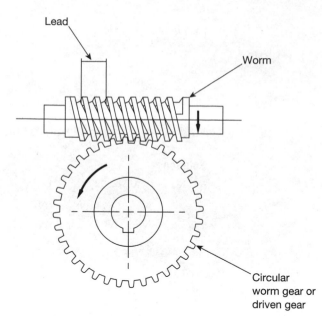

The results of Example 2–7 indicate that the compound gear affects the output and does not cancel out like the idler gears do in Example 2–6. Study these two examples until you see the difference between simple and compound gear trains.

The *worm gear,* pictured in Figure 2–27, is a second type of gear used in some automation machines. The worm gear has two components: the pinion (called the worm gear) and the driven gear. The worm gear is used to convert rotational motion in one axis to rotational motion in an axis rotated by 90 degrees.

Bevel gears (Figure 2–28) are often used in robot wrist mechanisms to transfer energy to an axis that must be rotated by 45 or 90 degrees. Bevel gears are classified into the following four categories: *straight-tooth, zerol, spiral-tooth,* and *hypoid.* The gear ratio for bevel gears used in robotics is usually 1:1 with a 90-degree shaft angle. Each type offers an application advantage. For example, the

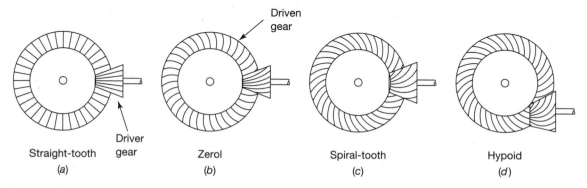

Figure 2-28 Bevel Gears.

zerol and spiral types have their teeth cut on arcs that make the gears capable of withstanding heavy loads. The hypoid offers the same load-carrying advantage but also provides an offset of the drive gear from the center line of the driven gear.

Gear Error

The primary cause of error in gear trains is *backlash*. Backlash is a measure of the movement that can occur in one gear when the meshing gear is held motionless. When a driver gear changes direction, it moves some small distance, called the backlash, before the driven gear starts to move. With precision gear machining and assembly adjustment, much of the backlash can be eliminated. However, some backlash is necessary for adequate lubrication and to eliminate excessive tooth wear. Some anti-backlash gear train designs place two spring-loaded driver gears side by side with springs that force the gears to maintain contact with the front and back tooth surfaces of the mating gear.

Ball-Screw Drives

The *ball screw,* a type of gear drive, is often used in robot drive systems. The ball-screw drive converts rotary motion into linear motion with a high degree of precision and in a relatively small package size. The ball-screw assembly, pictured in Figure 2–29, has a *rotating screw* shaft connected to the drive motor and held in place by screw bearings, a *block* or *nut* that rides on the screw and carries the load, and a *fixed guide shaft* that supports the block. Like a bolt and nut, when the screw (bolt) is rotated in place, the block (nut) moves in a linear fashion along the length of the screw.

Another ball-screw drive design technique frequently implemented uses ball bearings trapped in grooves in the block in place of the machined threads

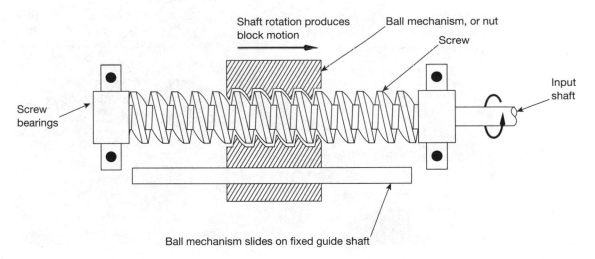

Figure 2–29   Ball-Screw Drive.

shown in Figure 2–29. Since the ball bearings can rotate, the friction between the block and screw are significantly reduced. The ball screw drive on the robot in Figure 2–30 uses the captured ball bearing nut just described, but the action of the ball screw is configured differently. In this application the block or nut mechanism rotates inside a housing and the nut is driven by a belt connected to the servomotor. The screw, which is not permitted to rotate, has one end in the end of the yoke. As a result, when the drive motor turns the ball mechanism or block, the screw forces the yoke to move in a precisely controlled vertical path.

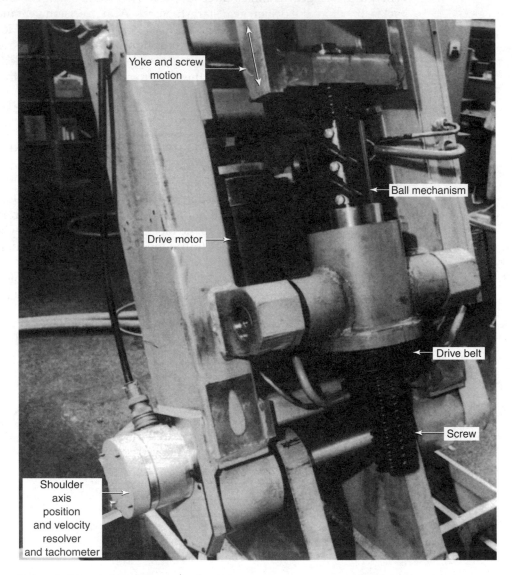

Figure 2–30 Ball Screw Drive Application.

Example 2–8

There are four threads per inch on a 36-in. screw from a ball screw drive. Determine how many turns of the screw are necessary to move the block 27 in.

Solution

$$\text{Distance per rev.} = \frac{1}{4} \text{ in.}$$

$$\text{Total rev.} = \frac{\text{distance moved}}{\text{distance per rev.}}$$

$$= \frac{27 \text{ in.}}{\frac{1}{4} \text{ in./rev.}} = 108 \text{ rev.}$$

Harmonic Drives

The *harmonic drive* is another type of gear drive frequently used in robot drive systems. An end view of a harmonic drive is illustrated in Figure 2–31 with all of the component parts in an exploded view. A picture of a harmonic drive inside a robot arm is shown in Figure 2–32. The harmonic drive contains three components: the *circular spline*, the *flexspline*, and the *wave generator*. The circular spline is a ridge cylinder with gear teeth machined on the inside surface, and it is typically the driven component in the gear system. The flexspline, as the name indicates, is a non-rigid cylindrical metal band with gear teeth machined on the outer surface. The flexspline slips over the elliptically shaped wave generator, and ball bearings are used to hold the flexspline away from the wave generator. The flexspline and

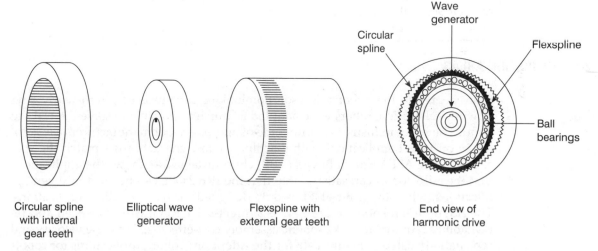

Circular spline with internal gear teeth

Elliptical wave generator

Flexspline with external gear teeth

End view of harmonic drive

Figure 2–31   Harmonic Drive Components.

Figure 2–32 Harmonic drive inside a robot arm.

wave generator assembly is placed inside the circular spline, and the flexspline gears make contact with circular spline gears at just two points 180 degrees apart. The number of teeth in the flexspline is less than the number found in the circular spline. Harmonic drives have no backlash and can provide a wide range of gear ratios in a small package size.

2-5 APPLICATION AREAS

It has been said that robotics is a solution looking for a problem. In the early days of work-cell design, robots were adapted as much as possible to solve the tasks at hand. The robot industry has now come of age, and robot manufacturers are zeroing in on target applications with machines whose characteristics match the specific job. The ASEA Brown Boveri (ABB) chart in Table 2–1 shows the applications that the ABB robots can satisfy. Note that the IRB 2400L can be used in seven applications, while the IRB 6400PE is only designed for poke welding applications. The AdeptOne robot in Figure 2–17 is a direct-drive robot with the speed and repeatability required to make robotic assembly cost-effective. Note the velocity and repeatability listed in Figure 1–16 for the AdeptOne robot. Applications for robots can be divided into two groups: *assembly* and *nonassembly*.

Table 2–1 ABB Robot Applications.

	Arc welding	Glueing/ sealing	Coating/ painting	Cleaning/ spraying	Cutting/ deburring	Grinding/ polishing	Assembly	Spot welding	Poke welding	Material handling	Machine tending	Palletizing packaging
IRB 1400	•	•			•		•			•	•	
IRB 1400H	•	•								•	•	•
IRB 2400L	•	•			•		•			•	•	•
IRB 2400/10	•	•			•		•			•	•	•
IRB 2400/16			•	•	•	•				•	•	•
IRB 4400/45		•	•	•	•	•					•	
IRB 4400/60			•	•	•	•						
IRB 510		•			•					•	•	•
IRB 6400/ 3.0-75						•	•	•		•	•	•
IRB 6400/ 2.4-120							•					
IRB 6400/ 2.4-150							•	•		•	•	•
IRB 6400PE									•			
IRB 6400S								•				
IRB 6400C								•				
IRB 8000L										•	•	
IRB 8000A										•	•	

Assembly

Assembly remains one of the applications most difficult to adapt to robot automation. The key for robot assembly is to start with a product designed for automated assembly. The design process used is called design for manufacturing and assembly (DFMA), and the finished product is easier to make and assemble. This can be a double-edged sword for robotics, however. A product designed for easy assembly is also much easier for humans to assemble. IBM, for example, redesigned its computer line for automated assembly. As a result, IBM found that it could improve profitability by taking many of the robots off the assembly line and replacing them with humans. Robot assembly is most effective when three assembly requirements exist: *high repeatability, small lot-size production,* and *varying part-size requirements.* If the assembly requires close tolerance but high-volume standard parts, then special-purpose high-speed assembly machines, called *fixed automation,* are the machines of choice. A carousel-type fixed automation machine with seven pneumatic robot positioners is illustrated in Figure 2–33. If the lot size is small or the parts to be assembled vary in size, then standard robots are a good choice.

Nonassembly

The primary applications in the nonassembly area have been *welding, spraying, coating, material handling,* and *machine loading* and *unloading.* These will continue to be the major areas for robot automation in the new millennium. The line drawn between assembly and nonassembly robots is certainly not a clear one, however, because assembly robots, like the AdeptOne, have been used in many nonassembly tasks. Nevertheless, robot classification by applications, assembly or nonassembly, is valid because there are tasks at which each machine excels (see Table 2–1).

Figure 2–33 Fixed Automation Assembly System for Radio Components Produced by METO-FER Corp. (Courtesy of METO-FER Corp., Pittsburgh, PA)

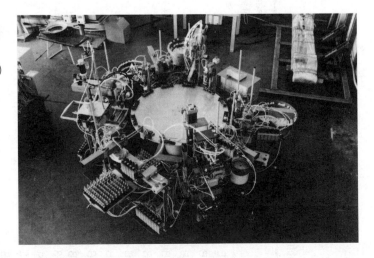

2-6 CONTROL TECHNIQUES

The type of control used to position the tooling separates robots into the categories of *servo* (closed-loop) or *nonservo* (open-loop) systems. An understanding of these two control techniques is fundamental to understanding the operation and programming of robots.

Closed-loop Systems

A closed-loop control system measures and controls the position and velocity of the robot tool center point (TCP) at every point in the robot's motion throughout the work envelope. The closed-loop, or servo, system illustrated in Figure 2–34 uses information provided by sensors attached to the robot's movable axes to measure the current position and velocity of the TCP. Study Figure 2–34 until you find the feedback sensor's signal path. Closed-loop controllers are much more complex and have a statistically higher chance of malfunctioning than open-loop systems; however, the advantages gained in the application of servo-controlled robots to certain manufacturing processes outweigh the problems created by the additional complexity inherent in the system. The advantages include these:

- Servo-controlled robots provide highly repeatable positioning anywhere inside the work envelope. This flexible, multiple-positioning characteristic is necessary for the implementation of robots in many production jobs.
- The servo type of controller with computer capability can provide system control for devices and machines that are external to the robot system. Many work cells require the synchronization of numerous peripheral

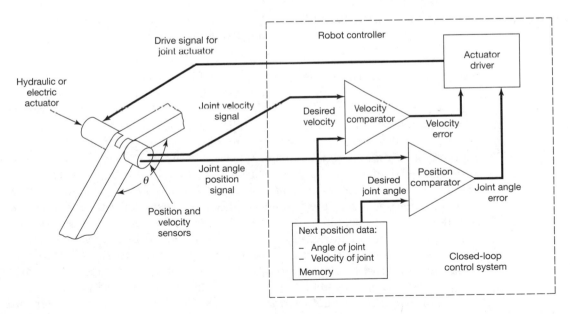

Figure 2–34   Closed-loop System.

devices, such as CNC machines, conveyors, gauges, sensors, and readouts with robot motion. The servo type of controller usually has the input/output capability and the programming power to control the robot arm and the other support equipment in the work cell. (Review Figure 1–15 to see some of the control capability of servo-type controllers.)

■ Powerful programming commands are available to perform complex manufacturing tasks with very simple program steps. For example, placing large numbers of parts in a rectangular pattern on a pallet takes only a few commands.

■ Interfacing robots to other computer-controlled systems, such as vision systems and host computers, is easier to accomplish with a servo-type controller.

Position Sensors

Internal positioning devices mounted on the robot arm mechanically measure the angle of the joints or the linear movement of joints and convert these measurements into proportional electrical signals. In addition, other feedback devices on the arm provide rate-of-change data to be used in controlling the velocity and acceleration/deceleration rates at the tool center point. The control diagram for a robot servo system with position and velocity feedback is illustrated in Figure 2–34.

The three primary positional feedback devices are the *potentiometer*, the *optical encoder*, and the *resolver*. The potentiometer, or pot, is a variable resistor with a linear resistance and a movable wiper. The resolver uses magnetic coupling between transformers to measure rotation, and the encoders use an interrupted beam of light to determine position. In every case, the device is attached to the axis drive motor either directly or through reduction gearing. The rotational displacement of the drive motor and the robot axes is measured by the positional feedback device. A positional and rate-of-change resolver attached to the elbow joint of a robot is pictured in Figure 2–22.

Potentiometers. The potentiometer is usually a single-turn type with gearing for reducing the multiple turns of the motor to a single turn of the potentiometer. The gearing is provided either by the robot manufacturer or is placed inside the potentiometer by its manufacturer, similar to standard ten-turn potentiometers. Figure 2–21 illustrates a simplified system with an ac servo, arm driveshaft, potentiometer gearing, and potentiometer. Assume that the arm axis in Figure 2–21 rotates through 180 degrees for every three turns of the motor, and the potentiometer has a rotation of 200 degrees for that same servo rotation. As a result, a relationship exists between the position of the arm and that of the wiper on the potentiometer. Zero degrees on the arm would equate to zero degrees on the pot, so that the wiper would be in position A in Figure 2–35. The pot output would be zero volts because the wiper is connected to the ground. If the robot arm moves to the 180-degree point, then the pot wiper is rotated through 200 degrees of its 230-degree maximum rotation and would be in position B. The output voltage would be 4.3 volts.

For arm positions between zero and 180 degrees, the potentiometer output voltage would be a value between zero and 4.3 volts.

Figure 2–35 Potentiometer.

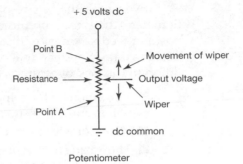

The transfer function of the potentiometer is expressed as follows:

$$TR = \frac{\text{change in output voltage}}{\text{change in shaft position}} = \frac{\Delta V}{\Delta \theta}$$

This equation represents the transfer function for the device because the voltage is the device output, and wiper rotation is the device input. The transfer function with units of volts per degree specifies the resolution for the device. The transfer function could also be expressed in ohms per degree, using the resistance of the potentiometer as the output parameter.

Example 2–9
Calculate the resolution for a potentiometer assuming 230-degrees of rotation for a 0 to 5 volt output.

Solution

$$TR = \frac{\Delta V}{\Delta \theta}$$

$$= \frac{5 \text{ V}}{230 \text{ deg.}} = 0.0217 \text{ volts/deg.}$$

Resolution is 0.0217 volts per degree change in the pot.

Example 2–10
Determine the input rotation required for the potentiometer in Example 2–9 when the output voltage is 3.73 volts.

Solution

$$\Delta V = 3.73 \text{ V and } TR = 0.0217 \text{ V/deg.}$$

$$\Delta \theta = \frac{\Delta V}{TR}$$

$$= \frac{3.73 \text{ V}}{0.0217 \text{ V/deg.}} = 171.89 \text{ deg.}$$

Even with the best resolution and accuracy possible in commercial potentiometers, the error introduced by the pot makes it difficult to use as the primary position feedback device. Instead, potentiometers are sometimes used for coarse position data in systems that combine potentiometers and optical encoders. In addition to the lack of accuracy, the potentiometer has the following shortcomings:

- Movement of the wiper contact against the resistance causes wear, which will eventually result in system failure.
- Potentiometer output is affected by temperature and humidity.
- The analog output of the potentiometer requires analog-to-digital conversion electronics to attain the necessary binary code (digital) for the controller computer.

Optical Encoders. *Absolute* and *incremental* types of optical encoders are frequently used as primary feedback devices to measure robot joint movement. In both types, a light source and light-sensitive receiver are separated by an encoder wheel, as illustrated in Figure 2–36. The beam of light from the source must pass through the encoder wheel to reach the light-sensitive receiver. The encoder in Figure 2–36 uses an opaque wheel with transparent regions, which pass the beam of light, and opaque sections, which interrupt the beam as the wheel is rotated. The disk can also be glass with opaque regions to block the light produced by depositing metal in a precise pattern. The light source is either a light-emitting diode (LED) or a neon or tungsten lamp, which produces greater signal strength as a result of greater light output. The receiver is usually a photodiode or phototransistor, but photo cells have been used in some encoders. The design illustrated in Figure 2–36 is called an *incremental-type* encoder because the sensor only indicates how much the axis has changed from the current position. The change is indicated by a series

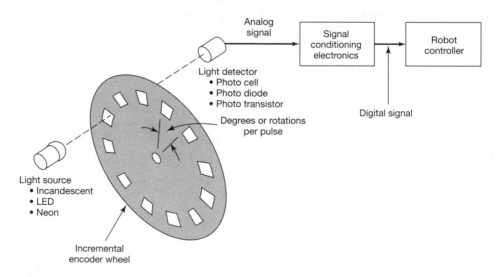

Figure 2–36 Incremental Encoder.

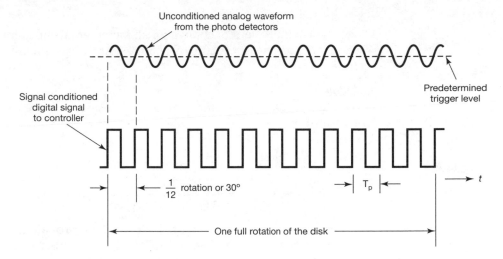

Figure 2–37 Encoder Waveforms.

of pulses produced as the light beam is interrupted. Note that the LED source is separated from the receiver by the disk. The pulses are produced when the beam, which is aligned with the transparent pattern, alternately passes through the wheel and then is blocked as rotation occurs. Each pulse represents the number of degrees or the increment of rotation necessary to produce it. The actual starting position of the encoder shaft is not known; only the number of increments changed from the previous position is provided. The term *incremental encoder* comes from the fact that an exact position is not provided, only the change from the previous position.

The typical pulse output waveforms from this type of encoder are illustrated in Figure 2–37. The waveform on the top in Figure 2–37 is the actual output from the receiver on the encoder, and the waveform on the bottom is the pulse train generated by the encoder electronics. Each time that the raw pulse crosses a predetermined trigger level, a single square pulse is produced electronically.

Example 2–11

The resolution of the incremental encoder is directly proportional to the number of marks (or holes) on (or in) the disk per degree of rotation. As the disk rotates, each hole or mark produces an output pulse, so the units will be pulses per degree. The transfer function for the incremental encoder will also have the units of pulses per degree. Determine the resolution for an incremental encoder disk that has 2400 marks evenly spaced around a disk.

Solution

$$\text{Resolution} = \frac{\text{total pulses}}{360°}$$

$$= \frac{2400 \text{ pulses}}{360°} = 6.67 \text{ pulses/degree}$$

Example 2–12

The smallest increment of change for all incremental encoders is 1 pulse. Determine the smallest rotational change that the system in Example 2–11 can detect.

Solution

The smallest rotational change will be degrees per pulse or the reciprical of the resolution.

$$\text{Smallest } \Delta\theta = \frac{1}{\text{resolution}}$$

$$= \frac{1}{6.67 \text{ pulses/degree}} = 0.15 \text{ degrees/pulse}$$

The incremental encoder in Figure 2–36 does not provide direction information, since the same pulse train (Figure 2–37) will be produced regardless of the direction of rotation. Direction information is included when a second light beam and hole patterns are added to the system. Figure 2–38 shows an incremental encoder with direction sensing added. The two rows of holes are slightly offset so that, for counterclockwise rotation, the first beam is interrupted before the second beam is broken by the inner track. When the disk is rotating clockwise, beam two is broken first, and then beam one. A digital circuit determines which beam is leading the other so that direction information is available. Using the time Tp (pulse period) in Figure 2–37, the controller electronics determine the rate of change so that joint velocity and acceleration/deceleration data are known.

Example 2–13

An encoder with a waveform like Figure 2–37 is turning at a rate of 10 rpm. Determine the time Tp for that rotation rate.

Solution

a) Determine the rotation rate in degrees per second.

$$\text{Degrees per second} = \frac{10 \text{ rev.}}{\text{min}} \times \frac{360 \text{ deg.}}{\text{rev.}} \times \frac{1 \text{ min}}{60 \text{ sec}}$$

$$= 60 \text{ deg. per sec.}$$

Figure 2–38 Incremental Encoder with Direction of Rotation Provided.

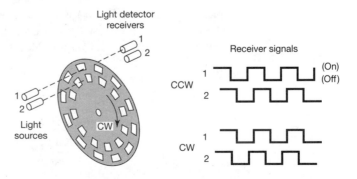

b) Determine circular degrees per Tp interval.

$$\text{Degrees per } Tp \text{ or pulse} = \frac{360 \text{ deg.}}{12 \text{ pulses}} = 30 \text{ deg. per } Tp \text{ or pulse}$$

c) Set up proportional relationship.

$$\frac{30 \text{ deg.}}{Tp} = \frac{60 \text{ deg.}}{\text{sec}}$$

$$Tp = \frac{30 \text{ deg.}}{60 \text{ deg.}} = 0.5 \text{ sec}$$

Example 2–14

If the time Tp for an incremental encoder is 20 microseconds and the resolution of the encoder is 20 pulses per degree, determine the rotational velocity in rpm.

Solution

a) Determine the number of degrees per mark.

$$\text{Degrees per mark} = \frac{1}{\text{resolution}}$$

$$= \frac{1}{20 \text{ pulses/deg.}} = 0.05 \text{ degrees per mark}$$

b) Determine rotation in degrees per second.

$$\frac{x \text{ deg.}}{s} = \frac{0.05 \text{ deg./mark}}{20 \text{ } \mu s/\text{mark}}$$

$$\text{Rotation} = 2500 \text{ degree/second}$$

c) Find rotation in rpm.

$$\frac{\text{rev.}}{\text{min}} = \frac{2500 \text{ deg.}}{s} \times \frac{1 \text{ rev.}}{360 \text{ deg.}} \times \frac{60 \text{ s}}{\text{min}}$$

$$\text{Rotation} = 416.67 \text{ rpm}$$

An *absolute-type* of optical encoder eliminates the primary shortcoming of the incremental system by providing actual rotational position information. This improvement is accomplished by producing a multidigit binary code for every incremental change in the disk position. Figure 2–39 shows a 4-bit absolute encoder. Note that each bit requires its own ring of transparent patterns and a separate light source and receiver for every bit of code or track on the disk.

Example 2–15

Determine the number of degrees of change per code for the absolute encoder in Figure 2–39.

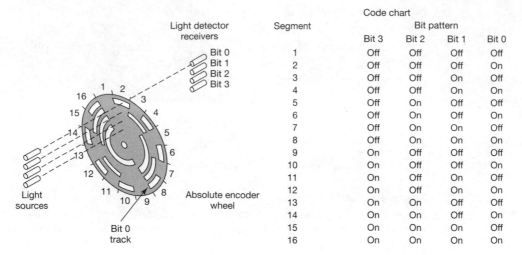

Code chart					
		Bit pattern			
Segment		Bit 3	Bit 2	Bit 1	Bit 0
1		Off	Off	Off	Off
2		Off	Off	Off	On
3		Off	Off	On	Off
4		Off	Off	On	On
5		Off	On	Off	Off
6		Off	On	Off	On
7		Off	On	On	Off
8		Off	On	On	On
9		On	Off	Off	Off
10		On	Off	Off	On
11		On	Off	On	Off
12		On	Off	On	On
13		On	On	Off	Off
14		On	On	Off	On
15		On	On	On	Off
16		On	On	On	On

Figure 2–39   Absolute Encoder with Code.

Solution

a) Determine the resolution for the 4-bit code.

$$\text{Resolution} = \frac{\text{codes}}{360 \text{ deg.}}$$

$$= \frac{16 \text{ codes}}{360 \text{ deg.}} = 0.0444 \text{ codes/deg.}$$

b) The reciprocal of the resolution yields the degrees per code.

$$\text{Degree change per code} = \frac{1}{\text{resolution}}$$

$$= \frac{1}{0.444 \text{ codes/deg.}} = 22.5 \text{ deg./code}$$

As the example problem demonstrates, detection of a small rotational change will require a disk with many tracks and a large number of bits in the code. The code produced by the disk in Figure 2–39 is outlined in the adjacent table. The rotational position of the shaft is indicated by the binary code on/off pattern, with the direction determined by the increasing or decreasing order of the patterns or binary values generated.

Both types of optical encoders are used in robot control systems because the encoders have a number of advantages over the potentiometer described earlier:

■ Greater resolution and accuracy

■ Noncontact measurements, which reduce wear and improve reliability

■ No loss of resolution in the conversion of encoder output to a controller-compatible code

Figure 2–40 Synchro and Resolver Schematics.

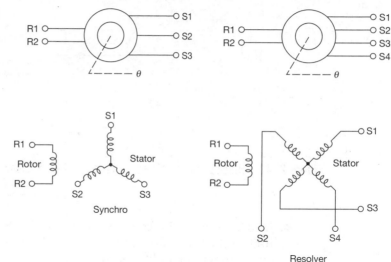

The two disadvantages encoders have when compared with the potentiometer type of positional measuring sensors are higher cost and larger physical size, especially absolute encoders with high resolution.

Resolvers and Synchros. A resolver is another type of device used for internal control of servo-driven robot arms. The word *synchro* is a generic term covering a range of ac electromechanical devices that are basically variable transformers. Figure 2–40 shows the schematic diagram and wiring of the two basic types frequently used. The synchro has three windings mounted at 120-degree intervals around a stationary stator; the resolver has two windings offset by 90 degrees around its stationary stator. Both types have either a single or a double winding, called the rotor, which rotates inside the stator windings. The degree of magnetic coupling between the rotor and stator windings varies with the shaft angle or amount of rotation present in the rotor. If an ac reference voltage is applied to the rotor winding, the stator windings have an ac output that is a function of the coupling coefficient or shaft angle of the rotor. The amplitude of the ac output present on the stator windings will vary as a function of the shaft angle and will be represented by a sine wave on one output and a cosine wave on the other. To interface the output of the resolver with the digital computer electronics in a robot controller, the analog position data must be converted into binary code. The three widely used methods of conversion are (1) time-phase shift, (2) sample and hold, and (3) tracking. The basic block diagram for a resolver-based position encoder is given in Figure 2–41. Resolver position sensors are identified in the robot pictures in Figures 2–22 and 2–30.

A comparison of the resolver-based position encoder to the optical encoder indicates the following advantages for the resolver-based system:

■ The optical encoder is more dependent on mechanical precision to achieve positional accuracy, whereas the resolver relies more on electronic circuits for the same degree of accuracy.

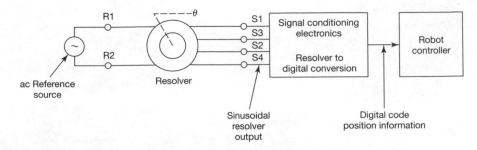

Figure 2–41 Block Diagram of Resolver-based Position System.

■ In the resolver, the conversion electronics are remote from the resolver unit itself. Thus, the electronics unit can receive greater protection from environmental conditions. In the optical encoder, the conversion electronics are inside the unit because signal strength from the optical unit does not permit long transmissions.

■ The resolver-based system is less susceptible to damage because there are no glass disks to break or light sources to burn out.

■ The resolver has smaller physical size at higher resolution values.

■ The resolver requires fewer output wires between the sensor and controller for absolute encoding.

■ The position sensing of the resolver is always absolute.

When compared with the optical encoders, the resolver-based system has the following disadvantages:

■ It is more expensive than an incremental encoded system, especially at higher resolutions.

■ It needs an ac reference voltage.

■ The conversion electronics are more expensive because the conversion unit must have 1 to 2 bits of resolution greater than that needed by the system as a whole.

Most servo-controlled robot systems use either optical or resolver-based positional encoders as feedback devices in the closed-loop servo systems to measure the axis joint angle. In general, optical encoders have greater application at lower resolution values, and resolvers are used when higher resolution is necessary.

Operation.   The operation of the closed-loop circuit was described earlier using the simplified system block diagram in Figure 2–34. However, a full understanding of the operation requires the use of the more complete circuit diagram in Figure 2–42. Study the figure until you are familiar with all the components. Every robot controller implements the closed-loop control differently. For example, some systems use the same sensor to measure both position and velocity, but the

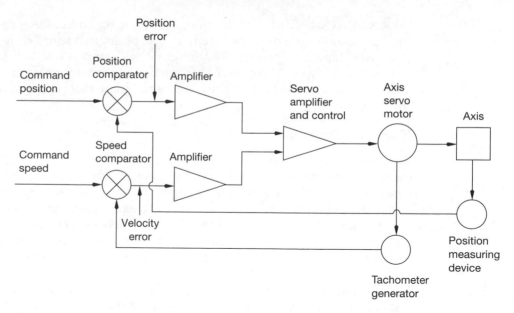

Figure 2–42 Closed-loop Control System.

configuration in Figure 2–42 is adequate for an overview. The feedback system receives position and speed information from the programmed path data. This data is compared with feedback information from the robot axis, and the difference, called the *error*, is amplified. For example, if the desired position data does not agree with the position information from the robot axis sensor, then a position error is generated and amplified. This amplified error voltage is used by the servo controller to apply the necessary voltage to the servomotor and move the axis toward the desired position. A similar process is used to keep the velocity at the programmed level. In the figure, joint angle information comes from a position sensor, which could be any of the devices covered in the last several sections. The velocity sensor could be the tachometer generator illustrated in the figure or velocity data extracted from the position information. Every axis on the robot has a circuit like the one shown to control position, velocity, and in some cases acceleration.

If a machine has 6 degrees of freedom or six axes, then the position of the joint in each degree of freedom is individually measured, and the electrical signals are sent by wire to the controller for analysis by the servo electronics. This sampling of joint positions and corresponding corrective action to the actuator occurs many times a second so that the visual effect is a smooth movement in the robot arm from one programmed position to another.

The servo, or closed-loop, system is used in any application in which path control is required, such as in welding, coating, and assembly operations. Operation of servo-driven robots is often more difficult for cell operators than that of their nonservo counterparts. System start-up involves a series of keystrokes from the keyboard and/or teach pendant, followed by the path programming process that satisfies the application. After testing and debugging the application program in the controller memory, the program is usually saved on disk or magnetic tape as a backup. The program will remain in the robot controller memory as long as the controller battery backup remains charged. The system is ready to execute the current program stored in memory whenever the robot is powered up.

Open-loop Systems

A *nonservo*, or *open-loop*, system (illustrated in Figure 2–43) provides no information to the controller concerning arm position or rate of change of the tooling. As the system drawing indicates, the arm receives only the drive signal, and the controller relies on the reliable operation of each axis actuator to move the arm to the programmed location. The reliability of the open-loop systems is based on the components used in their implementation. Most open-loop systems are pneumatic (Figure 2–4), with programmable logic controllers for control, the actuator is a pneumatic cylinder, and the driver is a pneumatic valve (Figure 2–20). All these components have a history of high reliability and long mean-time-between-failure (MTBF). Open-loop systems account for more than half of current robot applications in the United States and nearly 65 percent of those abroad. There are several advantages of an open-loop system:

- Lower initial investment in robot hardware
- Well-established controller technology in the form of programmable logic controllers
- Less sophisticated mechanical and electronic systems with fewer maintenance and service requirements

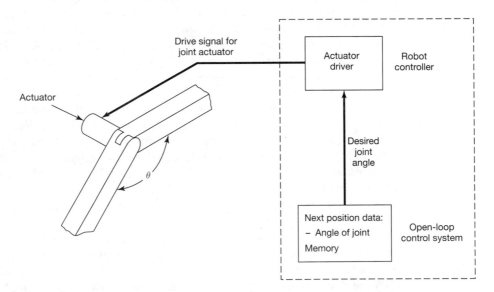

Figure 2–43 Open-loop System.

■ A larger pool of engineers and technicians familiar with the requirements of this type of system

The control of nonservo robots includes internal positioning mechanisms for establishing accurate positioning and external drive devices for ensuring the required sequence of movements.

Internal Positioning Devices. The elimination of feedback information on arm position in open-loop control requires that *internal positioning mechanisms* be used to position the robot arm accurately. The most frequently used position mechanisms for nonservo robots include the following four:

1. *Fixed hard stops* limit the movement of pneumatic or hydraulic actuators at each end of their travel. These stops can either coincide with the natural extension length of the actuator or can be added to the actuator to limit travel. The ends of the actuator cylinder illustrated in Figure 2–20 are excellent examples of hard stops. The motion of the cylinder stops when the piston reaches the end of the cylinder.

2. *Adjustable hard stops* limit the movement of pneumatic or hydraulic actuators at each end of their travel. These stops are identical to the fixed hard stops except that they are adjustable over part of the length of the actuator extension. Figure 2–44 shows a threaded shaft on a Mack modular robot used to adjust the stop on the Y transporter. The threaded adjustment screw positions the movable part of the Y transporter away from the hard end stop. Figure 2–45 illustrates how two different Z axis stops or positions are produced on a Festo pneumatic robot using an additional pneumatic cylinder. The multiple-stop cylinder has a white nylon stop on the end of the cylinder rod. When the multiple-stop cylinder is extended as shown in the figure, the Z axis slide block

Figure 2–44 Adjustable Hard Stops on B-A-S-E Robot.
(Courtesy of Mack Corporation)

Hard stop

Z axis
Slide block

Threaded stop

Figure 2–45 Limit Sensing and Multiple Stops on Festo Robot.

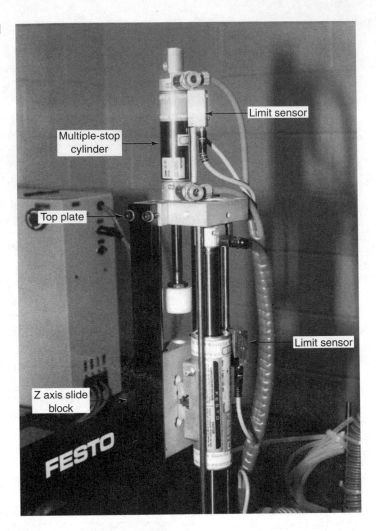

cannot move as high. When the multiple-stop cylinder is retracted (the white stop in Figure 2–45 is against the top plate), the Z axis slide block moves to the stop block.

3. *Limit switches* produce variable stops on nonservo actuators. The robot pictured in Figure 2–46 shows the limit switches used to produce variable degrees of rotation of the arm on the base. However, limit switches provide limited positioning accuracy, and they should be used only when precision stops are not required.

4. *Stepper motors* produce a fixed degree of rotation based on the number of pulses applied to the motor windings. Stepper motors are not often used on industrial robots, but they are found in peripheral hardware, such as positioning tables.

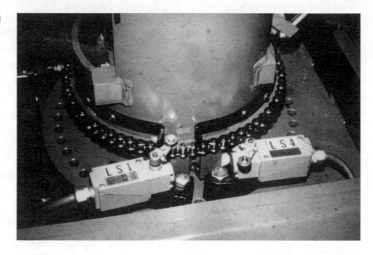

Figure 2–46 Multiple Stops on Prab Nonservo Feedback System. (Courtesy of Prab Robots, Inc.)

The rugged nature of the internal position control devices assures long operating life and high mean-time-between-failure (MTBF) for these open-loop control systems. The pneumatic systems require only clean dry air, line filters, and an oiler to provide trouble-free performance. Some pneumatic systems are prelubricated for life so that oilers are not necessary.

Of course, hydraulic systems require additional maintenance as a result of the pump and tank, which are part of all high-pressure oil systems. With regular filter changes and chemical tests of the hydraulic fluid, however, the open-loop control hydraulic robots provide a good MTBF.

Operation.   The operation of nonservo robot systems is similar to that of the servo type except that the number of programmed points is limited to the fixed or variable stops present on each actuator. In addition, the controller does not know the position of the arm and tool while the robot is moving from one point to another. On every axis, however, there is a *fixed stop,* or *limit,* at each extreme of travel that provides the positioning accuracy at that point. The joint will stop moving only when it reaches one of the two possible extremes of travel. A signal from the controller causes the power source to drive each axis actuator from the present limit or stop to the opposite limit.

In a technique known as *limit sensing,* a sensor is placed at the limit of arm travel to verify that the new position or arm limit has been reached. For example, if a nonservo robot is used to unload a die-casting machine, the system design would require verification that the robot gripper and the part held in the gripper are both clear of the dies before the die-casting machine would be cycled and the dies would be closed. Therefore, limit sensing would be required on the axes used to move away from the die-cast machine. After the controller drives the arm away from the machine, a verification from the limit switch signals the controller to continue in the program. A limit sensor, visible on the Festo robot in Figure 2–45, detects when the multiple-stop cylinder piston is in the fully retracted position. A second limit sensor is used to detect the position of the Z axis actuator. Although limit sensing is used in some applications, many robots use the integrity of the mechanical system and the time allowed to reach the new position to assure accurate operation.

The open-loop system is referred to in the industry as a *stop-to-stop* or *pick-and-place robot*. A pneumatically powered modular component system is pictured in Figure 2–4. The modular system is assembled by the user from a wide selection of linear and rotary actuators. The advantages of the nonservo robot are the simplicity of the system and the resulting reliability. Pneumatic robots or pneumatic-actuated devices are used successfully in every manufacturing area. The disadvantages for the open-loop control stop-to-stop type are its limited number of fixed-stop programmed locations and its inability to handle complex manufacturing tasks. Despite any disadvantages, the low initial cost makes this system an attractive choice for machine load-and-unload applications.

2-7 PATH CONTROL

The least ambiguous method for classifying robots is based on the type of path control that the controller provides. Path control is a way of defining the method that the robot controller uses to guide the tooling through the many points in the desired arm trajectory or program. The four types of path control from least complex to most complex are *stop-to-stop, continuous, point-to-point,* and *controlled path*. Some segments of the robotics industry identify stop-to-stop machines as point-to-point without servo feedback; however, except for this one deviation the industry is consistent.

Before describing the four types of path control, a clear understanding of how the controller stores the program points is important. The programmed points in a robot program are called *translations* or *position points*. Figure 2–47 shows a simplified robot with 4 degrees of freedom programmed to grasp a part. The robot is shown in the three positions through which it must pass to complete the pickup of the part. Point 1 is the starting point; point 2, often called the approach point, positions the gripper over the part; and point 3, called the pickup point, puts the gripper in a position where the parallel jaws can close to secure the part. For this operation the three *translation* points represent the minimum number of program steps that would have to be used. The robot operating system would record in the controller memory the necessary information about each arm joint to permit the robot to return to these three points as the program is executed. The minimum information necessary to permit the robot to return to any point in its work envelope is the position of each degree of freedom or of each axis when the robot is at the programmed point. In the case of joints that move through rotation, the angle is the critical value, and in the case of linear movement the distance extended would have to be preserved. The actual data stored in memory for each programmed point vary, depending on the type of path control present and the vendor's design. The data take three forms: joint angles in angular units, Cartesian space coordinate values, or pneumatic valve states, such as *on* or *off*.

Most often continuous-path-type machines store the actual joint angles for each programmed point. Figure 2–48 shows a portion of the controller memory

Figure 2–47 Point-to-point Path
Control.

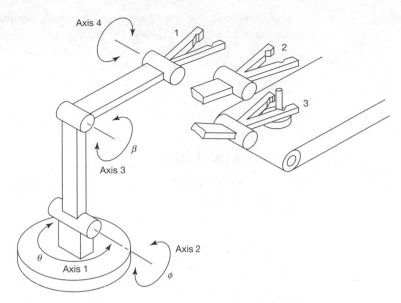

of a continuous-path system with joint angles for three recorded points. The angular values for each joint or axis are saved in a memory location as a binary value. For example, memory location 1003 would have the 90 degree angle saved as a binary number. For point-to-point servo machines, the system stores the Cartesian

Figure 2–48 Segment of Robot
Controller Memory.

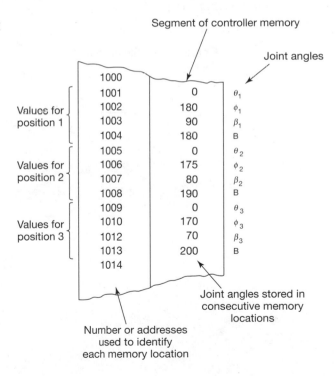

space coordinates of the programmed points. For example, the X, Y, Z, A, B, and C values would be saved. The coordinate values are used to calculate joint angles or actuator extensions during program execution. In actual practice the information or numerical values in memory would be recorded as binary (1s and 0s) numbers. Therefore, regardless of the path control in use by a robot system, the problem is reduced to recording the critical information about every location in the work envelope that the robot must pass through in the process of executing a series of moves. This basic process applies to each type of path control discussed in this chapter.

Stop-to-stop Path Control

In *stop-to-stop path control* the robot system operates open-loop, which means that when the axis moves, the position and velocity of the axis or tooling is not known to the controller. An example will demonstrate this. The robot in Figure 2–49 must go from point A to point B as part of a programmed move. Because the motion of the actuators is not sampled with a feedback system, the actual position of the axis is not known until all of the actuators are driven to the desired limit or mechanical stops in the actuator cylinders. As a result, the only information stored in memory is a sequential list of *on/off* commands for each actuator driver, or pneumatic valve in the case of air-driven actuators. The memory would have *on* for actuator one, followed by *on* for actuator two.

Point-to-point Control

The primary programming device in *point-to-point* controllers is the *handheld teach pendant*. The controls on the pendant for a servo robot controller are shown in

Figure 2–49 Stop-to-stop Path Control.

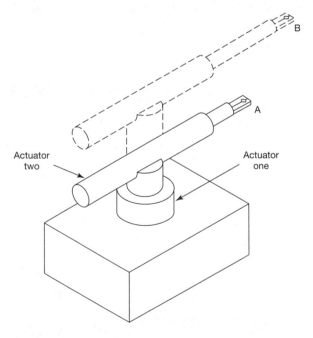

Figure 1–14. Teach pendants have two switches for each degree of freedom on the robot. The programmer can move each axis independently in either direction with these controls. In addition, the pendant provides a programming button that commands the controller to record in memory the current position of every joint or axis on the robot. A button is also provided for emergency stop and single stepping through a program. These functions are common among the teach pendants of most manufacturers; often, additional control switches are provided for special functions of a machine, such as programming and execution speed, gripper open and close, and programming mode. Several programming modes are often supported including:

- *Joint* or *axis*—Two buttons for each axis allow individual control of each degree of freedom.
- *Cartesian*—Two buttons cause the tooling to move in the positive and negative x direction, two cause movement in the y direction, and two in the z direction. The orientation joints are moved with three other sets of switches.
- *Tool*—Three sets of switches cause the arm to move along the x, y, and z coordinates associated with the tooling Cartesian coordinate system.

The tooling or gripper is moved into position by the human programmer using the teach pendant. Then the robot controller is commanded to record in memory the coordinate values for the position of the arm and tooling. This information represents one point in a programmed move. The same procedure is followed at the next desired point in the work cell. For every programmed point the X, Y, Z, A, B, and C values are recorded by the system when the record button is depressed. This process is repeated until all the desired positions of the end-of-arm tooling have been stored in the controller memory in the form of Cartesian space coordinates. It is important to note that the path that the robot takes as the programmer moves the tool from one desired position to another will have no effect on the final program path when the program is run. Figure 2–50 illustrates this concept with the positions that would be required for a robot controller to move a part from one conveyer to another. The following four points would be required: pick up the part at point 1; raise it to point 2; transfer it to point 3; and lower it to point 4. The solid lines represent the desired path of movement when the program is executed by the controller. The dotted line represents the path that the part might have taken when the programmer was teaching the robot the desired operation. *The translation points marked 1 to 4 are the only points that were recorded in the controller memory by the programmer*, so the deviation from the desired path as the programmer moved the tool from points 1 to 4 did not affect the final operation.

Point-to-point controllers can store thousands of program points for a complex application and can also store several different programs; thus a robot can adapt its movement to different manufactured parts or conditions in the work cell. For example, in an inspection operation the controller could drive the gripper through a programmed routine that moves parts through a laser gauge and onto an output conveyer. As soon as a part fails inspection, however, the process must

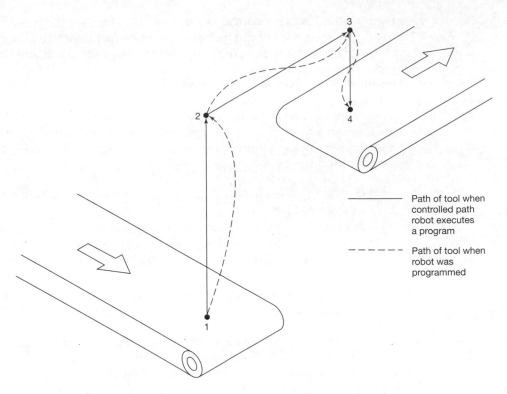

Figure 2–50   Point-to-point Programming Example.

Path of tool when controlled path robot executes a program

Path of tool when robot was programmed

be altered to drop the part into a rework bin. This is accomplished by activating a second stored program that has the set of points for the path to the rework bin. This process is called *branching*.

The point-to-point-type system provides feedback control of each axis over the entire path, with the controller driving each axis from its starting angle or position to the axis angle required for the next programmed point along the path. Figure 2–51 shows a symbolic spherical robot moving from point 1 to point 2, where the first and second points are the same distance from the robot, but point 2 is higher. The controller must change 4 degrees of freedom in this example. The base must rotate through angle θ; the shoulder must go from angle $\phi1$ to angle $\phi2$; the arm must extend along Z to reach the higher point; and the wrist must pitch from angle $\gamma1$ to $\gamma2$. The point-to-point controller will generally change each axis at its maximum rate; therefore, the shoulder and pitch actuators, with the least change, will complete the move well in advance of the base actuator, which requires the greatest travel distance. This causes the end effector to assume the non-straight-line path indicated by the dotted line in Figure 2–51. This lack of path control presents no problems with robots used in applications such as material handling or machine tending where a straight-line path between programmed points is not required; however, it could cause programming difficulty in arc welding situations in

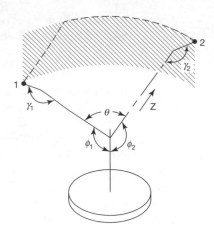

Figure 2–51 Example of Point-to-point Non-straight-line Motion.

which the robot is required to follow a seam between two metal parts. To achieve straight-line motion with a point-to-point controller in the welding example, many points must be programmed very close together to keep the changes in every axis small and about equal.

The advantage of point-to-point type control is that relatively large and complex programs can be obtained with a system that is moderate in cost yet has proven reliability. Although there are many applications in which path control is not a requirement, the lack of straight-line control does limit the number of applications of a point-to-point system and would thus be considered its primary disadvantage.

There is a good selection of servo-controlled point-to-point robots available with any desired arm geometry; however, the cylindrical type is the most popular. The lifting capacity varies from several kilograms to more than 900 kilograms with repeatabilities as precise as ±0.05 millimeters.

Controlled Path

The *controlled path* robot is a point-to-point system with added capability to provide control of the end effector or TCP as it moves from one program point to another. The system is programmed in the same manner as the standard point-to-point machines with each point in the path recorded using the teach pendant or an off-line system. The difference occurs when the program is executed; the primary distinction is that *straight-line motion* is produced between the programmed points. The axis actuators are driven in a proportional manner, with the axis requiring the most change driven faster than the axis requiring the least change. As a result, each axis motor starts turning at the start of a move and stops turning when the next programmed point is reached. As a result, a straight path is followed between programmed points without any additional programmed points. In addition, the *velocity* between points can be specified in the program, along with special tool moves such as an arc welding weave pattern. All this capability is a result of increased controller intelligence and has an added benefit of enhanced programming features such as program editing, trouble diagnostics, larger memory capability, and added end-effector control.

Continuous Path

The primary difference between point-to-point control described previously and *continuous-path* control is the number of programmed points saved in controller memory and the method used to save them. Study the programmed path in Figure 2–52 to see the difference. The point-to-point robot would require just four programmed points stored in memory to record this arm motion. The continuous-path machine would store hundreds of points for the same arm motion, however. The reason for the large difference in the number of points stored is the method used by each type of system to record a programmed path. In point-to-point programming, the programmer moves the robot to the desired location (points 1 through 4 in Figure 2–52) and presses the program button. A single point in the programmed path is saved. After a sequence of these, the program is complete. The point-to-point controller then drives the robot from one point to the next. By contrast, in continuous-path programming, the programmer teaches the desired

Figure 2–52 Comparison of Point-to-point and Continuous-path Control.

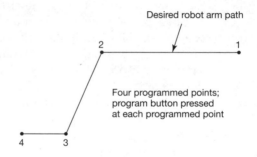

(a)   Point-to-point type control

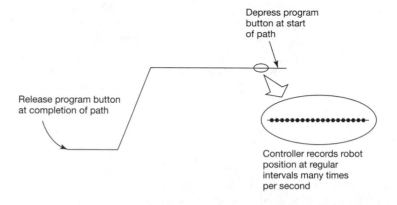

(b)   Continuous-path control

path using a teach stand or by physically moving the robot arm. The program button is depressed at the start of the move and is not released until the desired path is completed. While the program button is depressed, the continuous-path controller records program points in memory at the rate of six or more per second.

The continuous-path programming process records every move that the programmer makes as the arm or teach stand is moved. The point-to-point process, however, records only the location of the arm when the program button is pressed. Any motion in the arm when it is moved from one point to the next is not recorded. Continuous-path controllers are very useful in applications such as paint spraying in which the robot's motion must duplicate the skill of a human paint sprayer.

2-8 DESIGN GUIDELINES

A major step in the design process for a new automation system or work cell that includes robotics is the selection of the robot system. In this type of bottom-up design process, the designer of the work cell(s) has the liberty to set or adjust most of the manufacturing and production variables as part of the design process. As a result, the work cell can be configured: (1) to maximize the impact of automation components on the quality of finished products and the payback for the project; (2) to permit the lowest cost but most suitable robot model to be used in the work cell(s); and (3) to ensure that the likelihood of a successful automation project is close to 100 percent. In the bottom-up design of a new system, the robot is selected based on specifications developed for the overall integration and production requirements of the work cell.

Robot Selection Criteria

Development of this more general work-cell design process is covered later in Chapter 3; however, the robot selection criteria that are part of this cell design process include the following important specifications.

- *Positioning resolution, repeatability, accuracy:* These parameters are established by the product's parts, production requirements, and assembly demands.
- *Work envelope size:* This parameter is set by the size and layout of all the production equipment in the work cell.
- *Arm geometry:* The geometry required is dictated by the cell layout and the type of dexterity required by the production and assembly process.
- *Degrees of freedom:* The number of joints is also dictated by the machine locations and the types of moves the tooling must make during production.
- *Positioning flexibility:* This parameter, set by the production requirements, determines the type of wrist movement required and the number of axes necessary for the wrist.
- *Maximum and rated payload:* The payload or lifting limit for the robot is set by the maximum weight for the parts and gripper.

- *Maximum and payload-dependent velocity:* The robot velocity limits with the gripper loaded and empty are determined from the product cycle time requirements for the work cell.
- *Downward force:* The downward force capability of the robot is set by an analysis of the insertion requirements present in the production process.
- *Compliance requirements:* Robot compliance demands are usually determined from the work-cell parts-assembly analysis; however, in some machine loading applications compliance between the raw material and the fixtures needs to be considered.
- *Tool change requirements:* This specification is determined from a study of all the tooling needed to complete the production and assembly operations in the work cell.
- *Force/torque sensing requirements:* These specifications are usually determined from the work-cell parts-assembly analysis.
- *Programming (on- and off-line):* The flexibility and versatility needed in the robot programming language and operating system is dictated by an analysis of the data integration present in the production system into which the robot work cell will be integrated.
- *Cost:* The maximum cost is an important parameter that is determined in the justification analysis for the system and set by the allowed payback period.
- *Special options:* These requirements are dictated by special needs in the production and assembly process.
- *Vision integration:* This requirement is also dictated by special needs in the production and assembly process and by inspection and quality requirements.

The robot selection process starts with an analysis of the entire production cell. Based on that analysis, some of the robot selection criteria can be ignored and others become critically important. For example, in a machine-tending application, repeatability, work envelope size, arm geometry, degrees of freedom, positioning flexibility, cost, and maximum and rated payload would most likely be the critical parameters. However, if the parts needed to be palletized, then special programming options would also be critical. If cycle times were critical, then the velocity would also have to be checked.

Robot Survey

The designer of automation projects on existing production systems is often not permitted to change manufacturing and production variables. The current production system could be a manual operation that must be automated or an automated cell(s) that requires more current technology for improved productivity or a lower burden cost. Selection of robot automation for this type of automation project requires an additional step in the robot selection process. Before the design can start, a determination must be made on the feasibility of

Table 2–2 Robot Survey.

Robot applications—initial plant survey		
Answer the following questions for each workstation in the plant survey (the reason for the survey question is given in italics after each question.)		
1. Can inspection by operators be eliminated from this workstation? *It is difficult and expensive to include parts inspection in a robot work cell.*	Yes	No
2. Is the shortest machine cycle 3 seconds or longer? *The maximum velocity of servo robots is often a limiting factor. Human operators can work faster than robots when demanded by the process. Fast machine cycles often dictate the use of fixed automation type machines.*	Yes	No
3. Can the robot displace one or two people for three shifts? *If the average robot project costs $100,000, then it will be necessary to eliminate the cost of one or two operators for three shifts to get a 1- or 2-year payback.*	Yes	No
4. Can the parts be delivered in an oriented manner? *Picking parts from a tote bin is easy for humans but very difficult for robots. If the parts can come oriented for easy robot pickup, then robot automation is possible.*	Yes	No
5. Can a maximum of 6 degrees of freedom do the job? *Robots have one arm that moves through a restricted work space compared with the two-armed human. A single-armed robot must be able to do the job.*	Yes	No
6. Can a standard gripper be used or modified to lift the part or parts? *The tooling is a major part of the work-cell expense. The simpler the tooling, the greater the likelihood of a successful project. Also, the weight of the part plus gripper must be consistent with the robot's capability.*	Yes	No

using a robot in the production cell. One quick check to determine if robot automation is an option for a work cell(s) is the robot survey in Table 2-2. This survey is especially useful when robot automation is considered for a manually operated cell.

The survey technique has several steps: (1) usually a team of design or production engineers studies the targeted work cell(s) and completes the six-question survey; (2) the survey results of the team are combined; and (3) the introduction of robots into the work cells is prioritized based on the total number of positive responses awarded. The work cells receiving the maximum number of positive responses have the highest probability of success for a robot implementation. The survey quickly determines if a work cell could use robot automation to effectively improve production.

ISO has issued several standards documents to assist in the collection of valid robot data. The International Standard for Industrial Classification (ISIC) codes help target specific industries for data collection, while standard ISO/TR/8373 provides an extensive list of robotic terms and definitions. In addition, the standard classifies robots into four areas: *sequenced, trajectory, adaptive,* and *teleoperated*. The operation of the robot controller provides the primary difference in the classification categories named in the standard.

Sequenced

The nonservo-controlled pneumatic robot with stop-to-stop path control in either a Cartesian or cylindrical geometry best describes this category. The on/off, or binary, nature of the controller output drives the axes sequentially to well-defined end points. The trajectory or path, however, is not controlled or defined. The most frequently used controller for this category of robots is a programmable logic controller.

Trajectory

This category includes all robot geometries with servo-driven electric or hydraulic axes and controlled-path trajectory operation. Internally generated straight-line motion of the tool-center point between programmed points characterizes this classification.

Adaptive

This new category includes "thinking machines," and there are limited examples of machines in this category. Robots in this classification make heavy use of sensors, adaptive control, and learning functions.

Teleoperated

Teleoperated robots that extend the human sensory-motor functions to remote locations have been used for many years to handle radioactive material. This category includes a new class of teleoperated machines that can be programmed to respond to the operator's actions.

2-10 SUMMARY

The objective of this chapter was to introduce the general concept of robot classification and to provide an overview of all types of robot systems. Industrial robots can be grouped by several techniques that include arm geometry, power source for the joint actuators, intended applications, the presence or absence of feedback, the type of path control, and ISO standards.

The arm geometries currently available are Cartesian, cylindrical, spherical, and articulated. These mechanical configurations can use pneumatic, hydraulic, or electric drives as a source of power for the joints or axes that are moved. Application is not a good classification technique, but generally the areas of painting-coating, arc welding, assembly, and machine tending have received special attention from the robot industry. A much better classification scheme designates a robot system as either servo or nonservo controlled, which means that the system either knows the current position of every axis through the use of feedback signals or it does not sample the current position values and has no feedback. In the closed-loop, or servo, type of robot systems the joint angles and arm extensions are most often measured by potentiometers, optical encoders, or resolvers. The optical encoders used include both the incremental and absolute types. Most robot systems use either optical encoders or resolvers, because the accuracy and reliability of potentiometers are not adequate for industrial applications. Joint position sensors are part of the internal mechanisms that control the motion of the robot arm. In nonservo, or open-loop, systems, the limit of motion is set by mechanical stops on the arm. The internal limit devices are usually fixed hard stops, adjustable stops, or programmed adjustable stops. Another good classification technique uses path control to distinguish between current robots. The path control types include point-to-point, controlled path point-to-point, continuous, and stop-to-stop, with the last type using open-loop control in most situations. Standard ISO/TR/8373 uses the following four categories to classify robots: sequenced, trajectory, adaptive, and teleoperated.

One of the major design decisions in the creation of a robot work cell is the selection of robot hardware, software, and the vendor for the manufacturing operation. A robot selection criterion is provided in the chapter with fifteen of the most critical specifications described. The selection process starts with an analysis of the entire production cell so that this list of critical specifications can be prioritized. The analysis permits the most important specifications to be addressed first and some specifications to be ignored.

QUESTIONS

1. Compare the five basic robot geometries according to the advantages and disadvantages of each arm, the work envelope, typical applications, and power sources.
2. Compare the three types of drive power sources by preparing a table that illustrates the strengths and weaknesses of each.
3. What is the definition of a moment of force?
4. What is the difference between a direct-drive and reduction-drive robot system?
5. Name the five most commonly used reduction-drive systems used in robots.
6. Describe the differences among the v-groove, flat, and synchronous belt drive systems.
7. How do chain drives compare with belt drives?
8. Describe the three gear drive configurations used most frequently in robot drives.

9. What is the difference between a simple and compound gear train?

10. How is the direction of rotation of the driven gear affected by the number of idler gears present in a simple gear train?

11. What is the gear ratio and how does it relate to the transfer function of a gear train?

12. How does a ball-screw drive operate?

13. What characteristic(s) of ball-screw drives makes the use of these drives important in robot applications?

14. Describe the operation of harmonic drives.

15. What characteristic(s) of harmonic drives makes their use important in robot applications?

16. Make a list of the most important machine characteristics that should be incorporated into an assembly robot system.

17. What is the primary difference between a servo and nonservo robot system?

18. What are the three primary sensors used in servo feedback loops, and how do they measure the joint positions?

19. What two parameters are measured at every joint of most servo-controlled robots?

20. What are the advantages associated with nonservo robots?

21. Describe four techniques used on nonservo robot arms to achieve positioning.

22. What are the maintenance requirements on hydraulic and pneumatic nonservo robots?

23. Describe four advantages of servo-controlled robots.

24. Describe four limitations present when potentiometers are used for feedback sensors.

25. Describe how incremental and absolute optical encoders are constructed and how they measure shaft position.

26. Describe how the direction of the rotation is determined in each type of optical encoder.

27. What are the advantages and disadvantages of optical encoders compared with potentiometers?

28. What is a synchro?

29. What is the difference between a synchro and a resolver?

30. How do resolvers operate when used as shaft angle encoders?

31. What are the advantages and disadvantages of resolver position detection compared with optical type encoders?

32. Describe the three program arm-control techniques used in servo robot arms.

33. What types of manufacturing applications would be best served by a nonservo system?

34. What characteristics of manufacturing applications require that the robot used have a closed-loop system?

35. Why does the continuous-path system have a limitation on the length and number of programs that can be stored?

36. Why is the continuous-path system ideal for applications such as spraying and coating?

37. List three typical applications for each of the four types of path control. Try to match job tasks with the strengths of each type of path control.

38. How does the SCARA arm geometry differ from vertical articulated geometry?

PROBLEMS

1. Find the torque in foot-pounds on the shaft in Figure 2–23 when a force of 23 pounds is acting at a distance of 9 inches from the shaft.

2. A servomotor is connected to a forearm axis drive shaft without any reduction drive components. The maximum weight of the part and gripper at the end of the forearm is 55 pounds and the forearm length is 65 centimeters. Determine the motor torque in foot-pounds required to prevent the forearm from rotating under the gripper and part weight.

3. Assume that the lever arm in Examples 2–1 and 2–2 has a weight of 5 pounds and is 24 inches long. Recalculate both examples assuming that the lever arm weight acts in a downward direction at the midpoint of the lever.

4. Gear ratio relates the number of teeth in the driven gear to the number of teeth in the driver gear. Develop a similar concept for pulleys called the pulley ratio. The gear transfer function is the reciprocal of the gear ratio. Does the pulley transfer function have the same relationship with the pulley ratio?

5. The diameter of a v-groove driver pulley is 2 inches and the diameter of a driven pulley is 7 inches. Find the output pulley rpm if the input is turning at 5 rpm.

6. Show mathematically why the ratio of the drive pulley diameter over driven pulley diameter is used as the multiplier in Example 2–4 to find the torque of the system.

7. Calculate the torque in inch-pounds required at a driver pulley with a radius of 5 centimeters when the driven pulley's shaft is delivering 3 foot-pounds of torque to the load. The driven pulley has a diameter of 15 centimeters.

8. Find the gear ratio and transfer function for a simple spur gear system with a driven gear that has 36 teeth and a driver gear that has 9 teeth.

9. If an idler gear with 15 teeth is placed between the two gears in Problem 8, how does that affect the gear ratio, transfer function, and output of the gear system?

10. The gear train in Figure 2–21 has the following spur gear counts: driver gear B has 6 teeth, the large compound gear has 24 teeth, the small compound gear has 8 teeth, and the driven gear C has 36 teeth. Find the transfer function and gear ratio for the system.

11. Assume that the gear train in Problem 10 is placed between the servomotor and forearm drive shaft in Problem 2. Find the torque value for the motor.

12. Calculate the new number of spur gear teeth required for the driven gear in Problem 10 for the transfer function to remain the same when the compound gear is replaced with an idler gear.

13. Determine how far the block or nut of a ball screw drive will travel when the screw is rotated 75 revolutions. The pitch of the block is 6 threads per inch.

14. What rpm is required for the ball screw in Problem 13 to achieve a block velocity of 4 inches per second?

15. Find the transfer function and resolution for a 10-turn, 4.7 kΩ potentiometer with 3450 degrees of rotation and a 0 to 5 volt output. Calculate the resolution in ohms per volt.

16. What is the input rotation for the potentiometer in Problem 15 that will produce an output of 2.65 volts?

17. The B gear in Figure 2–21 has 6 teeth and the C gear has 30. Determine the appropriate number of teeth for the compound gear so that the motor rotates through 20 turns while the pot moves through its full rotation of 0 to 200 degrees.

18. Find the transfer function for an incremental encoder with 1800 marks evenly spaced around the disk. Determine the smallest rotational change in degrees that are permitted.

19. An encoder is used in place of the potentiometer in Problem 17. Determine the compound gear design needed to move the encoder through 360 degrees when the motor rotates through 8 turns. Determine the number of marks needed for the disk so that every 0.8 degree change in the motor shaft produces an encoder pulse.

20. Determine the Tp value for a motor turning at 35 rpm with an incremental encoder that has 500 marks.

21. Find the rotational velocity in rpm of an incremental encoder with a Tp time of 5 milliseconds and a resolution of 15 pulses per degree.

22. If the motor in Problem 19 rotates at 4 rpm, calculate the value of Tp.

23. Find the resolution for a 12-bit absolute encoder.

24. Determine the minimum number of bits that an absolute encoder must have to provide a maximum of 0.5 degrees per code.

PROJECTS AND CASE STUDY PROBLEMS

1. Use primary and secondary research to make a list of robot applications at the companies identified in Project 2 in Chapter 1. (If a single company has a large number of robot applications, just focus on that company.) Include the classification of each robot using the classification techniques (arm geometry, power source for the joint actuators, applications, servo/nonservo, path control) described in this chapter.

2. Use dBASE, Access, or another personal computer database software package to build a robot selection database. Include the following robot data: work

envelope limits, payload, speed, positioning accuracy, repeatability, geometry, program control techniques, actuator power source, special features, and any other parameter required by your instructor. Set up queries that permit searches for robots based on single parameters or combinations of parameters. Use the on-line resources for robot manufacturers listed in Appendixes A and B or other resources provided by your instructor to acquire the data.

3. Use the robot specification resources in Appendixes A and B, along with any other material available, to select four robots from a minimum of three vendors that would work in the die casting production cell illustrated in Figure 2–9. Use the same resources to obtain the approximate size for a 200-ton casting machine and a trim press. Draw a top view of the cell illustrating the footprint of all machines and the work envelope of the robot. Recommend one of the four models for the application. Document all assumptions; indicate which robot selection criteria were used in the selection process; and list all vendor names, robot model numbers, and descriptions.

4. Repeat Project 3 for the assembly cell illustrated in Figures 3–6 and 3–7, but identify a single robot from another vendor. Compare the specifications for the Adept robot used in the case study with the new robot selected as a replacement.

5. A company that manufactures kitchen products wants to add a set of plastic bowls to the product line. The set would have four bowls with diameters of 6, 8, 10, and 12 inches. Design a cell layout that would include a robot to unload finished bowls from a plastic injection molding machine, trim excess mold material from the bowls using a trim press, and place the finished bowls on an exit conveyer. A single mold die is used for all four bowls. Document all assumptions made; indicate which robot selection criteria were used in the selection process; and list vendor names, models, and descriptions for all machines. Draw a top view of the cell illustrating the footprint of all machines and the work envelope of the robot.

6. Use the Tol-O-Matic website in Appendix B or other resources to design a robot similar to the pneumatic system illustrated in Figure 2–4. Assume the following axes movement: X axis (horizontal) 12 inches, Z axis (vertical) 4 inches, and Y axis (in and out) 6 inches. The three-finger gripper has a 180-degree pitch and roll. Document all assumptions made, and list vendor names, models, and descriptions for all components.

7. Form teams of three persons and complete a robot feasibility survey (Table 2–2) for the manual work cells in the production videotape Forging from Genium Publishing Corp. Calculate a team average for each cell based on the number of positive responses, and prioritize a cell list with the cell most adaptable to robot automation at the top.

8. Develop an initial layout for the automation of the production of slugs (Step 2) in the West-Electric case study described in Chapter 12, Section 12-4. Bar stock is fed through a wraparound induction heater and then into a bar shear where different diameter bars are cut to different lengths. Document all assumptions made; indicate the robot survey results; indicate which robot selection criteria were used in the selection process; select two robot models that would satisfy

automation needs; and list vendor names, models, and descriptions for all machines. Draw a top view of the cell illustrating the footprint of all machines and the work envelope of the robot.

9. Develop an initial layout for the automation of the slug lubrication (Step 3) in the West-Electric case study described in Chapter 12, Section 12-4. Document all assumptions made; indicate the robot survey results; and list vendor names, models, and descriptions for all machines. Draw a top view of the cell illustrating the footprint of all machines and the work envelope of any robots used.

10. Develop a plan that would integrate the automated work cells designed in Projects 8 and 9 into a single automated cell.

11. Read the West-Electric case study (Sections 12-1 through 12-4 in Chapter 12) where robot selection is addressed. Use the Thomas Register link from Appendix B to select three robots from three different vendors for the upset forging cell design. Document all assumptions made and list all the devices selected, including vendor name, device name, description, and model/part number. Complete the questions and problems at the end of Chapter 12 that relate to Sections 12-1 through 12-4.

Automated Work Cells and CIM Systems

CHAPTER GOALS AND OBJECTIVES

The primary goal of this chapter is to step back from the specific robot information provided in the last chapter and take a look at the big picture called manufacturing automation. All automation, especially robots, must be integrated into the manufacturing system and linked to all elements of the enterprise. To reach this goal the concepts of computer integrated manufacturing (CIM) are introduced along with the ideas of flexible and fixed types of automated systems. Finally a design strategy for system automation is provided, together with systems level troubleshooting techniques. At the completion of this chapter you should be able to:

- Describe the three steps used to implement CIM.
- Identify value added and non–value added manufacturing operations.
- Describe and evaluate six manufacturing performance measures.
- Describe the difference between flexible automation and fixed or hard automation.
- Describe the difference between flexible manufacturing cells and systems.
- Apply the work-cell design checklist to the design of an automated system.
- Apply general systems-level troubleshooting techniques to a system block diagram to locate faults in linear, convergent, divergent, feedback, and switched types of signal flows.

3-1 INTRODUCTION

The ultimate goal of the enterprise is to develop an internal strategy that raises manufacturing performance to a level higher than that of the competition. The strategy often includes the development of production work cells with many forms of automation, including robots. In Chapter 1, order-winning criteria were used to identify and rank performance measures most critical for manufacturing success in a given market. To justify the investment in automation, the production cell must support some combination of the following performance measures: lower manufacturing cost, higher product quality, better production control, better customer responsiveness, reduced inventory, greater flexibility, higher productivity, and smaller lot-size production. In the 1970s the focus was on designing production work cells that were islands of automation. These stand-alone production cells were essentially isolated from other automated cells and the rest of the production system. Although the automated cells frequently improved the production performance, they fell short of the performance promised by the automation equipment vendors.

The disappointing performance was often traced to three sources. First, the automated cell was isolated from the rest of the production system. History has proved that complex, highly integrated production systems—called flexible manufacturing systems—are often difficult to maintain and operate; nevertheless, some integration of the production data and products of the automated cell with the rest of the production system is often beneficial. The degree of integration is a function of the end product and the production hardware and software implemented.

The second cause for the low performance of the automated cell resulted from deficiencies in the work-cell design process. The early automated cells were usually manual manufacturing operations that were changed to automated production. In the conversion process, the manual functions were just replaced by the automated equipment. For example, when a robot replaced a human operator who loaded and unloaded parts from a machine, the duplication of the human motion was the major focus of the cell design process. As a result, the automated cell incorporated many of the poor production practices present in the manual cell. The third cause of low performance in early automated production cells was insufficient training and preparation of people to set up and maintain the systems.

A study of the failures in early automation efforts led to the development of the process called computer integrater manufacturing or CIM to support effective integration of enterprise systems. Each company approaches the integration of the production systems, product data, and management of the organization differently. Even though every automation project is unique, agreement exists that a process like CIM should be used by organizations to respond to the external challenges present in the marketplace. The CIM process is effective for projects as small as a single automated cell with a robot and production machine or as large as an automotive assembly plant. This chapter describes the general process for integration of automated systems and the specific implementation process for robotic work cells. A good starting point for our study is the three-step CIM implementation process.

3-2 THE CIM IMPLEMENTATION PROCESS

The implementation of a successful CIM system follows a three-step process that includes *assessment, simplification,* and *implementation.* In the first step of the CIM design process, the organization is studied to determine strengths and weaknesses. After the initial assessment, a simplification process is applied to eliminate all waste from the manufacturing area(s) to be automated. The final step in the CIM design process is the acquisition and implementation of the hardware and software for the production cells or manufacturing systems. A detailed description of the three-step process follows.

Step 1: Assessment of Enterprise Technology, Human Resources, and Systems

Building a work cell or integrated production system that uses automation technology such as robotics requires extensive planning, many months of hard work, and a substantial investment in people, hardware, and software. The first step in the system design process must be assessment, because a thorough understanding of the current operation is critical for success in the simplification and implementation steps. The assessment of the enterprise's current *technology, human resources,* and *systems* includes a study to determine these factors:

- The current level of technology and process sophistication used in manufacturing
- The degree of employee readiness (readiness is a measure of the educational and psychological preparation of the general population of employees) for the adoption of CIM automation across the enterprise
- The reason *why* the production systems function as they do (a knowledge of the fundamental workings and operation of the production systems)

In each case the capability and strengths, along with weaknesses and limitations, are checked and documented.

The assessment process is an internal self-study with a large educational component. The critical nature of education is illustrated by the bar graph in Figure 3–1, based on a survey of 139 CEOs, presidents, and vice presidents of companies planning a CIM implementation. Note that 55 percent of the respondents listed *lack of in-house technical expertise* as a major obstacle to CIM implementation. To overcome this deterrent, the education component for every employee must focus on several elements:

- The necessity for enterprise change to remain competitive in a national and world marketplace
- The need to support a new order in enterprise operations that includes teamwork, total quality, improved productivity, reduced waste, continuous improvement, common databases, and respect and consideration for all ideas regardless of the level from which they are initiated
- The hardware and software necessary to implement a CIM system and the management strategy required to run the system successfully

CIM is not hardware and software; CIM is a way to manage the new technologies for improved market share and profitability. From the start of the implementation, all members of an organization must understand how CIM relates to their jobs; as a result, assessment and education must be first.

Figure 3–1 Obstacles to a CIM Implementation.
(*Source:* COMPUTER INTEGRATED MANU-FACTURING, 2nd ed. by Rehg and Kraebber©, p. 60. Reprinted by permission of Pearson Education, Inc., Upper Saddle River, NJ.)

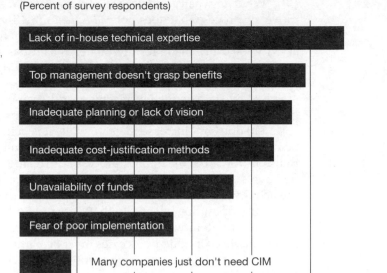

(Percent of survey respondents)

Step 2: Simplification—Elimination of Waste

In Step 1 the enterprise was studied department by department, process by process, and activity by activity to establish the starting benchmark for the current level of technology, employee readiness, and system operation. In addition, a comprehensive effort to educate employees about the company, the need for change, and the technology planned for the future was put into place. From this knowledge base, Step 2 is begun.

In most departments, CIM combines the manual operations and the islands of automation into an enterprise-wide, integrated solution. For example, the manual record-keeping process for tracking the location and quantity of inventory moves to a computerized system with all inventory information stored in an enterprise-wide computer database. If a company installs automation without first simplifying the process by eliminating poorly implemented operations and inefficient inventory processes, it just automates many of the poor practices present in the manual system. In other situations, manual production operations yield to automated cells and production systems that permit greater productivity and improvement in basic performance measures covered in Chapter 1. Building automated systems without eliminating operations that produce waste just automates the waste-production process. Therefore, *simplification* is necessary and is defined as follows:

> *Simplification is a process that removes waste from every operation or activity before that operation is implemented in the CIM solution.*

Automated Work Cells and CIM Systems   **123**

What is waste, and where is it found? Waste is every possible operation, move, or process that does not *add value* to the final product. Value is added if the material or part is worth more after the production process is performed. If an activity log were made for raw material as it passed through manufacturing, the state of the material would be described as moving, waiting in queue, waiting for process setup, being processed, and being inspected. Only one of the five, *being processed* by the production machine, adds value to the part, and that only if the production machine produces good parts. For example, an aluminum block that costs $10 does not increase in value while it is being moved, waiting in a queue, or being inspected. However, when the mill machines the block into a pump housing that sells for $100, the milling process adds value. The remaining states are classi-fied as *cost-added* operations because the part is *not* increasing in value as a result of the activity, and the enterprise continues to pay overhead cost during that time. In short, any operation or activity that does not add value to a product is cost-added and is waste. Removing all waste from operations is often not possible; however, all avoidable cost-added processes must be eliminated.

The search for waste often focuses on direct labor cost. Yet in most manufac-turing operations direct labor usually accounts for only 2 to 10 percent of the cost of goods sold. Therefore, eliminating waste only from direct labor cost would not provide significant savings. However, case studies show that cost-added opera-tions account for as much as 70 percent of all activities in the enterprise. Eliminat-ing as many of the cost-added operations as possible requires significant changes in the daily business operations and in the manufacturing process. The rules in Figure 3–2 offer a process designed to stimulate enterprise change and signifi-cantly reduce the level of cost-added operations. Armed with a comprehensive list of cost-added operations for an area, departments modify the processes so that half of the listed operations are eliminated.

Reducing waste in these significant amounts requires that the enterprise *attack waste fundamentally.* For example, a forklift was used originally to move ma-terial between production machines in a batch manufacturing operation. To reduce the transit time between machines—an identified cost-added activity—manage-ment links the production machines with a material-handling conveyer. Can you recognize the problem with this solution to waste reduction? Transit time is

Figure 3–2   Rules for Elimination of Cost-added Operations.

Three-Step Rule for Eliminating Waste	
Reduction	**Total (%)**
Reduce by 50%	50
Reduce by 50% again	75
Make it 10% of what it originally was	90

A world-class company works the three-step rule to arrive at a 90 percent reduction in waste.

reduced and the process is quicker; however, material is still moved, work-in-process (WIP) inventory is not reduced, and the batch sizes have not changed significantly. The conveyer is a *superficial improvement*. A *fundamental improvement*, elimination of the transportation, is achieved by moving the production machines closer together. Identifying fundamental waste reduction requires that every employee be trained in the fundamentals of the business during Step 1, assessment, in the CIM implementation process.

The reduction of waste by 50 percent in Figure 3–2 is the first step in the process to eliminate 90 percent of the identified cost-added processes. As the figure indicates, the waste still remaining is reduced by another 50 percent, so that the total waste present is only 25 percent of the original amount. The process continues until no more than 10 percent of the original waste is present. At this point, the manufacturing process is sufficiently stripped of unnecessary cost-added operations so that a successful CIM implementation is possible. The war on waste is never over, however, because employees use the *continuous improvement* technique to continue searching for ways to add value and to eliminate cost-added operations.

An aggressive simplification exercise on the shop floor to reduce cost-added activity assures that what gets automated by the CIM hardware and software is not bad production processes. When the simplification process is applied in every office and department in the enterprise, performance improves, and installed automation is effective.

Step 3: Implementation with Performance Measures

The first two steps prepared the enterprise, employees, and manufacturing area for the CIM implementation; Step 3 builds the system. Implementation with *performance measures* implies that the work cells and integrated systems are purchased and installed, but measurements of system performance must be included. In the past, companies made changes in the production system, such as installation of robot-automated work cells, but didn't measure effectively the production performance gains provided by the automated cell. The work cell's productivity must be measured before and after the automation project using the standards developed in Chapter 1. In addition, the efforts toward improvement in production cannot stop with the installation of the automated system. Step 3 focuses on the design and implementation of a system based on the results of Steps 1 and 2; however, the performance of the new system must be compared with past performance.

Implementation with performance measurement has two requirements: (1) measuring the success of the production process before the CIM implementation and at regular intervals afterward, and (2) recording the changes in key manufacturing and business parameters. The objective for the enterprise is to be competitive with the best in the world so that tracking success in world-class performance standards determines the level of improvement in the organization. The six performance standards frequently used for tracking progress include the three standards shown in Figure 1–3 (*setup time, quality,* and *inventory*) and three additional standards (*cycle time, productivity,* and *continuous improvement*). These six key measurement parameters are described next:

1. *Product cycle time:* Product cycle time is the actual amount of time from the release of a manufacturing order to its final completion. For this parameter to be an effective measure of improvement, the setup time, queue time, move and transportation time, run time, and lot size must be included in the total. In some applications the total should also include the time required for completion of the design of the product.

Example 3–1

A product with a lot size of 100 parts requires two work cells to complete the machining operation. Determine the cycle time in minutes for each part using the following data.

First machine: setup time is 2.5 hours per 100 parts; queue, move, and transport time is 0.9 hours per 10 part pallets; and run time is 22 minutes per part. Second machine: setup time is 1.25 hours per 100 parts; queue, move, and transport time is 0.7 hours per 10 part pallets; and run time is 13 minutes for every two parts.

Solution

a) Find time for first machine:

Setup time = 2.5 h (per 100 parts)

$$\text{Queue, move, and transport time} = 9.0\,\text{h}\left(\frac{0.9\,\text{h}}{10\,\text{parts}} \times 100\,\text{parts} = 9\,\text{h per 100 parts}\right)$$

$$\text{Run time} = 37.0\,\text{h}\left(\frac{22\,\text{min}}{\text{part}} \times \frac{1\,\text{h}}{60\,\text{min}} \times 100\,\text{parts} = 37\,\text{h per 100 parts}\right)$$

Total time = 48.5 h

$$\text{Cycle time} = \frac{\text{total production time for lot size}}{\text{lot size}}$$

$$= \frac{48.5\,\text{h}}{100} \times \frac{60\,\text{min}}{\text{h}} = 29.1\,\text{min}$$

b) Find time for second machine:

Setup time = 1.25 h (per 100 parts)

$$\text{Queue, move, and transport time} = 7.0\,\text{h}\left(\frac{0.7\,\text{h}}{10\,\text{parts}} \times 100\,\text{parts} = 7\,\text{h per 100 parts}\right)$$

$$\text{Run time} = 10.8\,\text{h}\left(\frac{13\,\text{min}}{2\,\text{parts}} \times \frac{1\,\text{h}}{60\,\text{min}} \times 100\,\text{parts} = 18.0\,\text{h per 100 parts}\right)$$

Total time = 19.05 h

$$\text{Cycle time} = \frac{19.05\,\text{h}}{100\,\text{parts}} \times \frac{60\,\text{min}}{\text{h}} = 11.43\,\text{min}$$

2. *Inventory:* The inventory is measured as either material resident time (the time raw material or parts spend in manufacturing) or product velocity (the number of inventory turns by product). Inventory cost must be clearly defined if the savings are to be used as a performance measure. Inventory turns are defined by the following expression:

$$\text{Inventory turns} = \frac{\text{annual cost of goods sold}}{\text{average annual inventory investment}}$$

If the cost of goods sold is $200,000 with an average inventory investment for the year of $50,000, then the product has 4 inventory turns. The $50,000 invested in inventory could have been invested in other financial instruments and generated cash returns; therefore, the inventory is a cost to the company. If the inventory turns for the product increased to 10 with the same annual sales, then the average inventory cost would be only $20,000: an annual savings of $30,000 for the company and a greater return on inventory investment. However, increased inventory turns create other costs such as order preparation costs.

3. *Setup times:* The setup time is part of the product cycle time; however, improvement in setup is a key factor in the struggle to become competitive. As a result, most companies use setup time improvement as a measurement parameter apart from the cycle time.

4. *Quality:* Quality is either an order-qualifying or order-winning criterion for most products, so it is no surprise that improvement in this area would be a measurement parameter. Quality takes on many forms; however, the two areas used as primary measurement parameters are *first-time good parts* and *reduction of scrap and rework.* First-time good parts is a measure of how often the first part produced in a production run is within specifications. The reduction of scrap and rework is a measure of production quality in two areas: scrap (unusable production that must be discarded) and rework (parts that are out of tolerance but that can be fixed with additional manufacturing operations).

5. *Employee output/productivity:* This term is a measure of the amount of output of goods and services per unit of input. Companies often use total employee hours worked divided into total units of output in this measurement parameter.

6. *Continuous improvement:* Of all the activities in CIM, it is hard to find one more important than continuous improvement. As a result, some measure of the activity in this area is always used to judge the success of a CIM implementation. In most cases the measurement is just the number of improvement suggestions per employee per week or month.

Initially, all six parameters are measured and recorded to establish a baseline before starting implementation of any automation or process improvement. In enterprises with successful CIM programs, the performance measurements are reviewed monthly to track changes in the performance. During the *initial* CIM implementation, companies often start by measuring and tracking just two or three key parameters. They focus first on the parameter(s) that would provide the greatest initial payback. For example, inventory is often selected first because most manufacturing operations have excess inventory and the dollars saved through reductions are easy to identify. The following example illustrates this type of savings.

Example 3–2

The inventory turns for a product with $2,000,000 in cost of goods changes from 10 to 150 due to improved production operations. What are the annual interest savings if the interest rate on the short-term loans used to finance the inventory is 7 percent?

Solution

a) Find average annual inventory investment (AAII) for both turns.

$$\text{AAII} = \frac{\text{annual cost of goods sold}}{\text{inventory turns}}$$

$$= \frac{\$2,000,000}{10} = \$200,000$$

$$= \frac{\$2,000,000}{150} = \$13,333$$

b) Find interest on each investment.

$$\text{Interest (10 turns)} = \text{annual interest} \times \text{principal}$$
$$= 0.07 \times \$200,000 = \$14,000$$
$$\text{Interest (150 turns)} = 0.07 \times \$13,333 = \$933$$
$$\text{Savings} = \$14,000 - \$933 = \$13,067$$

As control of the process is established, additional performance measures are tracked and improved.

The results listed in Figure 3–3 show the changes achieved over 18 months by a pump manufacturer with $25 million in sales. The remarkable improvements in *floor space* are a result of a shift from job shop (Figure 1–6b) and

Pump manufacturer's performance report card case history (pump housings)			
Measurement parameters	Baseline	12 months	18 months
Cycle time	18 weeks	6 weeks	1 week
Inventory turns	4	8	48
Scrap (percentage of lot size)	13%	5%	0.06%
First time good parts	45%	85%	93%
Floor space	1200 ft²	500 ft²	150 ft²

Figure 3–3 Results of Improved Business Operations.

repetitive production (Figure 1–8) environments to a *structured product-flow* system (Figure 1–6c), which groups machines by product. At the start of the improvement process, the company had two choices: (1) add automation to the existing production system, or (2) look at production alternatives that would improve order-qualifying and order-winning criteria. If a decision had been made to automate the existing process—a job shop manufacturing strategy with large batches or lot sizes of pump housings—then such dramatic results would not have been achieved. Instead, the major sources of waste in the production process were identified and attacked with fundamental changes in the way pump housings were manufactured. As a result, the automation was significantly more effective in improving productivity than would have been possible if the same automation had been added to the old production system. The investment for either approach would be about the same. However, the return on the investment in terms of competitiveness in the marketplace was much higher for the second choice because the company chose to adopt a new production strategy. The success achieved was a result of hard work in the assessment and simplification phase of the CIM system design. Compare the 18-month results with the standards listed in Figure 1–3.

The final step in the CIM implementation process includes the acquisition and installation of the hardware and software to the specifications developed in Steps 1 and 2. When assessment and simplification are completed and a performance measurement system is in place, then a successful implementation of hardware and software is assured. Unfortunately, many companies implement CIM automation beginning with Step 3. As a result, they automate all the disorder of the poor production processes and produce waste at a record rate.

3-3 MAKING THE CIM PROCESS WORK

The three-step process identifies three major tasks that must be performed for a successful CIM implementation. The specific process used in each step is a function of a number of variables. For example, the CIM process used by the pump manufacturer featured in Figure 3–3 was affected by the type of manufacturing systems used, the fact that management and labor had a good working relationship, current competitive position in the marketplace, earlier work in identifying training needs across the company, and experience gained in a small automation project completed previously. The specific process used by companies as they address the implementation of CIM principles will vary depending on the conditions present and the corporate culture. Some of the elements found in successful CIM implementation processes include these:

- Use of a program name different from computer-integrated manufacturing
- Support for the program beginning with the CEO
- Use of multifunctional employee teams at all levels in the process
- A willingness to look at all processes and products for potential productivity gains

- A willingness to accept recommendations for process and product improvement from every employee
- A willingness to accept a 3- to 5-year payback time for the investment

Most companies choose not to identify their project with the CIM acronym because they want a title that is unique to their operations. For example, the pump manufacturer in Figure 3–3 named its project the *Rapid Customer Response System (RCRS)*. The abbreviation avoided the use of the term *CIM* in the project title because the firm believed that the title should communicate the goal of the project to everyone in the company. Nevertheless, a study of the RCRS process indicates that it follows the philosophy and elements defined by CIM.

Several conditions guarantee the failure of a CIM implementation from the start, and lukewarm support from the top management is at the top of the list. The corporation has to change when CIM is implemented, and change will not occur without the support of top management. Other critical elements include the willingness to form teams with representatives from all corporate departments and all employee levels. Equally important is a belief that everyone in the organization has information and suggestions that are critical to a successful CIM implementation.

Most enterprises implementing CIM give the project a unique name and approach the assessment and simplification phase with a technique that fits their unique operations, products, and corporate culture. Regardless of the project name or the technique used to study the corporate operations, the end result is a CIM-based enterprise that is better equipped to compete in global markets.

3-4 AUTOMATED PRODUCTION

Reengineering of the manufacturing operation frequently requires the use of automation to improve production performance and to meet order-qualifying and order-winning criteria. The degree of automation is a function of manufacturing and assembly specifications, labor conditions, and competitive pressure. For example, some production operations and assembly tasks are performed better by human operators, whereas others require the precision, speed, and repeatability of automated machines. Labor costs and work requirements play a role in the degree of automation as well. Automation is necessary on some products to meet the competitive price demanded by the marketplace. In other situations, if human operators are used, the required work motions could cause a health hazard (e.g., carpal tunnel syndrome). In general, manufacturing is using more automation to meet the performance measures demanded by the competition.

The types of automation common in manufacturing are classified in Figure 3–4. The relationship among production capacity, production flexibility, manufacturing systems, and type of automation is illustrated in Figure 3–5. Study the two figures until the automation types and relationships are clear. Note that robots

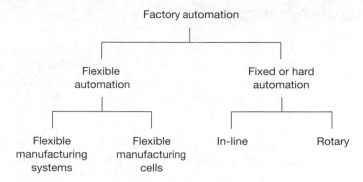

Figure 3–4 Types of Factory Automation.

are frequently used in all manufacturing systems except *project,* where only limited applications occur. Most robot applications occur in either flexible or fixed automation cells and systems, so a description of these areas is a good starting point.

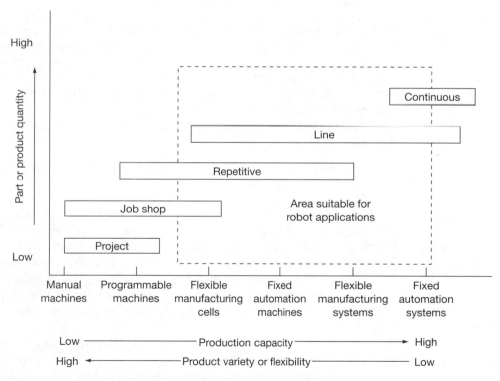

Figure 3–5 Production Systems.

(*Source:* COMPUTER INTEGRATED MANUFACTURING, 2nd ed. by Rehg and Kraebber©, p. 348. Reprinted by permission of Pearson Education, Inc., Upper Saddle River, NJ.)

3-5 FLEXIBLE AUTOMATION

Flexible automation enables agile manufacturing. Agile manufacturing is a measure of how fast a company can respond to changes in order-winning criteria. Flexibility and agility are often order-qualifying or order-winning criteria, and flexibility is a key performance measure listed in Figure 1–3. In manufacturing, flexibility is often described as (1) the ability to adapt to engineering changes in the part, (2) the increase in the number of similar parts produced on the system, (3) the ability to accommodate routing changes that allow a part to be produced on different types of machines, and (4) the ability to change the setup rapidly on the same machine or system from one type of production to another. In each case, the critical point is that properly designed production areas make production planning and scheduling easier and result in quicker response to customer needs. Flexible manufacturing takes two forms: the *flexible manufacturing system (FMS)* and the *flexible manufacturing cell (FMC)*.

An FMS is defined in the *Automation Encyclopedia* (Graham, 1988) as follows:

> *A flexible manufacturing system is one manufacturing machine or multiple machines which are integrated by an automated material handling system, whose operation is managed by a computerized control system.*

The definition implies features such as numerical control (NC) or smart production machine tools, automatic material handling, central computer control, production of various parts in random order, and data collection and enterprise integration. In general, an FMS is an integrated collection of production hardware linked together by computer software. The highly integrated nature of the FMS requires the design and implementation of the system as a unit.

A flexible manufacturing cell (FMC) is defined in the *Automation Encyclopedia* (Graham, 1988) as follows:

> *An FMC is a group of related machines which perform a particular process or step in a larger manufacturing process.*

Based on this definition, an FMC is a group of integrated product machines that perform part of a larger process. The FMC could be viewed as the production building blocks used to assemble an FMS. Figure 3–5 puts FMS applications and FMCs into a manufacturing perspective. Note that a single FMC would be used for medium-capacity production in job shop and repetitive manufacturing systems, whereas an FMS would more commonly be used to build a line-type manufacturing system. However, it is often difficult to differentiate between a small FMS and a large FMC.

For the purposes of this text, the study of flexible manufacturing will focus on two configurations of FMCs: (1) the FMC that is used as an integral part of a larger FMS, and (2) the FMC that results from automating a manual production machine or area. To obtain a clearer picture of the two configurations of FMCs, consider the Application Notes beginning on page 133. Read both cases studies at this time.

Designed by John Brown Automation, the electronic assembly system at Technophone (Figure 3–6) operates 24 hours a day assembling boards used in mobile cellular telephones.

The electronic assembly system is a vital part of an integrated production line that makes 8000 telephones per month.

Figure 3–6 Assembly of Cellular Phone Circuit Boards.
(Courtesy of Adept Technology, Inc.)

By automating its production line, Technophone has not only been able to handle the demand for cellular phones, but it has also reduced labor costs and improved product quality.

System Configuration

Fast, precise printed circuit board assembly is accomplished by an AdeptOne robot with integrated vision. The AdeptOne assembles all of the odd components at a single station, at the rate of one component every four seconds.

With its tool-changing capability, the robot can assemble ten devices on the two different Technophone boards: seven different connectors, a spacing washer, a filter and a coil. The robot has four grippers to choose from; two of the grippers use vacuum to pick up components. The other two are mechanical grippers—one with a flat jaw and the other with a curved jaw.

An AdeptVision XGS-II grayscale vision system is integrated with the robot system. The vision sys-

tem's unique recognition algorithms ensure component placement accuracy to ±0.2 mm.

Adept's patented AIM (Assembly Information Manager) software manages the printed circuit board assembly data—simplifying cell development and operation.

Work Cell Operation

The electronic assembly work cell (Figure 3–7) is one station in a larger production line. Before a board arrives at the Adept station, it has already passed through a screen printer, a solder paste-applicator and an automatic surface mount device placement machine.

The work cell builds boards in batches of 200–500, 24 hours a day. Two operators run the entire line—replenishing component supplies, handling batch changeover, and dealing with complications. All of the assembly programs are held in the robot controller, so batch changeover requires just one simple instruction.

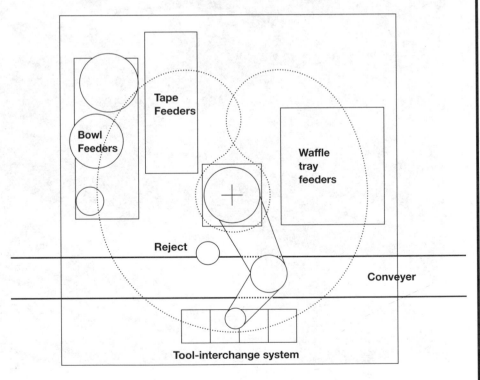

Figure 3–7 Layout of Technophone Flexible Manufacturing Cell.
(Courtesy of Adept Technology, Inc.)

Prior to assembly, the vision system verifies the board type. A camera mounted on the robot arm identifies four fiducial marks on each board, and the robot controller adapts the assembly program for the specific board type, location, and orientation. A collection of tape-feed devices, vibratory bowl feeders, and waffle trays presents the various components to the robot in the correct orientation. After the robot picks up a component, a second camera positioned just below the assembly area inspects the component.

If the part passes the inspection, assembly continues. However, if it is defective—for example, if it is the wrong part or if its leads are skewed—the robot drops it down a reject chute and selects a new component.

Including gripper changeover time, the robot achieves cycle times of 32 seconds for a 7-component board and 18 seconds for a 3-component board.

The system at Technophone can produce 900 parts per hour and has improved manufacturing performance in both quality and cost. The investment in the system was paid back in two and one-half years.

*(Reprinted with permission of Adept Technology, Inc.)

APPLICATION NOTE
Through-Hole Printed Circuit Board Assembly*

To maintain a competitive edge in the marketplace, French electronics manufacturer TRT Thomson needed to replace its labor-intensive board assembly operations with a solution offering higher productivity and better quality.

TRT required a solution that was flexible enough to handle a variety of components and several different board types—because TRT produced volumes as small as 10 boards per month of each design.

The company considered hard automation, but this solution would have required three different stations: one assembling axial components, a second assembling radial components, and a third assembling dual in-line packages (DIPs). Moreover, this solution would have required board-handling between machines.

In the end, TRT Thomson opted to replace its manual production with a flexible automation solution designed by French system integrator SCEMI.

System Configuration
The installation at TRT performs both component preparation and insertion at a single station. An AdeptOne robot with integrated vision works three shifts per day stuffing printed circuit boards at a rate of 800 components per hour. (See Figure 3–8.)

There are three different types of components to be assembled: axial, radial, and DIP. In the cell's feeding equipment, there is enough space for up to 50 reels of axial components, 2 of radial components, and 60 tubes of dual in-line packages.

In total, the robot has to choose among 112 different components. With a compliant wrist and an automatic gripper-change system, the AdeptOne can easily handle the number and variety of components—quickly and accurately.

Work Cell Operation
An automated handling system loads the boards onto the robotic assembly station. A bar code on the side of the board communicates to the robot which assembly program the robot must access.

An AdeptVision system verifies the location of a printed circuit board by "looking" at fiducial marks on the face of the board. An Adept MC robot controller then makes the necessary modifications to the assembly sequence.

With the help of an Adept vision system, the robot prepares and inserts the components into the board. If a component's leads are difficult to insert, the robot will use vision to "look" at the holes the leads should pass through. The robot controller will then modify the assembly program as necessary.

Figure 3–8 Assembly of Printer Circuit Boards.
(Courtesy of Adept Technology, Inc.)

To maintain accuracy over time, the vision system is used to recalibrate the robot's working envelope to within 0.02 of a micrometer.

Once a board has been assembled, it is automatically unloaded from the robotic assembly station.

The station at TRT Thomson requires minimal supervision—operators ensure only that the com-ponent feeders are full and available to sort out any issues. The system increased productivity and quality, and provided a four-year payback.

*(Reprinted with permission of Adept Technology, Inc.)

Both case studies describe similar FMCs used to assemble electronic circuit boards with either surface-mount devices or with conventional electronic components. In both cells, vision was used to check part location; multiple grippers were used to pick a wide variety of part geometries; and the robot was the primary production machine in the work cell. However, the two case studies exhibit one significant difference.

In the Technophone case study, the FMC was part of a larger FMS, with the cellular circuit boards and the production system designed simultaneously. The product was designed to be manufactured by the FMS, which was built for maximum compatibility with the product. In the TRT Thompson case study, manually produced product was converted to automated assembly. The FMC was not part of a larger automated line, and the automation hardware had to work with a product designed for manual assembly. The design process for the FMC in each case was similar; however, the TRT Thompson case, which added automation to a previously manual production system, created some unique problems. A process to automate manual production cells and a discussion of the problems encountered is the focus of an in-depth case study introduced in Chapter 12.

3-6 FIXED AUTOMATION

A fixed automation manufacturing system able to produce a large volume of parts or products is often necessary to meet market demands. An example of this type of product is the disposable razor or lighter. In the automotive industry, fixed automation is used to produce parts like fuel pumps and tires. Automation in this category has the following characteristics:

- A series of closely spaced production stations linked by material-handling devices to move the parts from one machine to the next
- A sequential production process with each station performing one of the process steps, and the cycle time at each machine usually about equal
- A system in which raw material and components are introduced at various points throughout the system, and finished parts or products exit at the end
- The number of stations in the system dictated by the complexity of the production process implemented
- A system that uses high-speed robots—either fast servo robots (Figures 2–5 and 2–17) or pneumatic robots (Figures 2–4 and 2–10)
- A tightly linked production system that produces either one or a few different products

Fixed automation systems are implemented in two configurations: *in-line* and *rotary.*

In-line Fixed Automation

An *in-line fixed automation* system (Figure 1–7) consists of a linear work flow through a series of workstations. Note that the system in the figure is designed to produce two different small products. These two products use the same machines for their first few production operations and then diverge into separate manufacturing paths. In this example, the production system uses thirty-seven pneumatic robots to move the product and parts between assembly machines and production stations.

Although most fixed automation systems are highly automated, it is possible to include manual stations along the line to handle operations that cannot be automated economically.

Rotary-type Fixed Automation

Rotary-type fixed automation locates the production stations around a circular table or dial. As a result, this type of production system is called an *indexing machine* or a *dial index machine*. The system pictured in Figure 2–33 has twelve production stations positioned on the rotary table and uses seven pneumatic robots to support the assembly. In this rotary system and in the in-line system in Figure 1–7, automatic parts feeders supply the pneumatic robots with components in the sequential assembly process. Rotary systems are limited to smaller workpieces and have fewer production stations than the in-line type.

3-7 WORK-CELL DESIGN CHECKLIST

The design of an automated work cell requires a detailed study of the attributes of the cell that influence the quality of the automation solution. The checklist in Table 3–1

Table 3–1   Work-cell Design Checklist.

Performance requirements	
Cycle times	Tolerance of parts
Part-handling specifications	Dwell time of tools
Feed rate of tools	Pressure on tools
Product mix	Maximum repair time
Equipment requirements	Malfunction routines
Human backup requirements	Allowable downtime
Future production requirements	
Layout requirements	
Geometry of the facility	Service availability
Environmental considerations	Floor loading
Accessibility for maintenance	Safety for machines and people
Equipment relocation requirements	
Product characteristics	
Part orientation requirements	Gripper specifications
Surface characteristics	Part size, weight and shape
Unique handling requirements	Inspection requirements
Equipment modifications	
Requirements for unattended operation	Maximum/minimum machine speed
Requirements for increased throughput	Requirements for automatic operation
Process modifications	
Lot-size changes	Routing variations
Process variable evaluation	Process data transfer
System integration	
Data interfaces and networks	Interface requirements
Hardware integration requirements	Software integrating requirements
Data integration requirements	

shows manufacturing variables that should be addressed during the design process. Depending on the function of the cell and type of automation proposed, the variables most critical to a successful design will change.

The list is divided into six major areas: *performance requirements, layout requirements, product characteristics, equipment modifications, process modifications,* and *system integration.* All parameters present in the cell should be addressed by the items in these six categories. To use the list, place the sentence fragment "What is/are the" in front of the checklist element, and place the fragment "in the work cell?" at the end. For example, the questions "What are the *cycle times* for the work cell?" or "What is the *allowable downtime* for the work cell?" illustrate how the checklist elements are used to cover design variables. The data gathered from checklist element questions are comprehensive in scope but contain sufficient detail to make design decisions. Study all six sections, and be sure that you understand what work-cell parameters are covered by each element.

3-8 IMPLEMENTING AUTOMATED WORK CELLS

The implementation of automated work cells follows a process similar to that defined for CIM. Two major issues must be addressed early in the design process: (1) the use of current production hardware versus the purchase of new process machines, and (2) the adoption of fixed versus flexible automation. An understanding of the design process starts with a look at the first issue.

New versus Existing Production Machines

Work-cell automation projects fall into two categories: *new automation* and *existing production cells.* In both types of projects, the automation can focus on a single production work cell or on the automation and integration of a group of cells. In the first category, new automation, both the production process machinery and the automation hardware are specified and purchased in the automation project.

In the second category, existing production cells, the current production machinery is integrated with new automation hardware and software to upgrade the production process. The decision to choose existing production machines over new ones is based on *market conditions* and *economics.* For example, if the order-winning criteria require a level of quality or price that current manual production machines cannot meet, then construction of a new automated production system is necessary to be competitive. If the cost of a new system prevents that course of action, however, then the company has two choices: (1) discontinue the product and abandon that market area, or (2) find a niche for the current product in the larger market. In the second case, the product could be designed to fill a special niche where quality is not as critical. With the quality issue satisfied, the current production equipment could be integrated and automated so that the cost is competitive in the niche area. In the pump manufacturing example (Figure 3–3), the process machines had the necessary precision but were not delivering order-winning values for cost, production flexibility, or cycle time due to poor work flow. Reorganizing the production machines and adding the appropriate automation to the reconfigured work cells was a cost-effective solution. When a new automation cell is designed and built with new process machines and automation hardware and software, the integration requirements for all of the equipment are specified in the purchase contract. As a result, the implementation of new automation cells often presents fewer integration and automation problems.

Fixed versus Flexible Automation

In the second issue, the use of fixed versus flexible automation is examined. Parts and products manufactured using fixed automation generally have the following characteristics:

- *Low product variability:* The product design is well defined, and no major changes are planned in size, shape, part count, and material. The production of electrical wall switches for houses is a good example of a product with low variability. The switch design is well tested, and the likelihood for a change in the design is very small.
- *Predictable demand:* The demand for the part or product is stable for a 2- to 5-year time frame. For example, the automotive industry frequently signs multiyear contracts with suppliers of parts like horns and water pumps. With this multiyear contract commitment, an investment in a fixed automation system is justified.
- *High production rates:* The number of units produced per hour or day is usually high. Consider the production rates necessary to match the number of disposable razors sold per day around the world. High-speed fixed automation is the logical choice for such a production system.
- *Cost pressures:* The prices of many products, like disposable razors, are set by market conditions, and the manufacturer must meet the market price to be competitive. Frequently, fixed automation is the system of choice when high volume and low cost must be achieved.

When one or more of these product characteristics are present, fixed automation is considered for the production system.

The Through-Hole Printed Circuit Board Assembly case study, covered earlier, describes a rationale for the selection of flexible automation over a fixed system. TRT Thomson chose a flexible system because the variation in components would require three different fixed automation work cells linked by material handling. With the cost and production speed of the fixed and flexible systems almost equal, the flexible system was chosen because of the large number of different products and relatively low quantities, and it offered the opportunity for *future cost avoidance.* Fixed manufacturing systems are often scrapped when the products produced on them are no longer sold. In contrast, flexible manufacturing cells and systems avoid future cost because they can be retooled to produce a different product with a minimum investment in new hardware. In general, flexible automation is used in these situations:

- The product mix requires a combination of different parts and products to be manufactured from the same production system.
- The product family has a history of unavoidable engineering changes that alter production requirements.
- The product or part family will expand with similar but not identical models.
- Production volumes are moderate, and demand is not as predictable.

CIM is not just the application of automation hardware and software; however, the full benefits of CIM require the judicious use of automation. Therefore, work cells with a wide variety of automation systems, including robotics, will be part of most manufacturing integration projects. Many cases of poorly designed systems indicate the need for a good design process.

3-9 SYSTEM TROUBLESHOOTING AND PROBLEM SOLVING

In every career field, technicians and engineers are called upon to solve problems. Oftentimes the problems are associated with the design of a new component or the creation of a machine or system. Another major problem-solving area is the troubleshooting of systems that are not functioning or of machines and systems operating outside the design limits. In every case, an organized approach to the solution of the problem is expected from a technically trained individual. While there are similarities, the problem-solving techniques used for design problems are different from those used in troubleshooting problems on operating systems; therefore, the following general introduction to troubleshooting is important.

Introduction to Troubleshooting

A good starting point is the definition of terms. According to *Merriam-Webster's* dictionary, a *troubleshooter* can be defined as

> *A skilled person employed to locate trouble or make repairs on machinery or technical equipment.*

That same dictionary defines *troubleshooting* as

> *The intricate process used to solve problems. While the process is predominantly mental, use of equipment and manipulation of mechanical objects is often necessary.*

The same source defines a *troubleshooting problem* as

> *A situation in which an answer, solution, or decision is not immediately apparent but may be found with a logical methodology that often has intuitive components.*

These definitions indicate some important concepts but don't dispel all of the common misconceptions with regard to the development of troubleshooting ability. The misconceptions include:

- *Troubleshooting is just a simple multiple-step process that can be memorized.* While there are techniques to learn that aid in troubleshooting, the solution process is as varied as the number of machines that can fail. Every new problem is different.

- *Copying the methods used by other troubleshooters is a good way to learn troubleshooting.* Rarely do people have identical troubleshooting styles, because everyone has a unique problem-solving technique. You can learn some standard solutions to specific problems by working with a skilled troubleshooter, but everyone must develop his or her own troubleshooting techniques through practice.

- *Good troubleshooters should be able to immediately spot the problem or trouble.* Experienced troubleshooters, who have solved many problems, often immediately know the solution after studying the symptoms because they have solved a similar problem earlier. When a problem is solved for the second time, it is merely an exercise, not true problem solving. However, when an experienced troubleshooter or a novice is presented with a problem never before seen, each one has to work through the logical process and analysis of the symptoms to arrive at the cause of the trouble. Experience makes you faster at the process, not better at guessing the fault.

- *Good troubleshooters are always busy measuring something while they search for the fault.* This could not be further from the truth. Good troubleshooters take time to *think* and *analyze* the system data before springing into action. Taking time to think is the mark of a good troubleshooter.

- *Good troubleshooting skills are inherited and cannot be learned.* Everyone is different, and some individuals approach problems in a more logical fashion than others. As a result, some people find troubleshooting an easy skill to master. However, anyone can learn to be a better troubleshooter by just learning some proven techniques.

- *Multiple system failures are common.* In most cases, system failure is a random event. For example, a failure could be caused by outside events like collisions between a forklift and system wiring, or voltage spikes as a result of lightning. Other failures result from component deterioration due to age, heat, vibration, or electrical and mechanical stress. Whatever the cause, it is not predictable in most cases. As a result, it is unlikely that two random events would occur at the same time to cause two unrelated failures in the system. So when the system is not functioning properly, a single failure is initially assumed. Multiple failures are more difficult to troubleshoot and are addressed later in this section.

Troubleshooting is a valuable skill to learn for engineers and technicians working with machines and systems because eventually something is going to fail. A good troubleshooter is then the man or woman of the hour.

Hardware versus Software Troubleshooting

It is hard to find a machine or industrial system today that is not a mix of electronic circuits, mechanical mechanisms, and computer controllers. Frequently the computer takes the form of a microprocessor in a single board microcontroller or in the form of a programmable logic controller. In either case, the presence of a computer indicates that a *program* or *machine code* exists that controls the operation of the machine. The presence of a program introduces a unique set of problems for the troubleshooter of the system. Troubleshooting of machine programs is discussed in later chapters in some detail; however, some general guidelines should be learned now.

- *If the program worked yesterday, then it should work today.* This statement implies that programs usually don't break or fail, so if the program controlled the machine correctly in the past it should continue to do so. To restate, if the machine worked yesterday, then the faulty operation today is probably not a problem in the computer program. There are two exceptions to this guideline:

  1. Data is frequently entered into a machine program to establish set points or operating conditions. If incorrect data is entered into the program, then faulty operation of the program and machine could result.

  2. It is not uncommon to have hundreds of signals in the interface between a complex system and the computer controlling the system. The computer programmer tries to consider all possible combinations of signal conditions in developing the control program; however, it is possible that an input signal condition might occur that was not considered. The program may operate correctly for months before this unique set of input conditions occurs. The malfunction is especially hard to find because the program appears to be correct based on recent operation.

- *Some knowledge of the programming language is necessary.* The engineer or technician troubleshooting the machine must be able to read, interpret, and analyze the program code used to control the machine.

- *Well-documented program code is critical for successful machine troubleshooting.* Machine control programs are developed by either the company supplying the machine or by the engineering department in the company where the machine is installed. In either case, a detailed and clearly documented program is essential for troubleshooting future machine failures.

The balance of this general introduction to troubleshooting focuses on the electronic and mechanical hardware components.

Troubleshooting Techniques

Troubleshooting techniques are proven procedures that provide a structure for the troubleshooting process. While the experienced troubleshooter and the beginner use the same techniques to find a faulty component, it is not always apparent because the experienced person often does many of the operations in his or her head. The beginner, on the other hand, is encouraged to write down the procedures until the techniques are mastered. The techniques most often used are *block diagrams, signal flow, bracketing,* and *information funneling.*

Block Diagrams

A block diagram is a set of rectangles used to describe all the parts of a system. A block diagram of a public address (PA) system is illustrated in Figure 3–9. Note that two microphones are used as inputs into the mixer in the figure. The mixer allows for the volume and tone levels for the microphones to be individually adjusted. The small signal amplifier, called the pre-amplifier, amplifies the combined output of both microphones. The power amplifier provides additional amplification to drive the two output speakers. A selector switch is present between the pre-amplifier and power amplifier to connect either the microphones or a CD player into the power amplifier. The CD player has an individual pre-amplifier and volume control included in the unit. Study the system in Figure 3–9 until the operation is fully understood.

The following characteristics of block diagrams can be extracted from this example.

■ *Complex systems are represented by series of simple rectangles.* The complex parts of the public address system, like the pre-amplifier and power amplifier, are reduced to single rectangles. As a result, very complex systems can be represented on a single page.

■ *Information flows from left to right through the rectangles.* Notice that the flow of information starts on the left side (sound into microphones) and moves from left to right where the speaker rectangles are located. Restated, inputs are on the left and outputs on the right.

■ *Systems and subsystems are represented.* The decision on what to include in a rectangle is usually based on how repairs are performed. If a failed

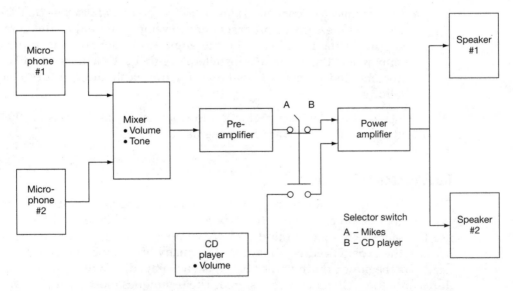

Figure 3–9 Public Address System Block Diagram.

unit were replaced with a spare to return the system to operation, then that entire unit would be a single block in the diagram. However, if components of the unit are replaced when a system fault is traced to the failure of a unit, then a subsystem block diagram of the unit must be created. For example, the power amplifier in Figure 3–9 has three internal modules that can be replaced. So the PA system has a block diagram as illustrated in Figure 3–9 and a subsystem block diagram for the power amplifier shown in Figure 3–10. If the fault is traced to the power amplifier, then troubleshooting continues at the subsystem level to locate the block inside the power amplifier (Figure 3–10) that caused the system failure.

The system block diagram is available from two sources. Ideally, the vendor of the equipment provides a system block diagram in the user manual. However, if one is not provided, then you have to create one for your troubleshooting requirements. A block diagram is easily generated as follows:

1. *Make a list of all of the system components that would be replaced in the advent of a failure.* For example, the selector switch and microphone are single components in the PA system in Figure 3–9 that would be replaced if they failed. The power amplifier has three circuit boards, shown in Figure 3–10, that could be individually replaced if that amplifier was the problem. However, the pre-amplifier, which is made up of many components, would be replaced as a unit if it failed. Entire units, like the pre-amplifier, are replaced with a spare because that is the fastest way to get the entire system back into operation. The decision to fix the bad unit after replacement is based on cost. In some cases, the cost of repair is too high and the unit is discarded. In other situations, a repair facility fixes the unit, and it is returned for use as a spare.

2. *Arrange the list of system components with inputs at the top, outputs at the bottom, and the remaining items in the order that signal or information flows through them.* The list for the PA system in Figure 3–9 would have microphones at the top and speakers at the bottom. The mixer would be listed under the microphones, and the pre-amplifier and CD player would be on the same line since they both feed into the selector switch. The selector switch and power amplifier would follow above the speakers.

Figure 3–10 Block Diagram of the Power Amplifier.

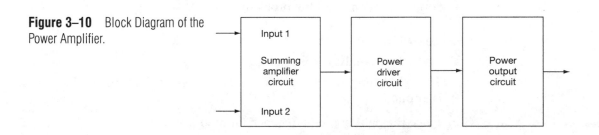

3. *Put all of the block diagram components into rectangles and apply signal flow techniques to link the rectangles in the diagram.* A major part of the block diagram is the signal flow representation and analysis that is described in the next section.

A troubleshooting technique called bracketing is used with the system block diagram to locate faults.

Bracketing

Bracketing is a technique that uses external markers to identify the portion of the system block diagram in which the fault exists. The brackets are initially placed on the system block diagram, but could be moved to a schematic diagram when the fault is narrowed down to a single circuit. A three-step process is used to establish the initial location for the brackets.

1. *Record and study all system symptoms listing all conditions that vary from normal operation.*
2. *Locate points on the system block diagram where abnormal operation is occurring and place a right bracket ( ]) after each abnormal block.* Remember, the signal flow on the block diagrams is from left (inputs) to the right (outputs). So in general, the right brackets are placed on the right side of output blocks that are not operating.
3. *Move to the left along the signal flow path from each bad bracket until normal operation is observed. Place a left bracket ([ ) to the right of the block where a normal output was detected.* In general, the left bracket is initially placed with a minimum of detailed testing, or only with easily performed tests.

The application of the brackets to the system block diagram may be an entirely mental process for experienced troubleshooters; however, the beginner should physically mark them on the block until all the troubleshooting techniques are fully internalized. Example 3–3 illustrates this process.

Example 3–3
The PA system in Figure 3–9 is needed in a conference room to use for a presentation. The system is set up by a technician and tested to verify correct operation. When the technician taps on both microphones there is no sound coming from either speaker. Locate the PA system fault. (Since this is an example of bracketing, let's assume the system power and volume setting on the mixer are not the problem.)

Solution
a) Record and analyze symptoms.

There is no output from either speaker when a valid input to the system is applied.

b) Place the right bracket after each faulty output.

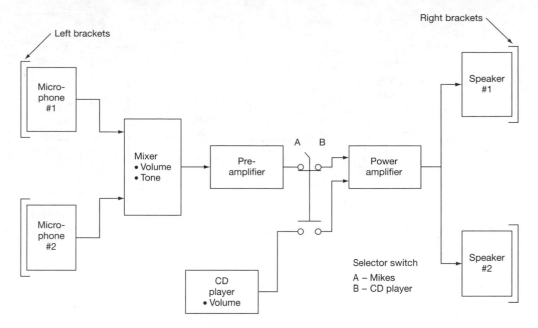

Figure 3–11 Starting to Troubleshoot the Public Address System.

The right brackets are placed on the right side of each speaker block as shown in Figure 3–11

c) Place the left bracket to the right of the last good or normal output. The symptoms indicate that the input to the microphones is valid, but the condition of the microphones is not known. Therefore, the left bracket is placed to the left of the microphone block as shown in Figure 3–11.

This example shows how the initial bracket positions are placed based on verified system symptoms. However, no new information is generated at this point, because all the brackets tell us is that the fault is somewhere in the total system. Before proceeding with additional checks on the faulty PA system, it is important to learn the concepts associated with signal flow in the block diagrams.

Signal Flow

Signal flow is generally divided into two groups called *power* and *information*. The *power flow* illustrates how power is delivered to all of the components in the system, and *information flow* indicates how information or data flows from source(s) to destination(s).

There are five distinct types of signal flow topologies: *linear, divergent, convergent, feedback,* and *switched.* The user must recognize each type of path because a unique troubleshooting approach is associated with each of the topologies. A description of each type is provided below.

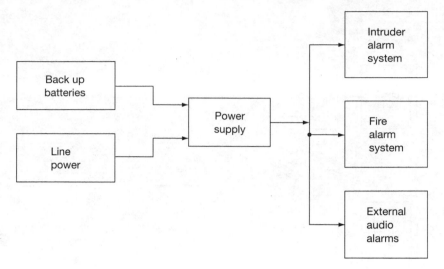

Figure 3–12   Power Flow Diagram for a Security System.

■ *Linear:* The linear signal path is a series connection of blocks as illustrated in Figure 3–10.

■ *Divergent:* A divergent signal path is present when a single block feeds two or more blocks. Figure 3–12 is a power flow block diagram for a security system. Can you find the divergent signal path in the figure? The divergent path in Figure 3–12 is the power flowing from the power supply to the three system blocks.

■ *Convergent:* A convergent signal path is present when signals from two or more blocks feed into a single block. The two microphones connected to the mixer in Figure 3–9 represent a convergent signal path. A second example is present in the security system in Figure 3–12 where the two power sources connect to the power supply.

■ *Feedback:* A feedback signal flow is created when part of the output signal is diverted back to the input of the system and added to the input signal. The block diagram for a feedback temperature control system is illustrated in Figure 3–13. The temperature of the process is measured and fed back to the input, where it is compared with the desired temperature called the *set point.* When the set point is higher than the actual temperature, a positive error signal is generated. The amplified error signal causes the mixer valve to increase the temperature of the liquid in the heating jacket. If the process is too hot, a negative error signal is generated, and the opposite action occurs.

■ *Switched paths:* Switched signal flow paths include linear, divergent, or convergent paths with switches present to change the flow of the signal. A

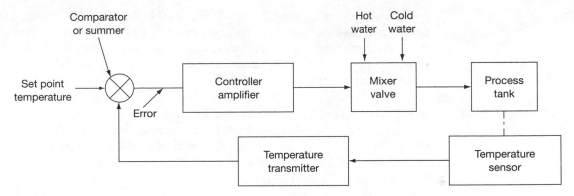

Figure 3–13 Feedback Circuit Block Diagram for a Temperature Controller.

switched convergent path is present in the PA system, Figure 3–9, where the CD player and pre-amplifier converge at the power amplifier through the selector switch.

Every system has a signal flow block diagram that is a combination of these five topologies. Analysis of the topologies is used to speed the troubleshooting process.

Signal Flow Analysis

The five topologies (linear, divergent, convergent, feedback, and switched), introduced in the previous section, have rules that can be used to speed the search for the faulty system component. The topology rules that cause bracket movement are described below.

- *Linear rule:* When brackets enclose a linear set of system blocks to be tested, the first test point should be at or just before the midpoint of the bracketed area. If the signal is faulty, then the right bracket ( ]) moves to that point because the fault is to the left of this point. However, if the signal is valid, then the left bracket ([ ) moves to that point because the fault is in the blocks to the right of this point. Application of this "*divide and conquer*" rule eliminates half of the components with a single check.

- *Divergent rule:* When brackets enclose system blocks with a divergent path, the stage before the divergence is fault free if any of the divergent paths are normal. For example, assume that brackets ([ ]) enclose the power supply and all three outputs in the power flow block diagram in Figure 3–12. If a test indicates that power is being delivered to the Fire Alarm system, then the power supply is not the problem. As a result, the left bracket ([ ) is moved to the output side of the power supply.

- *Convergent rule:* When brackets enclose system blocks with a convergent path, then two rules must be applied. Rule C1: If all convergent inputs are required to produce a valid output, then a valid output indicates all input paths are fault free. Rule C2: If only one convergent input is required to produce a valid output, then each input must be checked to verify that the input paths are fault free. For example, the convergent paths into the power supply in Figure 3–12 would use Rule C2 because either source could supply energy to the power supply.

- *Feedback rule:* When brackets enclose system blocks with a feedback path, a change or modification to the feedback path is used to indicate normal operation of the closed-loop system. In the temperature controller system shown in Figure 3–13, troubleshooting would start by changing the level of the signal back to the comparator.
- *Switched rule:* If brackets contain linear, divergent, or convergent topologies that are changed by a switch setting, then observe the system when the switch is moved to another position. If the trouble disappears, then the fault is in the path now switched out. If the trouble persists, then the fault is in the path common to both switch positions.

Information Funneling

The last general troubleshooting concept is called *information funneling.* This process is pictured graphically in Figure 3–14. As the figure indicates, when the brackets on the system block diagram are widely separated, the tests selected for identifying the fault tend to be of a more general nature. For example, tests at level 1 might include:

- Verifying that power is present at every unit.
- Verifying that switches required for normal operation are in the on position.
- Verifying that all cabling has been properly connected.

Figure 3–14 Information Funneling.

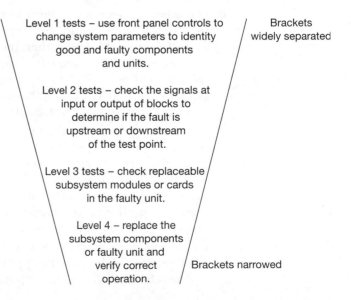

Level 1 tests – use front panel controls to change system parameters to identity good and faulty components and units.

Brackets widely separated

Level 2 tests – check the signals at input or output of blocks to determine if the fault is upstream or downstream of the test point.

Level 3 tests – check replaceable subsystem modules or cards in the faulty unit.

Level 4 – replace the subsystem components or faulty unit and verify correct operation.

Brackets narrowed

- Verifying that front panel controls are in the position for normal operation.
- Changing front panel controls or switch settings to see how the symptoms are affected.

Checks at this level can be performed quickly, usually don't require the use of any test equipment, don't require that equipment covers be removed, and eliminate some suspected blocks, so the separation between the brackets can be reduced. As a general guideline, you should exhaust all possible checks at each level, moving the brackets as close together as possible before dropping to the lower levels where tests are more specific.

At level 2, fewer suspect blocks are present between the brackets, and test equipment is used to move the brackets closer together. For example, cables might be removed from units and measurements made to verify that a valid input or output signal is present. Work at this level usually involves a number of tests with each test result changing the location of either a left or right bracket. The test equipment most often used includes meters, oscilloscopes, and data analyzers.

Level 3 tests in the funnel are usually performed when a single unit has been identified as the fault. In some cases, the troubleshooting continues inside the unit on subsystem modules or electronic cards using more precision measurements, but just as often the unit is replaced with a spare.

The lowest level, number 4, focuses on replacing faulty units or subsystem components with good spares and verifying that the exchange corrected the problem.

Bracketing and funneling work together to help eliminate:

- *Pointless and irrelevant system tests*—troubleshooters are required to focus only on the bracketed trouble area.
- *Redundant system tests*—if the separation between the brackets does not change, a measurement has not provided new information. When this occurs, it is an indication that additional tests at the current level may be redundant.
- *Premature system tests*—making detailed tests on a single block when the brackets are widely separated yields too precise information for the present situation. Funneling help focus on the most appropriate test for the current situation.

Example 3-4
Review Example 3–3 and Figure 3–11 to verify how the brackets were placed in the PA system based on the failure symptoms. Apply information funneling techniques to find the faulty unit.

Solution
Level 1 work in the information funnel

a) A check for power, correct operational switch settings, and proper cable connections indicates that none of these are causing the problem. The brackets do not move.

b) The mixer has a volume control for each microphone and a master volume control. Since neither microphone responds to an input all three volume controls are checked to verify that they are not at a zero volume setting. Volume controls are at a nominal value when checked, so the brackets do not move.

c) A study of the system block diagram, Figure 3–11, indicates that a second type of input could be applied if the selector switch was moved from the microphone to the CD player setting (see signal flow *switched rule*). The volume setting on the CD player is set to a nominal value, and the CD is selected with the switch. When these changes are made, both speakers have normal audio output, indicating that the power amplifier and speakers work with the CD. As a result, the single fault must be upstream from the switch in the pre-amplifier or mixer. The right bracket moves between the switch and the pre-amplifier. Since it is unlikely that both microphones would fail (single failure assumption), the left bracket moves between the microphones and the mixer.

The system block diagram with the new location for the brackets is shown in Figure 3–15. Using level 1 procedures, the location of the fault was narrowed from eight units to just two. However, no additional new information can be obtained from work on this level, so the test moves to level 2.

d) A study of the updated system block diagram, Figure 3–15, indicates that the fault must in the linear blocks labeled mixer and pre-amplifier. The "divide and conquer" strategy (see *linear rule* in signal flow analysis) is applied, and a check is made with an oscilloscope at the midpoint in

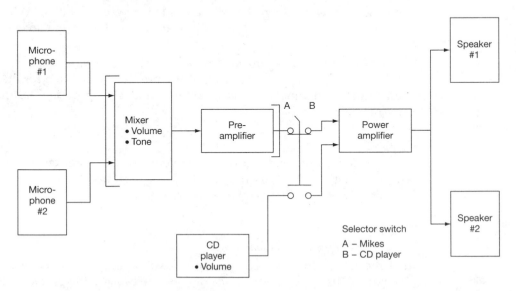

Figure 3–15   Further Troubleshooting of the Public Address System.

the linear signal between the mixer and pre-amplifier. The output of the mixer is correct, so the brackets move around the faulty unit, which is the pre-amplifier.

Level 3 in the information funnel is not used since the pre-amplifier does not have any subsystem components that are replaceable. If the power amplifier were the faulty unit, the tests would continue on the inside of the unit. The test moves to level 4.

e) The pre-amplifier is replaced with a spare, and the PA system is tested to verify correct operation. The test is successful; the system is ready for operation.

The example illustrates how bracketing, signal flow analysis, and information funneling can be used with a system block diagram to locate a bad component with the minimum number of tests and in the shortest time.

Symptoms and Use of System Data

A good troubleshooter uses a combination of tools and techniques that are related to problem solving. The troubleshooter is like a detective, the fault is the crime, and the symptoms are the clues. The following general guidelines should be applied to the solution process as necessary.

- Symptoms are a clear and detailed description of what the problem *is* and *is not*. Sometimes stating what is working helps to more clearly see what is not working. Use the following guidelines:
  - If possible, personally verify all symptoms. When this cannot be done, rely on written problem descriptions in maintenance logs and from reliable sources.
  - Verify that operating controls are in their nominal position before recording and analyzing symptoms.
- Keep an accurate and detailed troubleshooting log for each system that includes (1) conditions present when troubleshooting starts and (2) changes in the system as troubleshooting progresses.
- To recognize abnormal operation you must fully understand normal operation. You have to study normal system operation until every possible operational mode is understood.
- Abnormal operation has three sources:
  - *Adjustment*—improper adjustment of controls either on the front panel or inside the unit.
  - *Calibration*—the lack of calibration against some universal standard.
  - *Component or unit failure*—a component failure in some part of the system. Always look for evidence of a component failure such as noticeable wear, burn or excessive heat marks, or damage from impact.

Troubleshooting Sequence

With the tools in place to approach troubleshooting in a systematic and organized fashion, a general set of steps can be developed for use on system faults. The sequence includes:

1. *Define the problem.* Gather all the symptom data for the failure and information on the state of the system when the fault occurred.

2. *Decide what needs to be tested.* Put the brackets on the system block diagram (physically or mentally) to narrow the search for the fault to that part of the system where faulty operation has been verified.

3. *Decide what type of test to perform.* Apply the funneling criteria, which starts with changes in the signal flow path, to perform broad system tests that will eliminate some of the suspect units between the brackets. As the brackets narrow, the type of test performed becomes more precise with increased use of test equipment. When tests fail to provide any new information on the location of the fault, move to the next level of more precise tests.

4. *Correct the problem.* After the faulty unit or component is identified, the problem is corrected by adjustment or replacement of the unit or component.

5. *Verify correct operation.* After a problem is corrected, it is important to thoroughly test the system to verify that the applied fix corrected all of the system problems.

6. *Determine the cause of the failure.* Getting the system back into operation is only half of the solution. If possible, determine what caused the unit or component to fail and recommend system changes that would prevent a similar failure in the future.

The first three steps in the sequence are repeated as often as necessary until the fault is located. Symptoms are studied, possible faulty units are bracketed, and the easiest possible changes in the signal flow are applied to eliminate some suspected units. As the brackets move closer together, the test on suspected units gets more precise with test equipment introduced. Throughout the process, symptoms continue to be analyzed. The analysis includes time to *think* about the problem, reflecting on results of previous tests and possible future tests.

Multiple Failures

Systems with multiple failures provide the toughest troubleshooting problems. The first problem is to determine if a multiple failure is present. If a single unit or component failure could not cause the symptoms present, then multiple failures are assumed. Multiple failures can present two types of symptoms: *related* and *unrelated.* In the related symptom situation, failure of either unit or component would produce the symptom. For example, consider the case of a robot axis, which is driven by a servomotor with a servo control circuit. The robot axis would stop operating if either the servomotor or the servomotor control unit failed. So the symptom is related to both components. However, if a bad bearing on the robot axis caused the servomotor to stall, then the servomotor could fail because a winding overheated, and the motor failure could cause a failure in the servomotor drive. The result is multiple failures with related symptoms. When units and components have related symptoms, multiple failures could be present, so it is important to check each component before implementing a fix to the trouble.

The presence of unrelated symptoms indicates that multiple failures could be present in unrelated units or components. While this type of failure is more rare, the solution process is the same. Apply the troubleshooting principles to one of the symptom(s) until the fault is corrected, and then repeat the process for the unrelated symptom(s) until the second fault is fixed.

The tools and processes outlined in this section provide a general approach to troubleshooting system problems. Specific troubleshooting techniques are available for individual components and machines in the system. Engineers and technicians must be able to efficiently use both the general and specific techniques to solve system problems as quickly as possible.

3-10 SUMMARY

The use of stand-alone automated work cells did not deliver the productivity results needed by many enterprises to remain competitive. The poor results were traced to three sources: isolation of the automation from the production system, deficiencies in the work-cell design process, and insufficient training and preparation of people. A CIM implementation process that includes assessment, simplification, and implementation with performance measures will overcome most of the deficiencies of the earlier automation implementation.

The assessment step includes detailed self-study of the enterprise in three areas: technology, human resources, and process system. In each case, the assessment documents what the enterprise does well and what needs improvement. The second step, simplification, focuses on the elimination of waste. Waste is every operation, move, or process that does not *add value* to the final product. The first two steps prepare the enterprise for the CIM implementation; Step 3 builds the system. Implementation with *performance measures* implies that the work cells and integrated systems are purchased and installed, but measurements of system performance must also be included. The six key measurement parameters are product cycle time, inventory, setup times, quality, employee output/productivity, and continuous improvement. The specific process used by companies as they address the implementation of CIM principles will vary depending on the conditions present and the corporate culture. However, some of the elements are support from the top, multifunctional teams, willingness to study all processes, valuing every employee's input, and accepting a 3- to 5-year payback.

Flexibility in manufacturing is often described as (1) the ability to adapt to engineering changes in the part, (2) the increase in the number of similar parts produced on the system, (3) the ability to accommodate routing changes that allow a part to be produced on different types of machines, and (4) the ability to change the setup on the system rapidly from one type of production to another. Flexible manufacturing takes two forms: flexible manufacturing systems and flexible manufacturing cells. An FMS is defined as one manufacturing machine or multiple machines integrated by an automated material handling system, the operation of which is managed by a computerized control system. An FMC is defined as a group of related machines that perform a particular process or step in a larger manufacturing process.

Fixed automation systems and machines have little flexibility for changes in the manufacturing process or products. However, they do have the ability to manufacture large volumes of discrete parts to meet production demand economically for some products. Fixed automation systems are implemented in two configurations: in-line and rotary. The in-line fixed automation systems include a linear work flow through a series of workstations. Rotary-type fixed automation locates the production stations around a circular table or dial. As a result, this type of production system is called an indexing machine or a dial index machine. A rotary system is limited to smaller workpieces and fewer production stations than the in-line type.

The implementation of automated work cells forces two decisions early in the design process: (1) the use of current production hardware versus the purchase of new process machines, and (2) the adoption of fixed versus flexible automation. In the first instance, a choice must be made between adding automation to existing production machines or building a completely new production system. The second decision requires a choice between flexible and fixed automation.

Troubleshooting of hardware and software is a critical skill to learn for everyone working in manufacturing automation. A skilled troubleshooter uses a system block diagram and knowledge of the system operation with techniques of bracketing, signal flow analysis, and information funneling to effectively locate one or more faulty components.

QUESTIONS

1. Why were the islands of automation developed in the 1970s unable to perform to expectations?
2. How does the CIM development process correct the deficiencies present in the islands of automation?
3. What is assessment, and what areas of the enterprise are affected?
4. What is the focus of the educational component of assessment?
5. How is an enterprise changed by the simplification process?
6. What does implementation with performance measures imply?
7. Describe the six key measurement parameters used in CIM implementations.
8. Why do some companies avoid the CIM acronym when they start an improvement process?
9. What is usually considered the most critical element for the success of a CIM implementation?

10. Name the types of manufacturing systems in which robots are most likely to be used.
11. What does flexibility in manufacturing mean?
12. Compare and contrast FMCs and FMSs.
13. How are the automated systems installed at Technophone and TRT Thomson similar? How are they different?
14. What are the major differences between flexible and fixed automation systems?
15. Compare and contrast in-line and rotary-type fixed automation systems.
16. How do companies justify the use of existing production machines in an automation project?
17. What criteria are used in the selection of fixed automation over a flexible automation cell?
18. Why is it necessary to have the cycle times of each work station about equal in fixed automation systems?

PROBLEMS

1. Determine the cycle time in minutes per part for the total system and for each machine in a two-work-cell system with the following production data.

   First machine with a lot size of 200 parts: setup time is 8 hours per 200 parts; queue, move, and transport time is 1.4 hours per pallet of 50 parts; and run time is 10 minutes per part.

   Second machine with a lot size of 50 parts: setup time is 1.25 hours; queue, move, and transport time is 0.5 hours per pallet of 25 parts; and run time is 4.5 minutes per part.

2. The production of the parts in Problem 1 has to be increased by 50 percent. Identify the production constraints, and list some of the options to increase production capacity.

3. Determine the production inventory turns if cost of goods sold is $800,000 and average annual inventory cost is $12,000.

4. How low would the annual inventory cost drop in Problem 3 if the inventory turned over daily?

5. The inventory turns for a product with $1,500,000 in cost of goods changes from 4 to 125 due to improved production operations. What are the annual interest savings if the interest rate on the short-term loans used to finance the inventory is 9 percent?

6. The pump referenced in Figure 3–3 had cost of goods of $1.8 million, a lot size production of 200 parts, and overhead costs of $50 per square foot.

   a) Determine the annual average inventory cost at the start and end of the project. What was the percentage reduction in inventory investment?

   b) Express the quality in parts per million, and find the number of scrap parts for the lot size at the start and completion of the project.

c) Calculate the overhead cost saving as a result of floor space reduction.

7. What types of problems are created by the difference in cycle times for the parts in Example 3–1? How could these problems be eliminated?

The PA System in Figure 3–9 should be used for the following problems. In each case assume that the system was set up and the symptom occurred when power was initially applied.

8. Determine where the initial brackets would be placed for the following symptoms.
   a) No sound from speaker number one when the system is first tested.
   b) No response when microphone number one is tested, but both speakers respond to tests using microphone number two.
   c) No output from speaker number two, but speaker number one works with both microphones.

9. How could the single failure assumption be used in step b of Example 3–3 to justify putting the right bracket at the output of the power amplifier instead of after each speaker?

10. Why can't the single failure assumption be used in step a of Example 3–3 to start with the left bracket at the input to the mixer instead of at both microphones?

11. Determine how the brackets would move in step c of Example 3–4 if the speakers had no output when the selector switch was moved to the CD player setting.

12. Describe the testing sequence starting with step c of Example 3–4 if the power driver circuit (Figure 3–10) is faulty. What is the minimum number of tests to locate the bad component?

PROJECTS AND CASE STUDY PROBLEMS

1. Determine what elements of the work-cell design checklist (Table 3–1) would not affect the design of the injection molding production cell developed in Case Study Problem 3 in Chapter 2.

2. Complete the work-cell design checklist (Table 3–1) for the production cell design started in Case Study Problem 5 in Chapter 2. Document all assumptions made and justify why elements in the checklist were not used in the cell design.

3. Perform the initial design on a work cell to assemble one of the following kitchen products currently available in store's (rate is 2000 per week): set of knives, bottle cork and cap remover, hand-operated egg beater, or manual can opener. Complete the robot selection criteria and work-cell design checklist (Table 3–1), select robot model(s), determine the initial cell layout, and draw a top view of the cell illustrating the machines and robot work envelope(s). In addition, the design documentation should include the type of automation

selected (Figure 3–4), quantity and type of robots, number of stations, need for manual stations, drawings of the product parts, and required assembly cycle times.

4. Complete the work-cell design checklist (Table 3–1) for the West-Electric slug production cell design started in Case Study Problem 8 of Chapter 2.

5. Complete the work-cell design checklist (Table 3–1) for the West-Electric slug lubrication cell design started in Case Study Problem 9 of Chapter 2.

6. Determine how the work-cell design checklist would affect the integration of the slug production and slug lubrication cells (Case Study Problem 10, Chapter 2).

7. Calculate the five-day three-shift production maximums assuming 20 percent downtime per shift for the following work cells in the West-Electric case study in Chapter 12: extrusion, upset forging, block forging, and final forging.

8. Read the West-Electric case study in Sections 12-1 to 12-4 of Chapter 12 where the work-cell design checklist is introduced to the team. Complete the questions and problems at the end of Chapter 12 that relate to Sections 12-1 to 12-4.

End-of-Arm Tooling

CHAPTER GOALS AND OBJECTIVES

The goal of Chapter 4 is to provide a comprehensive explanation of the many different devices attached to the robot arm known as grippers or end-of-arm tooling. The treatment of the tooling includes a discussion of the gripping hardware and also includes numerous interface devices used to enhance the productivity of the tooling. At the completion of this chapter you should be able to:

- Name the five characteristics of good tooling design.
- Classify all tooling into one of the following five gripper categories: standard, vacuum, magnetic, air pressure, and special purpose.
- Differentiate standard grippers by closing motion, number of fingers, and power source.
- Calculate the maximum payload for a gripper given gripper parameters and the acceleration of the tooling.
- Design a vacuum gripper system using standard components for the vacuum cup and surface configuration given the degree of vacuum and device parameters.
- Describe the operation of the four grippers in the air pressure category.
- Calculate the gripping force for air pressure type grippers.
- Describe the difference between special purpose grippers and special purpose tools.
- Describe active and passive compliance and force torque sensing devices.
- Describe the operation of remote center compliance devices.
- Describe the difference between a multiple end-effector system and multiple gripper system.
- Describe the operation of wrist interface devices and tooling collision systems.
- Design end-of-arm tooling that satisfies the five tooling characteristics given case study data that includes a description of the part to be moved and work cell parameters.

End-of-arm tooling design is a good example of a career opportunity generated by the use of robot automation. Every robot has some type of tooling used to grip the parts. The tooling is usually a combination of off-the-shelf grippers and other mechanical and electrical hardware integrated into a tooling design that satisfies the production requirements.

The tooling design is usually performed on a computer aided design (CAD) system using one of the standard CAD software packages. In some cases a solid model of the tooling is developed to check for interference in moving parts and the degree of contact between the part and the gripper. In still other cases a finite element analysis (FEA) is performed to verify the structural integrity of the mechanical components. In many applications the tooling has sensors integrated into the design to determine part location. The tool designer can proficiently use one or more of the standard CAD software design systems like AutoCAD, Pro Engineer, Ideas, or Solid Works to create 2-D, 3-D wire-frame, and solid models. In addition, some designs require motion analysis using kinematics or other design verification software.

The designer must be well versed in classical mechanical design topics such as materials and material properties, static and dynamic force analysis, and design of standard mechanical mechanisms. However, in addition to classical design techniques, the robot-tooling designer must be familiar with the off-the-shelf gripper components that reduce design time and increase gripper reliability. Grippers use pneumatic or hydraulic power and incorporate electronic sensing, so knowledge of these technologies is also required.

The two- and four-year graduate who seeks a career in automation tooling has many options. All the major robot manufacturers who develop applications for customers need a variety of workers skilled in end-effector design, machining, assembly, and testing. The automation system design houses described in the Career Insights at the start of Chapter 3 also need employees with similar skills. Then there are a large number of manufacturers that supply standard grippers and vacuum grippers for use in end-of-arm tooling. These companies also need employees with the same skill background. The specific job title and responsibility again depend on the number of years of education and relevant experience. The creation of robot tooling offers the graduate in mechanical and electro-mechanical engineering and technology an exciting and rewarding career opportunity.

4-1 INTRODUCTION

Studies indicate that the process of joining two mechanical parts in an assembly operation uses an operator's sense of touch to a much greater degree than the sense of vision. This dominance of the tactile underscores the importance of human hands in all phases of manufacturing and assembly of production goods. It also emphasizes the demand placed on the *gripper* or *end-of-arm tooling* of an industrial robot if it is to perform many of the production duties of its human counterpart. Duplication of the human hand with its ability to grasp, sense, and manipulate objects remains one of the most difficult tasks facing the designer of end-of-arm tooling. Automation researchers are working on a robot hand with three fingers that can grasp irregularly shaped objects, but widespread application of this type of tooling is years away. This chapter investigates the types of robot tooling currently in use and covers the terminology involved in end-of-arm, or *end-effector,* tooling.

In general, the tooling or gripper mounted on the robot tool plate, called an end effector, has two functions. The function of the gripper is to hold the part as the robot presents it to the tool for work to be done, or to hold the tool as the robot moves it to work on the part. For example, a part held in a gripper could be positioned under a numerical control (NC) drill to produce a desired hole, or the end-of-arm tooling could hold a drilling mechanism, with the robot producing the holes in the same manner as a human operator using a hand drill.

The end-of-arm tooling used in a robot work cell should have all five of the following characteristics:

1. The tooling must be capable of gripping, lifting, and releasing the part or family of parts required by the manufacturing process.

2. The tooling must sense the presence of a part in the gripper, using sensors located either on the tooling or at a fixed position in the work cell.

3. Tooling weight must be kept to a minimum because it is added to part weight to determine maximum payload.

4. Containment of the part in the gripper must be ensured under conditions of maximum acceleration at the tool plate and loss of gripper power.

5. The simplest gripper that meets the first four criteria should be the one implemented.

As a result, one of the major design problems associated with robotic work cells is the selection and design of the end-of-arm tooling.

The end-of-arm tooling used on current robots can be classified in the following three ways: (1) according to the method used to hold the part in the gripper, (2) by the special-purpose process tools incorporated in the final gripper design, or (3) by the multiple-function capability of the gripper. The first category of gripping mechanisms includes standard mechanical pressure grippers, tooling using vacuum for holding or lifting, and magnetic devices. The second classification of tooling includes drills, welding guns and torches, paint sprayers, and grinders. The third type of gripper tooling includes special-purpose grippers and compliance devices currently in use.

4-2 STANDARD GRIPPERS

Standard grippers can have two different closing motions, *angular* or *parallel,* and can have *pneumatic, hydraulic, electric,* or *spring* power for closing and opening. The action of the angular and parallel devices is illustrated in Figure 4–1; standard off-the-shelf grippers are pictured in Figures 4–2 and 4–3. In most applications the gripper base is purchased with an actuator mechanism and jaw mounts for the fingers. The gripper in Figure 4–2 has each special-purpose finger attached to a jaw mount with two screws. The geometry necessary for the fingers was determined by the cell designer, and the finger parts were machined for the application. Some grippers are supplied with blank jaws that can be removed and machined into the configuration required for the application. For example, the gripper on the far left

Figure 4–1 Standard Angular and Parallel Grippers.

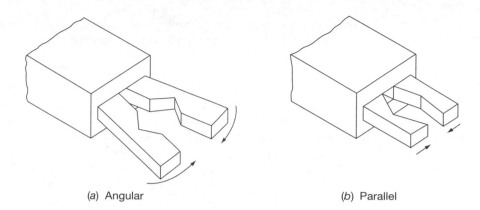

(a) Angular

(b) Parallel

Figure 4–2 Angular Gripper, Air Closed, Spring Opened.

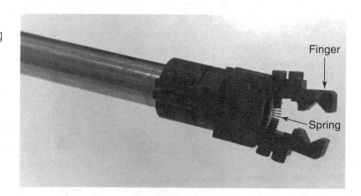

Finger

Spring

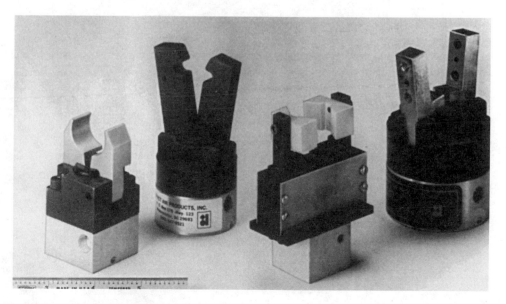

Figure 4–3 Typical Modifications to Blank Jaws.
(Courtesy of Compact Air Products of S.C., Inc.)

in Figure 4–3 was shipped with rectangular blanks that were machined as illustrated to fit the diameter of a round part. Starting with an off-the-shelf gripper and blank jaws that can be machined to satisfy the application is the most cost-effective approach to gripper design.

A standard gripper costs between 4 and 8 percent of the cost of the robot; small blank jaw pneumatic devices are priced under $1000. If a specially designed end-effector is required, then the cost for design and fabrication often exceeds 20 percent of the total costs for the robot system.

The gripper must be closed and opened by program commands as the robot moves through the production operation. The robot controller supplies the electrical signals that result in the gripper's action. Most grippers are opened and closed with a pneumatic actuator (Figure 4–3); however, in limited applications, hydraulic or spring power is used. In some cases, grippers are spring opened and power closed, as illustrated in Figure 4–2; less frequently, grippers are spring closed and power opened. Each of the types has advantages and disadvantages based on the specific application. The power-closed and spring-opened type, for example, require power only to close the gripper when a part is lifted and moved, which offers the advantages of simpler control and less power consumption. However, a part will be dropped or thrown if gripper power is lost while the robot is in the process of moving the part.

Grippers use both exterior features and interior geometry of the part to pick it up in an application. The two grippers in Figure 4–4 illustrate how the blank jaws of a standard gripper are machined to support external and internal part gripping. At the left in Figure 4–4, the jaw blanks were cut to grip the shaft around the external surface; however, in Figure 4–4 on the right the jaws were designed

Figure 4–4   Interior and Exterior Gripping.
(Courtesy of MACK Corporation)

to pick up the round part using the hole in the center of the part. The variation in gripper jaw or finger design is as different as the types of parts handled.

Another variation in standard grippers is the number of jaws or fingers used to grasp the part. Most applications use two-finger grippers. Three-finger grippers (Figure 4–5, far right) and four-finger grippers (Figure 4–6) are used when parts need to be centered by the gripping process. A three-fingered gripper, such as the one in Figure 4–5, would be fitted with fingers and would be used to handle large, flat circular parts by grasping the edge in three places. The four-finger grippers in Figure 4–6 center the part in both the horizontal and vertical directions when the jaws close on the part.

Figure 4–5   Two- and Three-finger Gripper.
(Courtesy of Compact Air Products of S.C., Inc.)

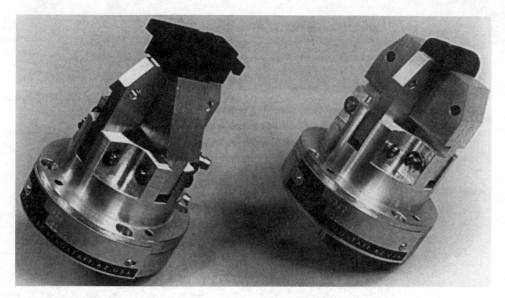

Figure 4–6 Four-finger Gripper.
(Courtesy of MACK Corporation)

Electrically powered jaws are also used in limited applications with a solenoid or dc servomotor providing the opening and closing action. Servo drives will be used more frequently when grippers are developed that can vary the pressure applied to the object as it is grasped. This would permit the same gripper to lift either a steel ball or an egg without dropping the ball or breaking the egg. Tactile or touch sensing must be developed beyond its present state, however, before this type of gripper will be practical for the factory.

Gripping Force

The first requirement for end-of-arm tooling is to grip and hold the part throughout the programmed motion. The ability to hold a part in the gripper is a function of the normal force transmitted to the gripping surface from the actuator in the gripper and the coefficient of friction between the gripper fingers and the part held. The relationship among a typical gripper actuator, jaw linkage, and jaws is shown in the picture of the grippers in Figure 4–4, and it is illustrated in the drawing in Figure 4–7a. Study these two figures to see how the picture and drawing are related.

The gripper is operated by the pneumatic cylinder. When the piston of the cylinder moves down, it provides a downward force (F_A) on the fingers at the points labeled A in Figure 4–7b. The force acts on each finger a fixed distance from the pivot point B in Figure 4–7b; as a result, a torque is applied to the finger

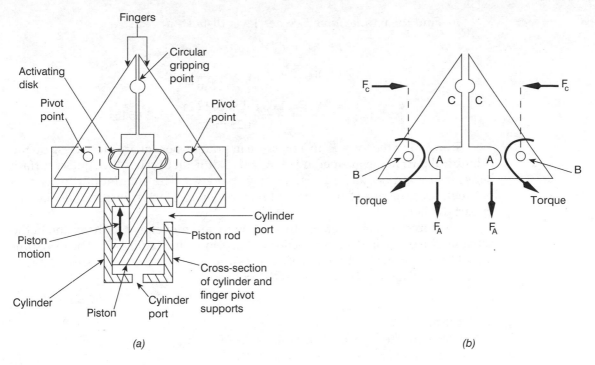

Fingers

Circular gripping point

Activating disk

Pivot point

Pivot point

Piston motion

Piston rod

Cylinder port

Piston

Cylinder port

Cross-section of cylinder and finger pivot supports

Cylinder

F_C

C C

A A

B

B

Torque

Torque

F_A

F_A

(a)

(b)

Figure 4–7   Gripper Forces.

around each pivot point. To find the force at point C (F_C) that is applied to the cylinder held in the gripper, the transformation of the torque to the gripping point is required. The torque produced by F_A at the pivot point and the vertical distance to the force acting on the part are used to calculate F_C. The following example illustrates these calculations.

Example 4–1

The force F_A in Figure 4–7b is 26 pounds and it acts at a distance of 0.75 inch from the pivot point. The vertical distance from the pivot point to the center line of the circular gripping point in the gripper is 1.25 inches. Calculate the force at F_C in pounds.

Solution

a) Find the torque resulting from force F_A.

$$F_A = 26 \text{ pounds and lever distance } d_1 = 0.75 \text{ in.}$$
$$T = d_1 \times F_A$$
$$= 0.75 \text{ in.} \times \frac{1 \text{ ft}}{12 \text{ in.}} \times 26 \text{ lbs} = 1.63 \text{ ft-lbs}$$

b) Find the normal force F_C when lever distance $d_2 = 1.25$ in.

$$T = d_2 \times F_C$$

$$1.63 \text{ ft-lbs} = 1.25 \text{ in.} \times \frac{1 \text{ ft}}{12 \text{ in.}} \times F_C$$

$$F_C = \frac{1.63 \text{ ft-lbs} \times 12 \text{ in.}}{1.25 \text{ in.}} = 15.65 \text{ lbs}$$

Translating the force F_C to a pressure in pounds per square inch acting on the cylindrical part in the gripper in Figure 4–7 is difficult because the contact surface between the gripper and part is not easy to predict. The amount of contact area between the gripper and the part is a function of many factors including variations in part diameter.

The force at F_A is generated by the downward movement of the piston in the actuator illustrated in the cross-section of the gripper in Figure 4–7a. The force produced by this action is a function of the pressure acting on the piston surface inside the cylinder. Example 4–2 illustrates this concept.

Example 4–2
Calculate the force F_A for a piston of 2 square inches and an applied pressure of 25 pounds per square inch.

Solution

$$F_A = A_p \text{ (piston area)} \times P_p \text{ (piston pressure)}$$

$$= 2 \text{ in.}^2 \times 25 \frac{\text{lb}}{\text{in.}^2}$$

$$= 50 \text{ lbs}$$

Minimum Gripper Force

In the previous examples, the force on the gripped parts was calculated from the air pressure applied to the gripper pneumatic cylinder without any consideration for the amount of force required to hold the part. In gripper design, the minimum force necessary to hold the part throughout the tool motion is the critical value. The ability to hold a part in the gripper is determined by the normal force (the force acting at 90 degrees to the part surface) and the coefficient of friction at the point of contact between the part and the gripper fingers. Two friction coefficient values are used, one for *static* and a second for *sliding* friction. The static value is defined as the friction coefficient that exists just before or at the point when the two surfaces start to slide. The sliding or dynamic coefficient of friction occurs after the two surfaces are sliding. The static coefficient is larger than the dynamic; stated differently, the amount of force required to cause a part to slide is greater than the

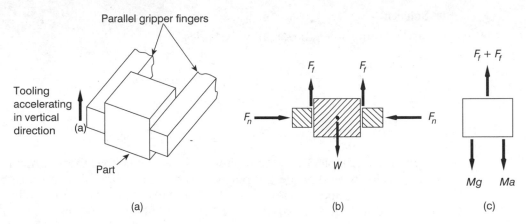

Figure 4–8 Force Diagrams for Grippers.

force necessary to keep it sliding. The coefficient of friction is a function of the materials used in the part and the surface of the gripper fingers in contact with the part. Coefficients of friction range from 0 to 1, with the larger values providing better resistance to slippage. In gripper design, the static coefficient is the most critical because the designer wants to make sure that motion between the part and the fingers never occurs during the motion cycle of the tooling. The relationship among the coefficient of static friction, the normal force, and the friction force are expressed in the following equation

$$F_f = \mu_s \times F_n$$

where

F_n is the force acting at 90 degrees to the sliding surfaces
F_f is the friction force that opposes the sliding motion of the surfaces
μ_s is the coefficient of static friction.

The parallel gripper illustrated in Figure 4–8a shows a block held in the fingers of a gripper. A cross-section of the fingers and part is illustrated in Figure 4–8b. The fingers apply a normal force F_n at the contact surfaces (Figure 4–8b). The part has weight represented by W and the opposing frictional forces represented by the forces F_f on each side of the part. As long as the sum of the two frictional forces is equal to or greater than the weight (Figure 4–8c), the part will not slip out of the gripper. The following example illustrates this concept.

Example 4–3

Determine the minimum normal force necessary to hold a 3-pound part if the coefficient of static friction between the part and fingers is 0.65.

Solution

Using Figure 4–8b, the equation becomes

$$W = F_f + F_f$$
$$W = F_n \times \mu_s + F_n \times \mu_s$$
$$3 \text{ lbs} = 2 \times F_n \times 0.65$$
$$F_n = \frac{3 \text{ lbs}}{1.3} = 2.31 \text{ lbs}$$

In the previous example, the robot tooling was assumed to be at rest or moving at a constant velocity so that the gripper had to overcome only the part weight to keep the part from slipping. However, if the robot was accelerating the tooling in a vertical direction (Figure 4–8a), then the part would be resisting that acceleration with a force equal to the mass of the part M times the acceleration of the tooling a (Figure 4–8c). In this situation, the normal force F_n would need to be greater to overcome the force due to gravity (weight) and the force due to the acceleration of the tooling. The following example illustrates this concept.

Example 4–4

Determine the minimum normal force for the gripper in Figure 4–8 to keep a 5-pound part from slipping when the robot tooling accelerates in an upward vertical direction at a rate of 2 feet per second squared. The coefficient of static friction is 0.75.

Solution

Using Figure 4–8c and treating the problem as a static friction problem, the equation becomes

$$0 = -M_p \times a + 2 \times F_n \times \mu_s - M_p \times g$$

where M_p is the mass of the part, a is the acceleration of the tooling in a vertical direction, and g is the acceleration due to gravity (32 ft per second squared).

a) Find part mass M_p.

$$W = M_p \times g$$
$$M_p = \frac{W}{g} = \frac{5 \text{ lbs}}{32 \dfrac{\text{ft}}{\text{s}^2}} = 0.156 \text{ slug}$$

b) Solve for the normal force.

$$2 \times F_n \times \mu_s = M_p \times a + M_p \times g$$
$$F_n = \frac{M_p(a + g)}{2 \times \mu_s}$$

$$= \frac{0.156 \frac{lb/s^2}{ft} \left(2 \frac{ft}{s^2} + 32 \frac{ft}{s^2} \right)}{2 \times 0.75} = 3.54 \text{ lbs}$$

The normal force must be equal to or greater than 3.54 pounds for the part to hold without slipping as the robot accelerates. If the tooling is accelerating at an angle upward, then the resultant force from the acceleration that is normal to the gripping force is used in the calculations.

4-3 VACUUM GRIPPERS

Vacuum is used as the gripping force in many tooling applications. The part or product is lifted by *vacuum cups,* by a *vacuum surface,* or by a *vacuum sucker gun* incorporated into the end-of-arm tooling. The lifting power is a function of the degree of vacuum achieved and the size of the area on the part where the vacuum is applied.

Most frequently, vacuum grippers use suction or vacuum cups to hold the desired part. Vacuum grippers can be as simple as a single vacuum cup, like the model illustrated in Figure 4–9, or as complex as the multiple-cup system illustrated in Figure 2–16 that uses four vacuum cups to assemble car windows. Figure 4–10 shows another vacuum application, palletizing of containers, where a Fanuc robot uses complex end-of-arm tooling with three vacuum cups to load boxes onto a pallet. In multiple-cup systems, the vacuum to each individual cup can be controlled so that patterns of cups can be activated to pick up material of varying sizes. This type of flexibility makes robots cost effective in manufacturing automation.

Lifting Capacity

The lifting capacity of the vacuum type of gripper is directly related to the pressure of the air surrounding the vacuum cups. Figure 4–11 provides additional background on the principles involved in this lifting technique. One *atmosphere of pressure* at sea level on the surface of an object is approximately 14.7 pounds per square

Figure 4–9 Vacuum Cup.
(Courtesy of Techno SOMMER Automatic)

without vacuum generator

SA 1/4

NS

Figure 4–10   Multiple-vacuum-cup System.
(Courtesy of Fanuc Robotics North American, Inc. All other logos are copyrights of their respective owners.)

inch. Vendor catalogs often use equivalent pressure units for one atmosphere, such as 76 cm of mercury or 1.0133 bars. Vacuum cups use this atmospheric pressure to lift objects, as Figure 4–11 (in which 14.7 pounds per square inch is rounded to 15) and the following example illustrate.

Example 4–5
Find the resultant force on the plate in Figure 4–11 if one square inch on the top of the plate has a perfect vacuum.

Solution
a) The plate in Figure 4–11b has 9 atmospheres of force acting on the bottom surface (14.7 pounds is rounded to 15 pounds in this example). The pressure (P_B) on the bottom of the plate in pounds per square inch is

$$P_B = 9 \text{ in.}^2 \times 15 \text{ lbs}$$
$$= 135 \text{ psi}$$

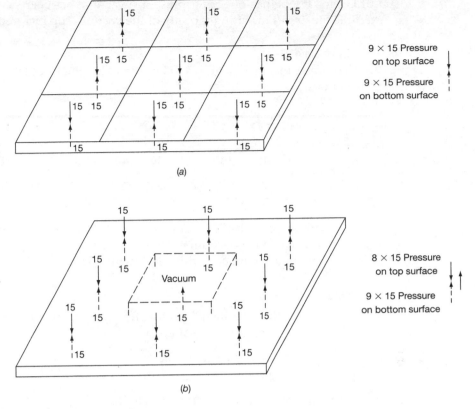

Figure 4–11 Atmospheric Pressure.
(*Note:* One atmosphere is represented by 15 lbs instead of 14.7 lbs.)

b) With only 8 atmospheres on the top surface, the pressure (P_T) is

$$P_T = 8 \text{ in.}^2 \times 15 \text{ lbs}$$
$$= 120 \text{ psi}$$

c) The missing atmosphere of pressure on the top is acting on the outer surface of the vacuum device. A force of 135 lbs acts against the bottom of the plate and only 120 lbs acts against the top. The plate has a net force of 15 pounds acting on the bottom surface. As a result, the plate and vacuum device have a force of 15 pounds holding them together.

As the previous example illustrates, a force of 15 pounds (or, more accurately, 14.7 pounds) holds the plate against the vacuum device as a result of the vacuum. This is the principle behind the vacuum gripper. However, a nearly perfect vacuum is not easy or economical to achieve; therefore, a partial vacuum (0.5 to 0.8 bars) is used with a larger vacuum cup area to obtain the equivalent lifting force. The term *bar* is a measure of how many atmospheres are present. One atmosphere is equal to

1.0133 bars. If less than one bar is present, a vacuum has been generated. A second example illustrates the lifting capacity of the vacuum cup pictured in Figure 4–9.

Example 4–6

The vacuum cup in Figure 4–9 has the following force table provided by the vendor.

Cup Diameter (mm)	10	15	18	24	30	40	50	60	70	85	100
Force (N)	4.7	10.6	15.2	27.1	42.4	75.4	118.0	170.0	230.0	340.0	471.1

The cup diameter is the outside diameter of the cup, and the force is in newtons when the vacuum is 0.8 bars. Calculate the lifting capacity in pounds for a 30-mm cup when the vacuum is 0.75 bars.

Solution

a) Convert the vacuum in bars to lifting capacity in pounds per square inch (assume 1 bar = 1 atmosphere).

$$\frac{\text{Holding capacity in the cup}}{\text{for a vacuum of 0.75 bar}} = 0.75 \times 14.7 \text{ psi} = 11.03 \text{ psi}$$

b) Convert cup diameter from mm to in.

$$d_c = \frac{30 \text{ mm}}{25.4 \ \dfrac{\text{mm}}{\text{in.}}} = 1.18 \text{ in}$$

c) Calculate the lifting capacity (F) of a 1.18-in. diameter cup.

$$F = \text{cup area} \times \text{holding pressure}$$
$$= \left(\frac{1.18 \text{ in.}}{2}\right)^2 \times \pi \times 11.03 \text{ psi} = 12.06 \text{ lbs}$$

The surface of the part determines the quality of the seal between the vacuum cup and the part; as a result, the surface of the material being lifted has the greatest effect on the total weight that can be raised. The smoother the finish on the surface, the closer the lifting force is to the ideal.

System Components

Vacuum gripper vendors provide several components to simplify the configuration of a system. For example, pneumatically powered vacuum generators use compressed air to generate a vacuum. In some cases the generator is integrated

Figure 4–12 Vacuum Cup with Sensor and Vacuum Generator.
(Courtesy of Techno SOMMER Automatic)

Outlet for proximity switch cable

Compressed air

NS-Suction cup

into the vacuum cup, as the model in Figure 4–12 illustrates. Study the specifications for this model in Appendix A. The generators require compressed air at 6 bars and a flow rate of 30 standard liters per minute to create the vacuum by the venturi principle. The specifications and drawings also indicate that the cup has an internal proximity sensor that is activated when a spring loaded pin in the center of the cup is depressed by the object to be lifted. Study Figure 4–13 to see how the internal proximity sensors operate. When the vacuum cup makes contact with the part the pin is depressed. The pin activates the proximity device, which closes electrical contacts to activate a compressed air valve connected to the vacuum generater. The part is acquired and the robot moves the part to the desired location. Other components include (1) a vacuum sensor with a variable trigger point that provides a contact closure if the vacuum falls below acceptable levels, (2) vacuum valves to control the application of the vacuum, (3) swivel ball joints and spring loaded cups (see Appendix A) that allow the cup to align with inclined

Figure 4–13. Internal Proximity Sensor.

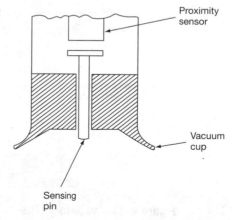

Proximity sensor

Vacuum cup

Sensing pin

surfaces, (4) solid cores at the center of the cups to cover holes in the part being lifting, and (5) bellows cups that provide compliance and a high degree of tolerance to surface orientation.

Vacuum Surfaces

Vacuum surfaces are just an extension of the vacuum cup principle. In some material-handling applications the product to be lifted is not ridged enough for vacuum cups to be effective. For example, robots are used to make composite material by building up multiple layers of graphite fiber cloth and resin. To lift the cloth into place, a vacuum surface such as the one illustrated in Figure 4–14 is used. Note the flat surface that forms one side of a vacuum chamber. Each hole in the vacuum surface provides a small lifting force so that the flexible cloth would be held against the vacuum surface from many points as the graphite fiber cloth is moved into place. Any flexible material that is not too porous can be lifted effectively in this manner.

Vacuum Suckers

The use of *sucker guns* in the production of nylon fiber is well established. The tool consists of a wand capable of sucking up a nylon end or a thread line as it is produced, thus permitting the operator to thread the machine. This type of end-of-arm tooling could be very useful for applications in which robots must handle material in the form of thread, line, or fine wire.

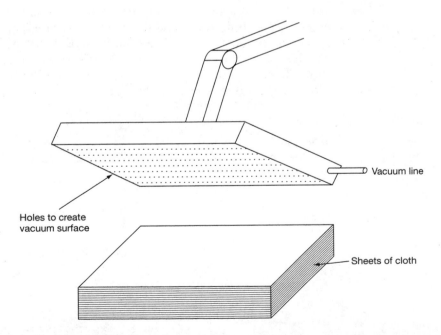

Figure 4–14 Vacuum Surface.

Figure 4–15 Magnetic Gripper.
(Courtesy of GCA Corporation)

4-4 MAGNETIC GRIPPERS

Parts that contain ferromagnetic material can be lifted with an electromagnet mounted on the robot tool plate. Figure 4–15 is a picture of a *magnetic gripper*. An electromagnet has a center core of iron (circular metal in the center of the gripper in Figure 4–15) with a coil of wire around the core. When a dc voltage source is placed across the coil, the current creates a magnetic pole at the end of the iron core. This pole will attract material with ferromagnetic characteristics and hold them to the core.

4-5 AIR-PRESSURE GRIPPERS

Fingers, mandrel grippers, pin grippers, and *bellows* form a group that uses air pressure to grip parts. The fingers, also called *pneumatic fingers,* have a hollow rubberlike, cylindrical body with a smooth surface on one side and a ribbed surface on the opposite side. With pressure applied to the inside of the hollow body, the finger deflects in the direction of the smooth side. Figure 4–16 illustrates this process, and Figure 4–17 shows a pair of parts being lifted by robot grippers with two pneumatic fingers. The pneumatic fingers do not apply a large force for gripping; therefore, they are usually used under a flange, as illustrated in Figure 4–17.

Mandrel grippers are inside grippers with an airtight flexible diaphragm mounted to a mandrel. Figure 4–18 shows a section view of a circular mandrel gripper inside a part. When air under low pressure (25 to 35 psi) is forced into the port, the diaphragm expands and traps the part. Figure 4–19 shows a mandrel gripper (diaphragm is light gray) on robot end-of-arm tooling moving into a part.

Figure 4–16 Pneumatic Finger.

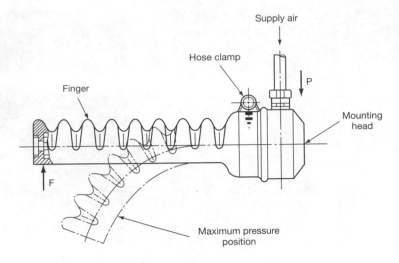

Figure 4–17 Finger Gripper.
(Courtesy of GCA Corporation)

Figure 4–18 Mandrel Gripper.

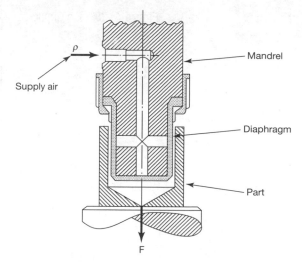

Supply air

ρ

Mandrel

Diaphragm

Part

F

Pin grippers (Figure 4–20) are similar to mandrel grippers except that the part is gripped from the outside. The gripper fits down over a round pin, then the diaphragm is expanded inside the ridged housing, and the part is trapped. The primary advantage offered by air pressure grippers is the gentle variable force applied to the part being held.

Figure 4–21 illustrates a *bellows* type of air pressure gripper. The part to be grasped is located between the two gripper fingers and air is applied to the bellows on the inside surface of each finger. The expansion of the bellows traps the part between the fingers for transport.

Figure 4–19 Mandrel Gripper.

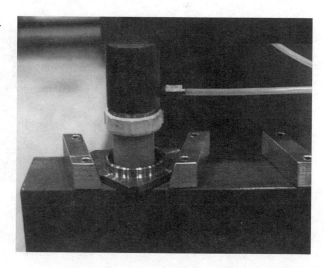

Figure 4–20 Pin Gripper.

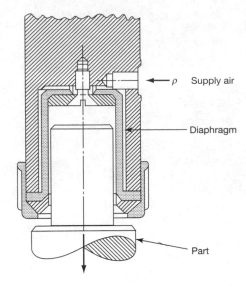

← ρ Supply air

Diaphragm

Part

Gripping Force

The gripping force for the air pressure type of grippers is calculated like the gripping force for the standard grippers. The compressed air used to expand the gripper diaphragm acts directly on the part wherever the diaphragm contacts the part. The gripping force is thus a product of the air pressure in psi and the diaphragm surface in contact with the part. On circular parts, the force acts uniformly around the part circumference. The following examples illustrate this concept.

Example 4–7
The mandrel gripper in Figure 4–18 extends 0.75 inch into a part that has a 1-inch diameter hole. If 30 psi of air are used, calculate the gripping force applied to the part by the gripper.

Figure 4–21 Bellows-type Parallel Gripper.

Solution

a) Calculate the area of contact (A_C) between the part and the gripper. Part diameter (d_1) = 1 in. and penetration (h) = 0.75 in.

$$A_C = \text{part hole circumference} \times \text{depth of contact}$$
$$= d_1 \times \pi \times h$$
$$= 1 \text{ in.} \times \pi \times 0.75 \text{ in.} = 2.36 \text{ in.}^2$$

b) Calculate the gripping force (F).

$$F = A_C \times \text{air pressure applied}$$
$$= 2.36 \text{ in.}^2 \times 30 \frac{\text{lb}}{\text{in.}^2} = 70.8 \text{ lbs}$$

Example 4–8

The pin gripper in Figure 4–20 overlaps a 0.5-inch diameter pin by 1 inch. If 25 psi of air are used, calculate the gripping force applied to the part by the gripper.

Solution

a) Calculate the area of contact (A_C) between the part and the gripper.

$$A_C = d_1 \times \pi \times h$$
$$= 0.5 \text{ in.} \times \pi \times 1 \text{ in.} = 1.57 \text{ in.}^2$$

b) Calculate the gripping force (F).

$$F = A_C \times \text{air pressure applied}$$
$$= 1.57 \text{ in.}^2 \times 25 \frac{\text{lbs}}{\text{in.}^2} = 39.25 \text{ lbs}$$

Example 4–9

The bellows gripper in Figure 4–21 has a bellows geometry that includes a center rectangle 0.5 inch by 3 inches and semicircular ends. Calculate the gripping force on a rectangular part when 35 psi of air are used.

Solution

a) Find the bellows area (A_C). Height (h) = 0.5 inch, length (l) = 3 inches, and the ends are equivalent to a circle.

$$A_C = \text{area of center rectangle} + \text{area of round ends}$$
$$= h \times l + \left(\frac{h}{2}\right)^2 \pi$$
$$= 0.5 \text{ in.} \times 3 \text{ in.} + \left(\frac{0.5}{2}\right)^2 \times \pi = 1.7 \text{ in.}^2$$

b) Calculate the gripping force (F).

$$F = A_C \times \text{air pressure applied}$$

$$= 1.7 \text{ in.}^2 \times 35 \frac{\text{lb}}{\text{in.}^2} = 59.5 \text{ lbs}$$

4-6 SPECIAL-PURPOSE GRIPPERS

The expanding robot application base supports the development of off-the-shelf grippers to fill special applications. As a result, the largest classification of grippers now in use falls into the category of special-purpose tooling. In addition to those available off the shelf, many more are fabricated by the robot end user. Often the user starts with a basic pneumatic or electric mechanism available in the industry and modifies the gripper to do the special job. In almost every case special gripper fingers must be fabricated to hold parts for specific jobs. Figure 4–22 shows two internal grippers mounted together and turned 90 degrees so the robot can unload

Figure 4–22 Robot Loading a Cinturn Turning Center.
(Courtesy of Cincinnati Milacron Corp.)

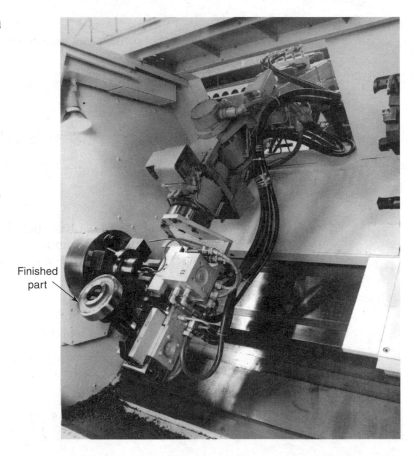

Finished
part

a finished part from the chuck and then load raw material in one trip to the turning center. In the figure, a finished circular part is visible in the gripper farthest from the tool plate. The two expanding fingers can be seen inside the hole of the circular part.

Again, robot flexibility is responsible for the large variety of gripper designs. It is often a simple matter to identify a secondary task that the gripper can perform during the production cycle in addition to the primary parts-handling or assembly responsibility. Examples of secondary tasks that can be incorporated into a standard gripper include hooks for lifting or turning parts over, air guns for cleaning, and sprayers for lubrication of parts and dies.

4-7 GRIPPER SELECTION AND SYSTEM INTELLIGENCE

In setting up any manufacturing system, some criteria must be used in the selection of equipment. When robots are included in the automation, an analysis of each production sequence will establish the requirements of the gripper for each step of the manufacturing process. Every sequence in the manufacturing process should be examined and the relative difficulty established. The robot system capability, including the end-of-arm tooling, must be equal to or greater than the most demanding sequence in the process. In this way, the robot system intelligence level will always match the need of the production requirement.

There is another advantage to analyzing the requirements of each sequence in the manufacturing process. Often only one sequence is classified as difficult and that sequence dictates a robot system or end-of-arm tooling with far greater capability than is required for the rest of the operations. As a result of such an analysis, the demanding sequence can be identified and possibly modified to permit a lower cost implementation.

4-8 SPECIAL-PURPOSE TOOLS

Grippers are also designed to hold power tools in the same way that human hands do. Those most frequently used include drills, welding guns, glue and sealer dispensers, spray guns, grinders, and sand blasters. Robots are often used for jobs that fulfill the three Ds: dirty, dangerous, and dull. A human operator using any of these tools must wear equipment to protect his or her eyes, eardrums, or lungs from harm and often requires counterbalance devices to overcome the weight of heavy tools. As a result, these jobs are ideally suited for robots.

4-9 ROBOT ASSEMBLY

Assembly applications for robots will grow at a rate faster than all other current or future application areas simply because assembly requirements occur in every segment of manufacturing. As gripper dexterity and part manipulation capability

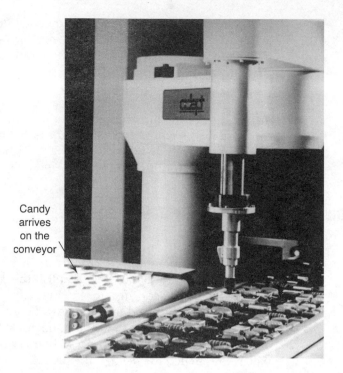

Figure 4–23 AdeptOne Robot at Chocolat Frey of Buchs, Switzerland. (Courtesy of Adept Technology, Inc.)

Candy arrives on the conveyor

increase with developing technology, and as microprocessor systems are designed to control more complex grippers, the number of assembly applications will increase. Figure 4–23 shows an AdeptOne robot inserting chocolates into candy box liners; note that the candy is delivered on a conveyer in random locations. A vision sensor interfaced to the robot is used to locate the chocolates on the conveyer, then a vacuum gripper is used to move the candy to the box.

Grippers designed to be used in assembly operations have special needs that must be satisfied. For example, parts must be moved into place for the assembly operation, fasteners must be put into place, and assembly tools such as screw drivers must be applied. In all these operations, the compliance between the parts must be considered.

4-10 COMPLIANCE

Compliance deals with the relationship between mating parts in an assembly operation. For example, if a pin must be fitted into a hole, alignment between the hole and the pin must be achieved. Because it is impossible to guarantee perfect alignment between the robot gripper that holds the pin and the fixture holding the part, compliance techniques must solve the slight misalignment problems between mating parts during assembly. Compliance is defined as follows:

Compliance means initiated or allowed part movement for the purpose of alignment between mating parts.

When a pin is inserted into a hole, three types of contact can occur during the insertion process: *chamfer, sliding,* and *two point.* The chamfer contact occurs when the pin is not perfectly aligned with the hole and hits the chamfered edge of the hole (Figure 4–24a). If the assembly system is not perfectly rigid, the pin will rotate slightly and start to slide along one side into the hole (Figure 4–24b). If the misalignment is severe, the pin will make a two-point contact, with the base of the pin hitting the far wall of the hole (Figure 4–24b). Two-point contact on insertion often results in damage to both the parts and the assembly system. Problems with mating part alignment in assembly and other applications are resolved using *active* and *passive* compliance techniques.

The initiated or allowed movement during insertion will be lateral, rotational, or axial about a center of rotation. Figure 4–24 shows the three basic

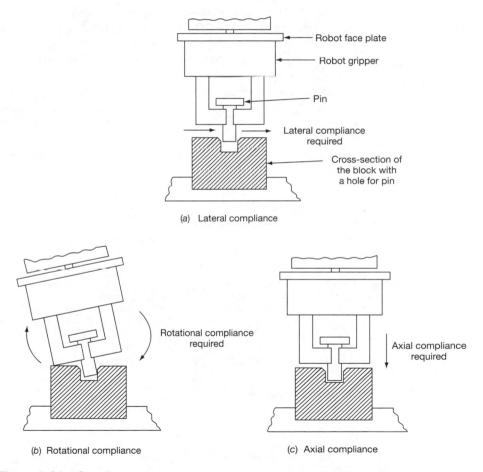

(a) Lateral compliance

(b) Rotational compliance

(c) Axial compliance

Figure 4–24 Compliance.

types of movements and the misalignment that must be eliminated by each. In Figure 4–24a, lateral part movement is required for mating. A rotational correction is necessary for proper part insertion in Figure 4–24b. Devices that provide lateral and rotational compliance are commercially available. The axial compliance illustrated in Figure 4–24c is not necessary for proper alignment but is added when the pin must be flush with the bottom of the hole. Using axial compliance, the robot could move in the axial direction, slightly past the point where the pin is seated. In these drawings, each compliance condition was considered separately, but in operation they all may be present at one time, and the compliance technique employed must continue to function. Without compliance the entire system is rigid, so that force applied by the robot when two-point contact is present will result in damage to the two mating parts.

Active Compliance

Active compliance systems measure the *force* and *torque* (*F/T*) present when the robot performs the programmed task; as a result, they are often called *F/T sensing* systems. Force sensing allows the robot to detect changes and variations in the workpiece or tooling and adapt the program to correct them. In F/T sensing systems, the robot moves are based on a required or measured force instead of a programmed point.

F/T sensing uses an adaptor placed between the gripper and the robot tool plate to measure the force and torque caused by contact between mating parts. The adaptor has an outside ring or wall (Figure 4–25a) that attaches to the robot tool plate. A center hub is connected to the ring by three webs or beams (Figure 4–25a) that have strain sensors attached. With the gripper attached to the hub, any force

Figure 4–25 Force/Torque Transducer (a) and Vectors (b).
(Courtesy of Assurance Technologies, Inc.)

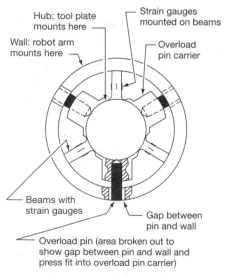

or torque on the gripper as a result of a programmed action causes the webs connecting the hub to the ring to deflect. The deflections are measured by the strain sensors and converted into three force vectors and three torque vectors (Figure 4–25b) that are used by the robot to adjust the programmed motion. Note that overload pins attached to the hub prevent damage to the sensor if too great a deflection is experienced.

Active Compliance Applications

The gear assembly application, pictured in Figure 4–26, shows the F/T sensor active compliance device between the gripper and tool plate of a robot. Adept Technology, Inc. has integrated the control of an F/T sensor into the robot controller; as a result, the robot can read and react to contact with the tooling in less than 3 milliseconds. In the Adept integrated system, the following force strategies are available: move until the specified force threshold is exceeded, move in a defined spiral search pattern to find a hole, move to verify assembly, and move to cause a snap-fit assembly. F/T sensors are available from 3 pounds of force and 1 inch-pound of torque to 600 pounds of force and 3600 inch-pounds of torque.

Figure 4–26   Force/Torque Sensor in Robotic Assembly.
(Courtesy of Assurance Technologies, Inc.)

Passive Compliance

The second technique, *passive compliance,* provides compliance mechanically by means of a *remote center compliance* (RCC) device. The concept was originally developed in the 1930s to reduce vibrational stresses in radial aircraft engines by permitting deflections in the engine mounts that coincided with rotations about the engine's center of rotation. The theory of operation of RCC devices is a little abstract, but understanding the principle of operation is a minimum requirement for anyone using these devices in a work cell. The *center of compliance* (remember, compliance implies movement) is that point at which the entire compliance system, including the gripper and part, is considered to be concentrated and acting. The location of the center of compliance is determined by the design of the RCC device and depends on the location of the compliance elements and their orientation to the RCC device. As the name indicates, the center is usually remote from the device itself. Figure 4–27 illustrates this concept.

As we discuss the operation of the RCC device, consult the design in Figure 4–27, which came from research work at the Charles Stark Draper Laboratory at MIT. The RCC device consists of three plates; the center plate is connected to the top plate with four rods (only 2 of the 4 are visible) and to the bottom plate with four additional rods (only 2 of the 4 are visible). In operation, four rods, one on each corner, are used for lateral compliance (only two rods are shown in Figure 4–27, for clarity), and four angled rods, one on each corner, are used for rotational compliance (again, only two rods are shown). The pivoted rods allow the plates to move relative to each other and provide a combination of lateral and

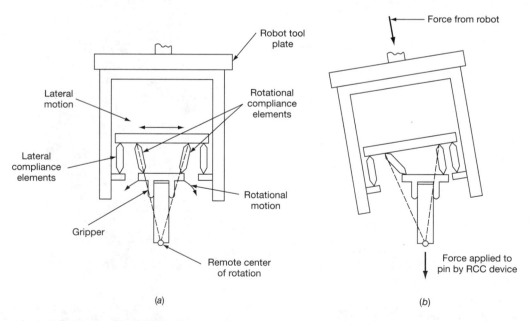

Figure 4–27   RCC with Lateral and Rotational Compliance Links.

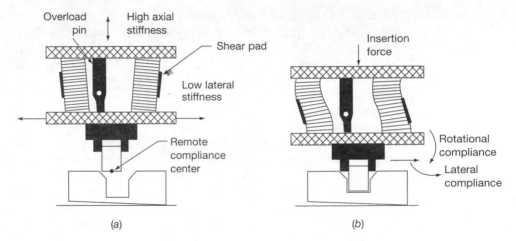

Figure 4–28 Operation of Remote Center Compliance Device.
(Courtesy of Assurance Technologies, Inc.)

rotational compliance; however, this device is rigid in the axial direction (in the direction of the robot force in Figure 2–27b) with no compliance provided.

Modern RCC devices, like the model illustrated in Figure 4–28 and pictured in Figure 4–29, consist of three to six shear pads sandwiched between two

Figure 4–29 Remote Center Compliance Device.
(Courtesy of Assurance Technologies, Inc.)

aluminum plates and shear pins for protection against excessive movement. This type of RCC design is called the *single-stage shear pad* device. The key to the performance is the construction of the shear pads, which are made from 15 to 30 alternate layers of neoprene or nitrile elastomer and metal shims. The upper plate is attached to the robot tool plate and the gripper is attached to the lower plate. The shear pads are stiff in the axial directions but highly compliant in the lateral and rotational directions. The slight angle of the shear pad causes the rotational compliant effect illustrated in the model in Figure 4–27b. Note in the model and in the actual device (Figure 4–28) that the lower plate rotates because the shear pads are placed at slight angles; as a result, the increased angle on one shear pad reduces the distance between the upper and lower plates. Additional rotation is caused by the other shear pad, which moves to a more vertical position, thereby increasing the distance between the two plates. The combination of these two actions straightens the pin in Figure 4–27b. Study the action in Figure 4–28 until you understand the lateral and rotational compliance.

Two additional features, *shear pins* and *lockup,* are available on current RCC devices and improve the performance of the original design. The shear pins provide protection for the shear pads against the occasional and inevitable overloads that occur. The shear pin is attached to the top plant and is centered inside a hole in the bottom plate. If the lower plate tries to move an excessive distance, then the shear pin comes in contact with the lower plate and prevents further movement. The RCC device in Figure 4–29 has a shear pin visible on the front left side of the device. The lockup feature removes the compliance from the RCC device by locking the top and bottom plates through rigid pins whenever the device is not performing an assembly. The weight of the part and gripper could cause the RCC device to oscillate after a quick move to the assembly. The insertion operation would have to be delayed until the oscillation died out naturally. To overcome this problem, some RCC devices have a lockup feature that makes the device rigid for moves or for operations where compliance is not desired. The lockup capability reduces assembly cycle time and increases the operational life of the shear pads. Figure 4–30 shows an RCC device used in a pneumatic deburring application. The soft lateral compliance permits the burr to have a constant pressure against the part.

Selective Compliance Articulated Robot Arm (SCARA).   Another form of passive compliance is found in the SCARA configuration. The AdeptOne robot in Figure 2–17 uses SCARA technology to provide a variable tool movement for insertion compliance at programmed points. After the tool reaches the programmed point, the controller frees the servo system so the gripper can move freely over a selected distance in the X and Y directions. If the gripper attempts to move beyond the selected range, the servo system stiffens, and the gripper position is maintained.

In addition to the SCARA arms that initiated the selective compliance technique, some non-SCARA arms—for example, the Seiko RT-3000—have language commands that permit any of the axes to have compliance. The command for the Seiko is **FREE**<*axis*>where the argument (*axis*) is the letter for the axis where compliance is desired.

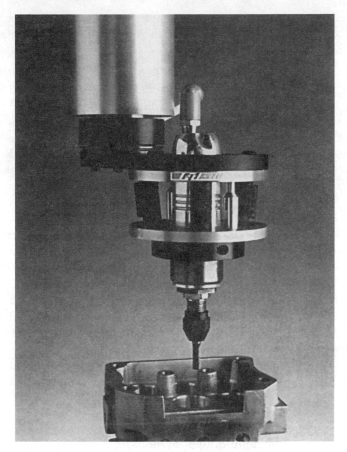

Figure 4–30 Remote Center Compliance Device in a Deburring Application.
(Courtesy of Assurance Technologies, Inc.)

4-11 MULTIPLE END-EFFECTOR SYSTEMS

In most applications currently using robots, the machines have a single function to perform in the work cell, and only one end effector is necessary. As work cells become more complex and robots more sophisticated, however, the application of multiple gripper systems increases. Multiple gripper systems are implemented using two techniques. One approach is to mount physically two or more tools to a single tool plate, and the other approach uses tool changing technology to swap end effectors. The introduction of tool changing technology demands a standard interface at the robot wrist.

Wrist Interface

The wrist interface for flexible manufacturing must satisfy the following interface requirements:

- *Mechanical interface:* The robot must be able to change tooling under program control, and the integrity of the mechanical linkage must be as good as that experienced with threaded fasteners. The interface must provide both registration and orientation control from one tool to the next.

■ *Electrical interface:* The electrical signals used for control of the tooling or the signals coming from sensors mounted on the gripper must be separated automatically when the tooling is changed. For example, the power to electrically powered tools must be disconnected and then reconnected as the tooling is changed.

■ *Pneumatic interface:* The same rationale for a quick-breaking interface that was developed in the electrical area also applies here.

■ *Replaceable or quick-change capability:* Future applications must provide for rapid tool change by the user and the robot.

The advantages of a standard wrist interface include these three: (1) the same tooling can be used on every robot in a manufacturing facility without special adapters for each robot; (2) no program editing or reprogramming is required when a worn or damaged gripper is replaced; and (3) off-the-shelf tooling for general applications can be developed and sold.

The Robotic Industries Association (RIA) has several standards that affect the mechanical interface area. One of those, ANSI/RIA R15.03, for industrial robots and robot systems—mechanical interfaces, has two components. The first is the circular mechanical flange interface standard and the second is the shaft mechanical interface standard. Development of standards will help move the robot industry toward interchangeable tooling.

Multiple-gripper Systems

A multiple-gripper system is one that has a single robot arm but two or more grippers or end-of-arm tools, mounted on the tool plate, which can be used interchangeably on the manufacturing process in the cell. These multiple tooling systems can be completely separate grippers or tools mounted to the fixture on the end of the robot arm. The tooling pictured in Figure 4–31 shows six pneumatic grippers with a combination of two, three, and four jaws to handle many tasks.

In other applications, the robot tooling is designed to permit two or more grippers to be used on the same robot arm on an interchangeable basis. The discussion of the wrist interface indicated that interchangeable tooling would require an interface for mechanical, electrical, and pneumatic requirements. A quick change mechanism, developed by Assurance Technologies, Inc., satisfies the mechanical, electrical, and pneumatic requirements. The quick change device, shown in Figure 4–32, consists of a master plate that connects to the robot tool plate and a tool plate that attaches to the gripper. Study the figure and notice the parallel rows of fifteen electrical contacts and the eight pneumatic interfaces. The external connections for the electrical and pneumatic ports are visible as well. The interface

Figure 4–31   Multiple End Effector.
(Courtesy of MACK Corporation)

sequence is illustrated in Figure 4–33. Device models have payloads that range from 5 to 150 kg, have from ten to seventy-four electrical connections, and have from six to fourteen pneumatic ports. The current for the electrical interface is typically

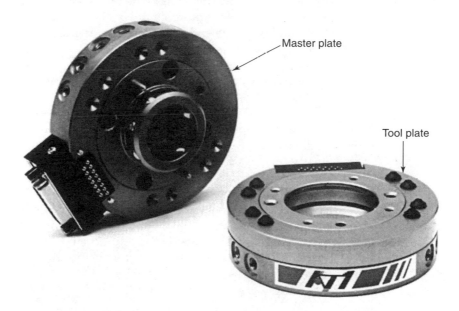

Master plate

Tool plate

Figure 4–32   Quick Change Device.
(Courtesy of Assurance Technologies, Inc.)

Figure 4–33 Quick Change Tooling System.

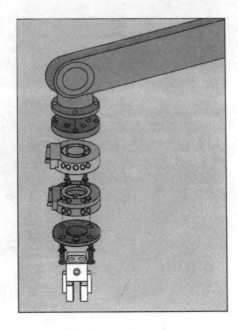

3 amperes at 50 volts (higher capability is available on larger models). The pneumatic interface can operate from a vacuum up to 100 psi. The repeatability of the interface is ±0.010 millimeters and would add to the repeatability present in the robot.

As work cells become more complex and costly, the concept of increased flexibility through multiple end-of-arm tooling makes economic sense. Cost justification on this basis alone is difficult to establish in terms of person hours saved, but in many production situations the use of robot automation would not be possible if multiple grippers were not employed. For example, in a production situation in which the range of part sizes is great, the only way to avoid installing several robots is through multiple grippers. In an assembly application in which the number and variety of operations to be performed is numerous, it would not be economically feasible to have a robot and separate gripper for each operation.

The advantages of using multiple grippers include (1) increasing the production capability of the work cell, (2) reducing work-in-process time for the part because it must be moved through fewer workstations, and (3) using the robot arm and controls more efficiently. In addition, many assembly applications would not be possible without the development of multigripper work cells. The primary disadvantage is the additional complexity added to the tool design problem. This also results in greater cost for the design and construction of the devices.

The multiple-gripper system pictured in Figure 4–34 is used in an assembly work cell. The tools, from left to right, include a parallel gripper, automatic electric screwdriver, and a vacuum cup. Each tool would have a different TCP offset value that would be changed in the robot program when the tool is changed.

Figure 4–34 Multiple-Gripper System.

(Courtesy of Fared Robot Systems)

4-12 COLLISION SYSTEMS

The tooling for a $200,000 robot work-cell application can cost $40,000, so the protection of that tooling from a damaging collision with other work-cell hardware is critical. Collision protection hardware falls into three categories: breakable link devices, spring loaded couplers, and pneumatic controlled couplers. All the collision protection devices are placed between the gripper or tooling and the robot tool plate, as illustrated in Figure 4–35. Each type of device detects a collision by sensing that the robot tool plate tooling is continuing on the programmed path, but the tooling has stopped moving. The difference in motion between the tool plate and tooling creates forces on the collision device that joins them. If these forces exceed an allowed lower limit, the collision device gives way and also initiates an emergency stop signal to the robot.

Breakable Link Devices

In the breakable link devices, replaceable metal links are severed by the force of collision so that the tooling can move, and a switch in the emergency stop circuit is activated. The system is serviced by replacing the broken links with new ones. The accuracy and repeatability of the robot program is affected because the alignment between the tool plate and the tooling is changed by the replacement of the links. In applications where programmed points have a tolerance of less than 0.001 inch, the points would need to be reprogrammed. Other types of breakaway mechanical collision protection devices do not have breakable links, but the separation mechanism must be reset manually after a collision.

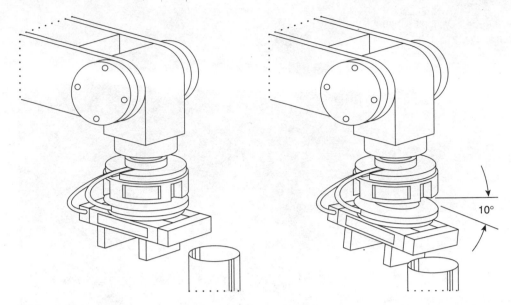

Figure 4–35 Collision Protection Coupling.
(Courtesy of Robotic Accessories)

Spring and Pneumatic Collision Devices

The spring and pneumatic collision devices also give way to overload forces in the pitch, yaw, roll, lateral, and insertion motion directions and provide a limit switch for use in the robot's emergency stop circuit. However, both have operational features that make them superior to the breakable link units. For example, both types have the following features:

- An automatic reset that returns the tooling to within 0.002 inch of the original tooling orientation after a collision.

- A compliance option that allows for up to 0.04 inch of movement in the coupling before the load sensing switch is tripped. When noncompliance devices are used, the load sensing switch is tripped with less than 1 degree of deflection. Other options include only vertical travel in the noncompliance devices.

- An over-travel feature (Figure 4–35) that allows from 10 to 15 degrees of deflection in the tooling for continued robot motion after the emergency stop signal.

- A wide range of resistance, from 2 inch-pounds to over 7000 inch-pounds.

The spring devices use stiff springs to link the production tooling to the robot tool plate, while the pneumatic devices use a pneumatic coupling mechanism. In addition, the pneumatic collision couplers have the following operational features:

- A pneumatic control system that allows the tripping force to be set pneumatically using external valves and program signals from the robot or cell controller.
- A spring set lower trip limit used for collisions with a pneumatically controlled high limit used for rapid movements of the arm. The spring-pneumatic combination also ensures that the tooling will not sag in the case of accidental air loss.

The low cost of collision devices makes them a good investment for protecting expensive robot tooling.

4-13 SUMMARY

The objective of this chapter was to introduce the different types of tooling that are currently available for use on robot arms. Grippers must be capable of holding the required part so that safety is assured under all operating conditions. In addition, grippers should conform to the lightest design using the simplest technique possible while holding and sensing the presence of the manufactured part.

In general, end-of-arm tooling groups include standard grippers, vacuum grippers, and special-purpose grippers and tools. Assembly fixtures require that special attention be paid to the problem of compliance. The alignment problem between mating parts can be solved by either active or passive compliance. In active compliance, or force/torque sensing, the compensation for part alignment error occurs in the commands the controller sends to the arm. The adjusted coordinate values are generated as a result of signals received from the gripper indicating a force and torque resulted from an attempt to mate two parts. The second type of compliance is called passive and uses an RCC device. The required movement of the part to permit alignment and mating is produced in the RCC, which is mounted between the gripper and the robot wrist-mounting plate. The corrective movement results from forces acting on the part in the gripper as it is inserted into the assembly without perfect alignment.

Multiple end-effector systems have the advantage of permitting robot automation to be implemented in work cells involving complex operations. In addition, they permit increased use of expensive robots on a wider range of manufacturing problems. These advantages overcome the primary disadvantage, which is increased cost because of greater complexity.

The rate at which robots are introduced into new production situations is directly related to the rate at which the tooling develops. Many of the new applications require grippers that approximate the dexterity of the human hand. The development of a gripper with three or more fingers that can grasp a variety of irregularly shaped objects is critical for many future applications. Increased robot use is directly proportional to technical breakthroughs in end-of-arm tooling design.

End-of-arm tooling is protected by collision couplings. Three types are commonly used: breakable links, spring loaded devices, and pneumatic systems.

QUESTIONS

1. What is end-of-arm tooling and what function does it serve?
2. What are the five characteristics that all end-of-arm tooling must satisfy?
3. What are the three categories for classifying end effectors, and what is included in each category?
4. Draw a classification tree that includes all the types of end-of-arm tooling described in this chapter.
5. What is the estimated cost for off-the-shelf and in-house designed tooling as a percentage of total system cost?
6. Write a procedure for calculating the lifting power of round vacuum cup grippers.
7. How does the complexity of the production job affect the intelligence level of the robot and tooling used for the application? How can that be changed?
8. What does the term *compliance* mean?
9. Describe the operation of an F/T sensor used for active compliance.
10. Describe the two methods used to achieve passive compliance.
11. Describe the three compliance conditions normally found in robot applications.
12. What is an RCC device?
13. What are the advantages and disadvantages of multiple end-effector systems?

PROBLEMS

1. Calculate the force at F_C in Figure 4–7b when the force F_A is 50 pounds. The lever arm between F_A and the pivot point is 1 inch, and the force on the part acts at a vertical distance of 2 inches from the pivot point.
2. How much pressure does a parallel gripper with 0.75 square inch of contact surface between a finger and the side of a square part produce if the force acting on the contact surface is 35 pounds?
3. Calculate the force generated on the piston rod by a 1.5-inch diameter piston for a pneumatic cylinder with 30 pounds per square inch of applied air pressure.
4. Determine the diameter of the piston (Figure 4–7a) required for the pneumatic cylinder in Problem 1 if air pressure is 18 psi and the piston rod has a 0.375 diameter.
5. What percentage change is necessary in the linkage distance between the force F_A and the pivot point to create a 25 percent increase in the force F_c in Problem 1?
6. Determine the minimum vertical stroke required on the piston rod in Figure 4–7 that would produce a distance of 1 inch between the gripping surfaces when the gripper is open. Use dimensions from Example 4–1.

7. Determine the minimum normal force (Figure 4–8) necessary to hold a 5-pound part if the coefficient of static friction between the part and fingers is 0.7.

8. The fingers of a parallel gripper apply a normal force of 54 newtons to each side of a plastic part. If the coefficient of static friction is 0.85, calculate the maximum weight, in pounds, of the part that can be lifted without slipping.

9. Determine the minimum normal force for the gripper in Figure 4–8 to keep a 15-pound part from slipping when the robot tooling accelerates in an upward vertical direction at a rate of 0.75 feet per second squared. The coefficient of static friction is 0.6.

10. What is the percentage change in the normal force for the gripper in Problem 9 if the robot accelerates upward at a 45-degree angle?

11. What is the minimum normal force for the gripper in Problem 9 if the robot accelerates downward at 5 feet per second squared?

12. Design a standard gripper that would have the following specifications: overall gripper diameter equal to or less than 3 inches; overall gripper length equal to or less than 6 inches; and the ability to grip a one-quarter-inch thick, 9-inch diameter, solid aluminum part by using a 1.5-inch diameter hole in the center of the part and 20 psi shop air. The documentation should include all design assumptions, required normal force when fingers are fabricated from aluminum and for aluminum fingers covered with rubber, drawings of gripper parts with full dimensions, and calculations to verify compliance with the specifications.

13. Determine the contact size in square inches for the fingers of a parallel gripper to meet the following specifications: rubber coated aluminum contact surfaces with a μ_s of 0.85, part that is a cardboard box weighing 179 newtons and capable of withstanding a maximum pressure of 1 psi, application of a 25 percent lifting safety factor, and a maximum vertical acceleration of the robot arm of 35 centimeters per second squared.

14. Convert the vendor force table in Example 4–6 from millimeters and newtons to inches and pounds.

15. Determine the diameter of a vacuum cup that would produce 14.7 pounds of force when the vacuum is 0.7 bars. How does the area compare with that in Example 4–5 where a perfect vacuum was used? What is the new cup size if the vacuum is 65 centimeters of mercury?

16. Design vacuum gripper tooling like the system pictured in Figure 2–16 for window glass that weighs 22 pounds. Assume a vacuum of 0.75 atmospheres and a 50 percent safety factor. Use vendor data in Appendix A to select components and to document the number, size, and layout of the cups.

17. Redesign the vacuum system pictured in Figure 4–10 so that it picks up either one or two boxes. A vacuum of 0.8 bars is provided, each box weighs 9 pounds, and the safety factor is 25 percent. Use vendor data in Appendix A to select components and to document the number, size, and layout of the cups. Also indicate how the system is configured with vacuum valves to control what cups have a vacuum.

18. The gripper in Figure 4–14 picks up 4-mil polyvinyl sheets that weigh 3 ounces per square foot. If the vacuum is 0.95 bars behind the gripping surface, how many $\frac{1}{8}$-inch holes will be required per square foot to lift the material with a 50 percent safety factor?

19. Design a vacuum gripper that would pick up 18-gauge sheet metal, 10 inches on a side, from a stack with a random number of pieces present. Assume a vacuum of 0.7 bars and a safety factor of 10 percent. Use vendor data in Appendix A to select components and to document the number, size, and layout of the cups. Also indicate how the system is configured with a proximity sensor to signal the robot that the top of the stack was reached and to turn on the vacuum valves.

20. A mandrel gripper extends 1.25 inches into a part that has a 0.4-inch diameter hole. If 20 psi of air are used, calculate the lifting force applied to the part by the gripper.

21. A pin gripper overlaps a 1.5-inch diameter pin by 0.5 inch. If 15 psi of air are used, calculate the lifting force applied to the part by the gripper.

22. If a safety factor of 25 percent is added to the gripping force in Problems 20 and 21, what change in pressure would be required?

23. A bellows gripper has a bellows geometry that includes a center rectangle 0.4 inch by 2.75 inches and semicircular ends. Calculate the holding force on a rectangular part when 25 psi of air are used.

24. Design a rubber bellows gripper that will grasp a 3 inch by 4 inch rectangular part, 1 inch thick, with a gripping force of 20 pounds. Specify the bellows geometry and air pressure required.

25. Determine the maximum part weight in Problem 23 if the coefficient of static friction is 0.85 and the maximum upward vertical acceleration is 5 feet per second squared.

26. Calculate the maximum acceleration that a robot using the gripper from Problem 1 could have in the upward vertical direction if the coefficient of static friction is 0.65 between a 25-pound part and the gripper surface.

PROJECTS AND CASE STUDY PROBLEMS

1. Design a decision tree that will select the best end-of-arm tooling for a manufacturing problem. Use questions at each branch of the tree that require a yes or no response.

2. Write a computer program in Visual Basic that will execute the design tree from Project 1.

3. Write a computer program in Visual Basic that will input the size and number of vacuum cups in a gripper along with the degree of vacuum achieved and output the lifting capability of the gripper.

4. Design a gripper for use in the work-cell design started in Case Study Problem 5 of Chapter 2. The tooling should be capable of removing the parts from the

mold die, retrieving the bowls from the trim press in a single move, and placing the bowls on the exit conveyer. Document all assumptions and vendor parts selected, include all calculations associated with the sizing of the tooling, and complete a three-view drawing of the parts in fabricated tooling.

5. Design a gripper for use in the work-cell design started in Case Study Problem 8 of Chapter 2. The tooling should satisfy all part transportation needs identified in the initial cell design. Document all assumptions and vendor parts selected, include all calculations associated with the sizing of the tooling, and complete a three-view drawing of the parts in the fabricated tooling.

6. Design a gripper for use in the work-cell design started in Case Study Problem 9 of Chapter 2. The tooling should satisfy all part transportation needs identified in the initial cell design. Document all assumptions and vendor parts selected, include all calculations associated with the sizing of the tooling, and complete a three-view drawing of the parts in the fabricated tooling.

7. Read the West-Electric case study (Section 12–7) where the gripper design is discussed. Determine the normal force required on the part for the following specifications: (1) a 50 percent safety factor is present, (2) the gripper force is determined using the heaviest part, and (3) maximum robot acceleration of 3 feet per second squared occurs upward at a 60 degree angle off the vertical. Use available references to find the coefficient of static friction for the gripper and part material. Document all assumptions and include all calculations. Complete the questions and problems at the end of Chapter 12 that relate to the gripper design.

Automation Sensors

CHAPTER GOALS AND OBJECTIVES

The goal of the automation sensor chapter is to provide a comprehensive overview of the three main types of sensors used in automated manufacturing. In addition to a complete description of contact and noncontact sensing devices, the chapter provides guidelines for incorporating automation sensors in work-cell designs. At the completion of this chapter you should be able to:

- Name and describe the six reasons sensors are used in automated manufacturing.
- Classify sensors into either the discrete or analog category.
- Classify sensors as either contact or noncontact type of operation.
- Select the appropriate limit switch given the parameters for a sensing application and design a trip dog consistent with the switch selected and the motion present.
- Describe the difference between simple touch and tactile sensing.
- Describe the operation of the noncontact sensor group including inductive and capacitive proximity type and photoelectric type.
- Select the appropriate noncontact sensor given the parameters for a sensing application and integrate the sensor using good design practices.
- Match the output of a sensor to the input of the controller using current sourcing and current sinking techniques.
- Describe the operation of smart sensors, including their operation in networks such as DeviceNet.
- Use sensor troubleshooting techniques to locate system faults.

5-1 INTRODUCTION

Even the most unskilled production worker can tell when a part has fallen to the floor or when a finished part is not ejected from a machine. The most sophisticated robot available today cannot perform these routine tasks without the help of sensors. Strategically mounted sensors provide the robot system with the same data that an operator gathers using the five human senses. For example, in Figure 5–1 a sensor checks for the presence of a part in the die of a forming press. A light source and light-sensitive receiver are positioned so that the beam of light from the source is aimed directly at the part in the die on its way to the receiver. Figure 5–1 shows this type of sensor detection system with the part in the die. A robot system would use a broken light beam or an *off* condition at the receiver to warn that the part had not been rejected and another part should not be loaded into the die.

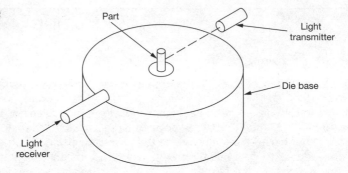

Figure 5–1 Part Ejector Failure Sensing.

Part

Light transmitter

Die base

Light receiver

Sensors are used in work cells for the following six reasons:

1. To detect a condition where an operator or some other human worker could be harmed by the robot or other manufacturing equipment
2. To detect a condition where the robot or other machines could be harmed by some other manufacturing equipment
3. To collect data on the production system to ensure that system parameters are at the correct setting
4. To monitor the work-cell operation to detect and analyze system malfunctions
5. To measure production parts to determine the current level of product quality
6. To monitor production parts for identification, location, and orientation data that can be used by other production systems in the work cell

Sensors are grouped into two categories, called *contact* and *noncontact*. As the name implies, a contact sensor must physically touch an object before the sensor is activated. In contrast, a noncontact sensor measures its parameter values without touching the object of interest. In each of these two categories, the sensors can have *discrete* or *analog* output signals.

Discrete Sensors

Discrete sensors have a single trigger point, for example, a temperature sensor that changes output states when the temperature of the sensor crosses 68°F. A good example of this type of sensor is the thermostat that controls the heating systems in homes. A single temperature level is set, and the heating system cycles *on* and *off* as the house temperature moves below and above the thermostat temperature setting.

The two most distinctive characteristics of discrete sensors are that (1) a single-input condition triggers a change in sensor output state, and (2) the output of the sensor swings between two conditions, *on* and *off*. The light sensor in Figure 5–1 has both of these characteristics. The single input condition is the presence or absence of a part in the die, and the receiver output cycles between 0 and 24 volts dc. Many discrete sensors used in manufacturing signal the change in output states by the opening or closing of a set of electrical contacts. Other discrete sensors use voltage levels to indicate the *on* and *off* state of the output. Often the voltage levels are 0 and 24 volts dc or 0 and 115 volts ac for *off* and *on* conditions, respectively. The trigger point for the sensor is usually adjustable so that the change in output states can be set to an input condition within a range dictated by the device. For example, the sensitivity of the receiver in Figure 5–1 can be adjusted for different light levels based on the distance separating the transmitter and receiver.

Analog Sensors

In contrast to the discrete sensors, the analog type measures a range of input conditions and generates a range of output states. In analog temperature measurement, for example, the sensor system could produce an output that varies from 0 to 5 volts, or 4 to 20 milliamperes, when exposed to a temperature range of 0 to 100 degrees Centigrade (°C). A direct and linear relationship exists between the input condition and the output response. For example, 0°C produces 0 volts output, 100°C produces 5 volts output, and 50°C produces 2.5 volts output. Other sensors have inverse input and output relationships, and some types of sensors (e.g., a thermocouple) produce a nonlinear output from a linear change in the input. Analog sensors include all sensors that measure process parameters, such as temperature, pressure, level, and flow. This type of sensor is found most frequently in process industries such as food, chemical, and petroleum.

The description of robot controllers in Chapter 2 indicated that robots use analog sensors to measure the position of axes and the rate of change of the joints as the robot moves. Although some analog sensors measure unique conditions in the work cell, more are found on the robot itself than in the automated cell where the robot works. The primary sensor used to automate manufacturing cells is the discrete type; therefore, the sensors discussed in this chapter will be primarily contact and noncontact discrete-type devices.

5-2 CONTACT SENSORS

Contact sensors include *limit switches* and all *tactile-sensing devices.* Limit switches, which have been used in flexible automation for many years, are reliable and easy devices to interface in an automated work cell. Tactile sensing is in the development stage; a high level of research activity is occurring in both industrial and university laboratories.

Discrete Devices

The most frequently used discrete contact sensor is the *limit switch.* The selection and application of limit switches requires an understanding of the *physical* and *electrical* properties of the devices along with their *operational characteristics.* As these

Figure 5–2 Limit Switches: General Specifications; Electrical Contacts and Operating Specifications; Trip Mechanisms.
(Courtesy of Omron Electronics, LLC)

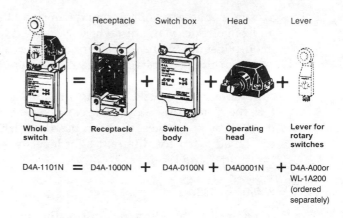

properties and characteristics are described, refer to the Omron Model D4A device data sheet in Appendix A. Take time now to become familiar with the switch specifications.

The *physical* properties of a limit switch include the fact that it is a mechanically actuated electrical switch consisting of a receptacle, switch box, operating head, and lever contacting device illustrated in Figure 5–2.

The receptacle provides the electrical contacts for interfacing with the control circuit. The switch box holds the mechanical switch and contacts, as well as providing a base for attaching the operating head. The different types of side levers are illustrated on the second page of the specifications in Appendix A.

On the Omron Model D4A switch data sheet two types of switch bodies are listed and pictured on the first page. There are a number of *side rotary* switches listed and three types of switches with *built-in actuators*. The five switches pictured on the data sheet include three side rotary (adjustable roller lever, standard lever, and adjustable lever rod) and two with built-in actuators coming out of the top of the switch (top plunger and wobble lever). The rotary type requires the contacting lever to rotate to activate the switch, whereas the built-in actuators require a linear motion of the plunger (either into the side or top of the head) or the deflection of the wobble lever to trigger a change in the switch.

The contact configurations include *single-pole double-throw* (SPDT) and *double-pole double-throw* (DPDT) with double break action illustrated on page 7 of the specifications. In addition, *sequential* operation and *center neutral* or *center off* operation are also available. In the sequential operation pole 1 operates before pole 2, and in the center neutral pole 1 operates for clockwise rotation and pole 2 operates for counterclockwise rotation. The two different indicator options include light emitting diode (LED) and neon lamp. The switching operation and indicator circuits are described on page 7 of the specifications.

The *electrical* properties start on page 4 of the specifications and include the current ratings of the contacts for different applied voltages. The specifications are for a heavy duty or *pilot duty* type of switch that is expected to handle relatively large currents. The current values listed would be normal service conditions for controlling motors, motor starters, lamps, and other high-power loads. The current values differ for contact closure (make) and contact opening (break) operation. Electronic duty covers the operation at low values of voltage, 5 to 30 volts, and low values of current, 0.2 to 100 milliamperes, respectively. The measure of a switch's ability to handle low currents is called the *dry circuit rating.* An input module of a programmable logic controller requires less than 150 milliamperes when 28 volts are applied. Therefore, limit switches interfaced directly to programmable logic controllers should have contacts with good low-current switching characteristics. The D4A has an electronic duty model available but not shown on the specifications.

The *characteristics* and *operating characteristics* of the model D4A switch (page 5 on the data sheet) include the operating force and travel distance for the switches. Review the characteristics listed on the data sheet and note the excellent service life of these devices. Figure 5–3 shows a listing and illustration of the movements and forces for the plunger and lever-type switch.

Dogs

Limit switches are frequently used to detect the movement of parts or part carriers in an automated work cell. The switch is tripped by a specially designed part called a *dog* that is attached to the part carrier. The dog is used to turn the lever or push down the plunger on mechanical limit switches. For example, limit switch LS1 in Figure 2–46 is engaging the trip dog and the switch lever is rotating. The speed at

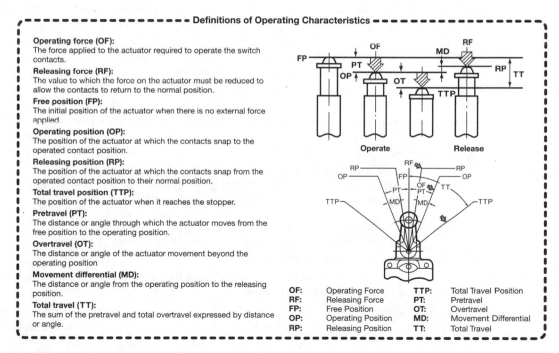

Figure 5–3 Definition of Switch Movement.
(Courtesy of Omron Electronics, LLC)

(1) Non-overtravel dog		(2) Overtravel dog

(a) V ≤ 1.31 ft/s

φ	Vmax. (ft/s)	Y
30°	1.31	0.8 (TT)
45°	.82	Dog stroke is
60°	.33	allowed up to
(60°–90°)	.16	80% of TT.

(b) 1.31 ft/s ≤ V ≤ 6.56 ft/s

θ	φ	Vmax. (ft/s)	Y
45°	45°	1.64	0.5 to 0.8(TT)
50°	40°	1.97	0.5 to 0.8(TT)
60°	30°	4.26	0.5 to 0.7(TT)
75°	15°	6.56	0.5 to 0.7(TT)

(a) V ≤ 1.31 ft/s

φ	Vmax. (ft/s)	Y
30°	1.31	0.8 (T.T)
45°	.82	Dog stroke is
60°	.33	allowed up to
(60°–90°)	.16	80% of TT.

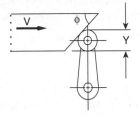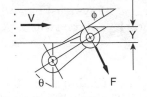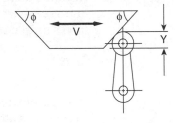

Figure 5–4 Trip Dog Design Data.

which the part is moving influences the shape of the trip dog used to move the lever. Figure 5–4 illustrates the type of design data provided by Omron Electronics for their limit switches. This data is used to determine the three critical parameters—the angle of the dog (ϕ), the angle of the lever (θ), and the height of the roller relative to the base of the dog—in the design of dogs for lever- and plunger-actuated switches. Trip dogs are designed to minimize the velocity of the lever on the switch as the switch is tripped by a passing part. Study Figure 5–4 and note that as dog (and carrier) velocity increases, the angle of the dog striking surface (ϕ) decreases, and the angle of the lever (θ) increases. Two types of dog operation are used: overtravel and non-overtravel. The non-overtravel dog keeps the switch in the actuated position after the switch is engaged, and the overtravel dog allows the switch to return to the *off* state after the dog has passed. The following steps are used in the design of trip dogs:

1. Using the mechanical and electrical specifications for the application, select a receptacle, switch body, and lever or plunger configuration that meet the requirements.

2. Determine the velocity of the trip dog at the time of impact with the lever or plunger of the limit switch.

3. Determine if an overtravel (switch signal should return to initial state after switch is activated) or non-overtravel (switch signal should remain in switched state after switch is activated) type dog is required.

4. Using the data from Steps 1, 2, and 3, locate the appropriate velocity table provided by the switch vendor (Figure 5–4).

5. Locate the dog velocity values in the table that are above and below the value identified in Step 2. Use the values for (ϕ) and (θ) associated with the larger of the two velocity values. Find the equation for Y in this table.

6. Locate the pretravel (PT) and overtravel (OT) values for the switch selected in Step 1 from the operating characteristics table on the data sheet for the limit switch.

7. Determine the TT in degrees (PT + OT) and the total travel (TT) for the selected switch as follows.

   For lever-type limit switches:

   TT (in.) = arm radius (in.) $\times$ (1 − cos (TT in degrees))

   For plunger-type limit switches:

   TT (in.) = value listed in switch data sheet

8. Solve for the trip dog stroke value Y using the value of TT from Step 7 and the equation from Step 5.

9. Draw the trip dog and locate the limit switch using the values obtained in Steps 8 and 6.

Example 5-1

A signal is needed in a robot welding application to initiate the welding cycle on the part located on an indexing table. Design a trip dog for the edge of the table to actuate a limit switch as the table is turned into position. The velocity of the dog is 0.5 feet per second, and the limit switch selected for the application is model D4A-1101N with a D4A-A00 lever. Non-overtravel operation is desired.

Solution

1. The switch specified is a D4A-1101N with a D4A-A00 lever. The switch is a standard SPDT model with side rotary. Specifications for the switch are given in the operating characteristics table under the row marked D4A-☐ ☐01N. The lever specifications are given with the outline drawing for the levers.

2. The velocity is 0.5 feet per second.

3. Non-overtravel operation is desired for this application.

4. The velocity range for table 1a in Figure 5–4 is 0.16 to 1.31 feet per second, and 0.5 falls within that range.

5. The trip dog velocity of 0.5 feet per second falls between 0.33 and 0.82 in the table, so the larger value of 0.82 is used.

6. The values include: ϕ = 45 degrees, PT = 12 degrees, and OT = 70 degrees. The equation for Y is:

   Y = 0.8 $\times$ total travel (in.)

Figure 5–5 Results of Trip Dog Design.

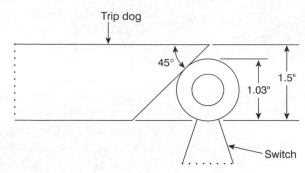

Trip dog

45°

1.03"

1.5"

Switch

7. TT (total travel) = PT (pretravel) + OT (Overtravel)

$$TT = 12 \text{ degrees} + 70 \text{ degrees} = 82 \text{ degrees}$$

$$TT \text{ (in.)} = \text{arm radius} \times (1 - \cos (TT \text{ in degrees}))$$

$$TT = 1.5 \text{ in.} \times (1 - \cos (82))$$

$$TT = 1.5 \text{ in.} \times 0.860$$

$$TT = 1.29 \text{ in.}$$

8. $Y = 0.8 \times TT$ in.

   $Y = 0.8 \times 1.29$ in.

   $Y = 1.03$ in.

9. The trip dog and limit switch position to satisfy the problem are drawn in Figure 5–5 using the values obtained.

Artificial Skin

Effective tactile sensors and systems are needed for robot grippers used in assembly applications. Research on small and medium-sized parts-assembly tasks indicates that the development of a basic form of artificial skin would permit robots to perform about 50 percent of the generic tasks found in industrial assembly. Currently, most robot interaction with sensors is of the open-loop type. For example, gripping devices close on all parts with the same force. A signal indicating that the gripper is closed is generated by the sensor on the robot end-of-arm tooling, but feedback to the robot controller of gripping characteristics, such as closing force, is not provided. In grippers with artificial skin, the fingers are closed under servo control with finger pressure and other tactile parameters measured. There is an important distinction between *tactile sensing* and *simple touch*. Simple touch includes simple contact or force sensing at one or just a few points on the gripper surface. For example, the gripper in Figure 5–6 uses two switches to determine that the part is centered in the jaws. The switches provide binary data in the form of either an *on* or *off* signal depending on the location of the part. If the two switches were replaced with analog pressure pads, the data would be continuously variable, and the signal value would become a function of the pressure applied by the fingers. The sensing would still be classified as simple touch, however, because only two contact points were monitored.

Figure 5–6 Simple Touch-Sensing Gripper.

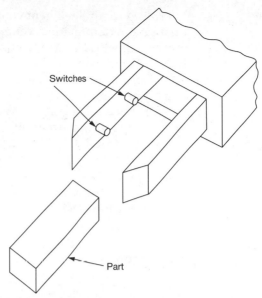

Switches

Part

Tactile sensing requires a group of sensors arranged in a rectangular or square pattern called an array. Figure 5–7 shows gripper fingers with an 8×8 tactile-sensing array. The array has sixty-four sensing elements, each capable of measuring the continuously variable force applied to the element. A tactile sensor

Figure 5–7 Tactile-Sensing Array.

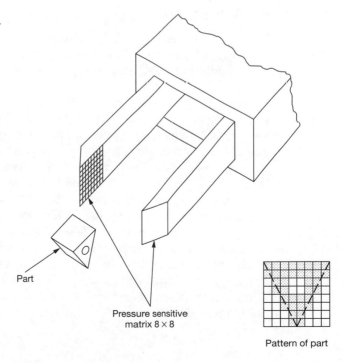

Part

Pressure sensitive matrix 8×8

Pattern of part

interfaced to an intelligent controller can determine the shape, texture, position, orientation, deformation, center of mass, and presence of torque and slippage of any object held. The array output pictured in Figure 5–7 shows the results of holding a triangular-shaped object with a hole in the center in the gripper.

The three steps required for intelligent acquisition of parts by either a robot or a human operator are first *vision,* followed by *proximity sensing,* and finally *tactile sensing.* For a robot system to retrieve randomly placed parts or to assemble products at any scale automatically, further development of all three of these sensing areas is required. In addition, if machines are to develop tactile sensing like that found in humans, then a second element, *haptic perception,* must be incorporated. Haptic perception is the sensing information that comes from the joints and muscles. Thus, robot software must combine the data from the artificial skin with the joint torque information if machines are to achieve what humans call "touch."

5-3 NONCONTACT SENSORS

As the name implies, noncontact sensors measure work-cell conditions without physically touching the part. In robot work cells the most frequently used noncontact sensors are *proximity* and *photoelectric* devices and *vision* systems. All three types of sensors are available from many commercial vendors. Proximity and photoelectric sensors are discussed in the following sections, and vision systems are described in Chapter 6.

Proximity Sensors

Proximity sensors detect the presence of a part when the part comes within a specified range of the sensor. Proximity sensors are available in three package shapes and several different sizes. The package shapes are *cylindrical, rectangular,* and *through-head type.* The shape of the part to be sensed and sensing application dictate the type of package shape for best operation. Figures 5–8 and 5–9 list a sample of proximity sensors available from Omron Electronics LLC illustrating the different geometries for varying automation applications. These examples are representative of the devices available from numerous sensor manufacturers. Two types are shown, *inductive high frequency* oscillation and *electrostatic capacitance.* The operation of each type is described in this section. As the *features* heading indicates, each model has distinguishing characteristics designed for specific applications. The sensing distance for a sensor is usually fixed; for example, the E2EC model comes in six sensing distances from 0. 5 mm to 4 mm. However, some models have a variable sensing distance that is set by an adjustment in the sensor amplifier. Note in the figure that the sensing distance is specified for shielded and unshielded types of sensors. Figure 5–10 shows the outline drawings of shielded (top) and unshielded (bottom) E2E2 sensor. The shielded version has a flush sensor end and is design to mount flush with the mounting surface. Study the end of the unshielded model on the bottom and notice that the sensing coil

Type	 High frequency oscillation	 High frequency oscillation	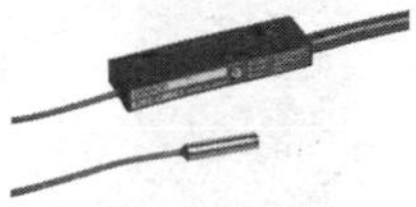 High frequency oscillation	 High frequency oscillation
Model	E2EM	E2E2	E2EC	E2S
Features	Extended range sensing distances about 2X farther then other conventional sensors. Extended range reduces work piece collision.	Long metal cylindrical inductive with full length threaded body and strong durable housing.	Shielded two-wire sensors with in-line amplifier for robotic and space-confined applications. Robot cable lasts ten times longer than standard cable. Operation and stability indicators on in-line amplifier allows easy set-up and monitoring.	World's smallest square sensor with a built-in amplifier and 10 micron repeatability. High frequency response (1kHz) for fast machine processes and low power consumption.
Shielded Sensing Distance	2, 4, 8, 15 mm	2, 3, 5, 7, 10 mm	0.5, 0.8, 1.5, 2.5, 3, 4 mm	
Unshielded Sensing Distance	16, 30 mm	5, 8, 10, 14, 18, 20 mm		1.6, 2.5 mm
2 wire DC Output	NO, NC – 100 ma max.	NO, NC – 100 ma max.	NO, NC – 100 ma max.	NO, NC – 50 ma max.
3 wire DC Output	NPN-NO, NPN-NC, PNP-NO, PNP-NC – 100 ma max.	NPN-NO, NPN-NC, PNP-NO, PNP-NC – 200 ma max.	NPN-NO, NPN-NC, PNP-NO, PNP-NC – 100 ma max.	NPN-NO, NPN-NC, PNP-NO, PNP-NC – 50 ma max.
Frequency Response	0.1 kHz to 1.5kHz	0.8 kHz to 1 kHz	1 kHz to 1.5 kHz	1 kHz

Figure 5–8 Proximity Sensors.
(Courtesy of Omron Electronics, LLC)

Type	 High frequency oscillation	 High frequency oscillation	 Electrostatic capacitive	 Electrostatic capacitive
Model	TL-L	F2LP-W	E2K-C	E2K-F
Features	Detects all types of metal (ferrous and non-ferrous) and is a long distance sensor. Submersible and current output models available.	Inductive ring sensing head detects small moving metal objects anywhere in the ring. Separate amplifier provides relay and transistor outputs with a 40 ms OFF delay.	Capacitive sensor with AC or DC power supply and adjustable detecting distance. Ideal for noncontact detection of glass, wood, water, oil, plastic, and metal. Allows indirect detection of materials in nonmetallic containers.	Flat, thin capacitive sensor with DC power supply. Ideal for mounting directly to metal. Detects glass, plastic, wood, water, oil, and metals.
Shielded Sensing Distance		0.3, 2, 2.5, 3 mm min.		
Unshielded Sensing Distance	100 mm		5 to 25 mm adjustable	10 mm
2 wire DC Output	NO, NC – 100 ma max.	AC output SPDT relay – 3 amp max. and NPN-NO – 100 ma		NO, NC – 50 ma max.
3 wire DC Output	NPN-NO, NPN-NC, PNP-NO, PNP-NC – 100 ma max.		NPN-NO, NPN-NC, PNP-NO, PNP-NC – 200 ma max.	NPN-NO, NPN-NC – 100 ma max.
Frequency Response	0.1 kHz to 1.5 kHz	75 to 125 millisec between objects	10 Hz to 70 Hz	100 Hz

Figure 5–9 Proximity Sensors.
(Courtesy of Omron Electronics, LLC)

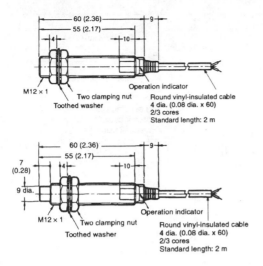

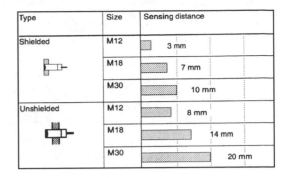

Figure 5–10   Shielded and Unshielded Proximity Sensors.
(Courtesy of Omron Electronics, LLC)

extends beyond the threaded body of the sensor. The unshielded types are mounted with the sensor end beyond the mounting surface. Two-wire or three-wire outputs are provided with *normally open* (NO) or *normally closed* (NC) contacts options for the two-wire operation. The output options for the three-wire devices are either NPN (current sinking) or PNP (current sourcing) transistors. The operation of the three-wire outputs is described later in this section. The maximum output current is listed for each type of output. The frequency response for the different models is specified as well. The ambient operating temperature range for most models is –13 to 158°F. The sensors are mounted to supporting brackets using either a threaded part of the sensor body or mounting holes. The operating parameters for proximity sensors are defined as follows:

- *Reference plane:* The plane of reference on the proximity sensor from which all measurements are made.
- *Reference axis:* An axis through the sensor from which measurements are made.
- *Standard object:* A definition of the object to be sensed in terms of a specified shape, size, and material composition.
- *Sensing distance:* The distance from the reference plane to the standard object that causes the output of the sensor to change to the *on* state.
- *Vertical sensing distance:* The sensing distance measured by bringing the standard object toward the reference plane with the standard object normal to and centered about the reference axis. (See Figure 5–11.)
- *Horizontal sensing distance:* The sensing distance measured by bringing the standard object along a plane parallel with and at a fixed distance from the reference plane. (See Figure 5–12.)

Type of sensor	Illustration	Explanation
Column head & square pillar head types	**Vertical sensing distance** 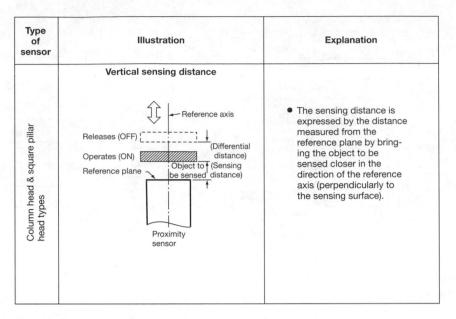	• The sensing distance is expressed by the distance measured from the reference plane by bringing the object to be sensed closer in the direction of the reference axis (perpendicularly to the sensing surface).

Figure 5–11 Vertical Sensing Distance.
(Courtesy of Omron Electronics, LLC)

Type of sensor	Illustration	Explanation
Column head & square pillar head types	**Horizontal sensing distance** 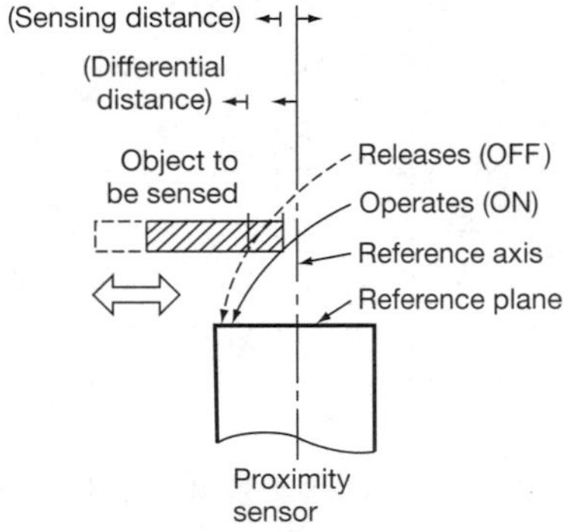 (Sensing distance) (Differential distance) Object to be sensed Releases (OFF) Operates (ON) Reference axis Reference plane Proximity sensor	• The sensing distance is expressed by the distance measured from the reference axis by moving the object to be sensed parallel to the reference plane (i.e., sensing surface). This distance can be expressed as the locus of operating points, since it varies with the passing position of the object (distance from the reference plane).

Figure 5–12 Horizontal Sensing Distance.
(Courtesy of Omron Electronics, LLC)

- *Resetting distance:* The distance from the reference plane to the standard object that causes the output of the sensor to change from *on* to *off* as the standard object is withdrawn from the sensor. (See Figures 5–11 and 5–12.)
- *Differential distance:* A measure of the hysteresis present in the system. The difference between the resetting distance and the sensing distance for the type of sensor used. (See Figures 5–11 and 5–12.)
- *Setting distance:* The maximum sensing distance when worst-case ambient temperature and supply voltage variations are assumed. (See Figures 5–11 and 5–12.)
- *Response time:* The time required for the output to change states after the standard object's position triggered a change of state.
- *Frequency response:* The maximum rate at which standard objects can cause the output to change states.
- *Leakage current:* The maximum *off* current that will flow from the output terminals with the output stage in the *off* state.
- *Temperature variation:* The variation in sensing distance as a result of variations in ambient temperature.
- *Voltage variations:* The variation in sensing distance as a result of variations in supply voltage.

The *release off* point in Figures 5–11 and 5–12 is the same as the resetting distance, and the *operates on* point in the figures is the same as the sensing distance.

The selection and the application of proximity devices requires an understanding of the *physical* and *electrical* properties of the sensors along with their *operational characteristics.* All models have solid-state circuits enclosed in the sensing head to generate the fields necessary for remote sensing of objects and generation of output signals. Proximity sensors come in several sensing styles: inductive all metals, inductive ferris metals, inductive nonferris metals, and capacitive.

Inductive Sensor Operation

Inductive sensors use a property of tuned circuits wherein the quality factor of the circuit changes as eddy current losses in a conductive material increase. A detailed description of this effect starts with a typical *RLC* oscillator circuit like the one shown in Figure 5–13. This type of circuit is used in the inductive type sensors. The circuit is a noninverting operational amplifier with a positive feedback resistor R_1 back to the input where an *RLC* tuned circuit is placed. The gain of the amplifier is set by resistors R_3 and R_4, and the frequency of oscillation is set by the capacitor C, inductor L, and variable resistor R_2. Resistor R_2 represents the unavoidable loss portion of the *LC* parallel oscillator circuit. The circuits are set up to oscillate in the frequency range of 100 kHz to 1 MHz by adjusting the values of L and C. Oscillations will occur only if the condition in the equation

$$\frac{R_1}{R_2} \leq \frac{R_3}{R_4}$$

Figure 5–13 Oscillator Circuit for an Inductive Sensor.

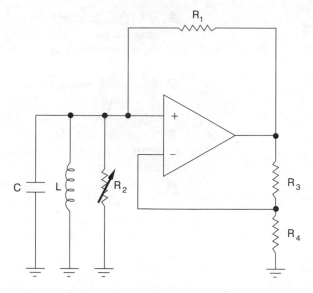

is met. The inductance L is constructed so that the oscillating current sets up the electromagnetic field illustrated in Figure 5–14. The shape of this field is determined by the core of the coil, which is made from highly permeable ferrite material. If a conductor is placed in the field, as illustrated in Figure 5–14, then eddy currents are

Figure 5–14 Electromagnetic Field of an Inductive Sensor.

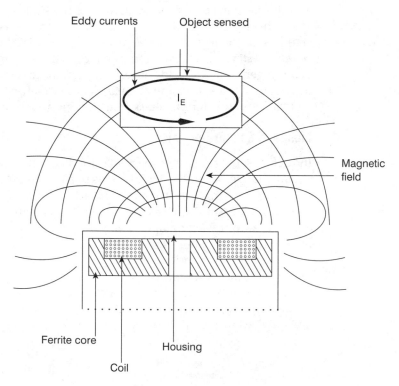

induced in the conductor. The energy of the field is reduced in proportion to the level of eddy currents created. The level of eddy currents depends on: (1) the distance and position of the conductor in the field, (2) the size and shape of the conductor, and (3) the conductivity and permeability of the conductor. If any of these three conditions cause the eddy currents to increase to a level not sustainable by the oscillator, then oscillations will stop. For example, if the object in Figure 5–14 is moved closer to the coil or inductive sensor, the eddy currents increase and the amplitude of the oscillations decreases. At some point close to the sensor, all oscillations will stop, as shown in Figure 5–15. Note the relationship among the sensor output signal, the position of the target, and the oscillation amplitude illustrated in the figure.

Based on the operation of the inductive sensor circuit, the performance characteristics of the sensor include:

- Detection of all materials that are electrical conductors.
- Detection not limited to magnetic materials or metals.
- Detection of stationary and moving objects.
- Preferred target is a flat, smooth object.
- Low levels of energy so they do not create radio interference or generate heat in the target.

Figure 5–15 Inductive Sensor Oscillations as a Function of Conductor Distance.

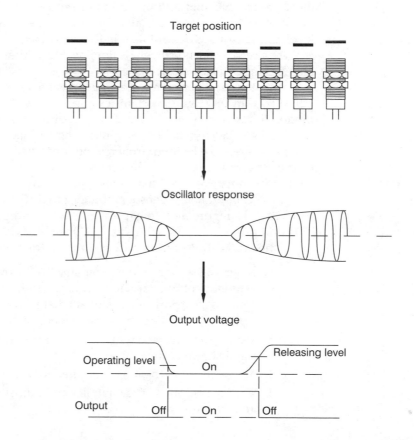

Some manufacturers provide three different output circuits for interfacing to other work-cell hardware. The output circuits in Figure 5–16 include NPN and PNP transistor outputs in the dc switching type and an ac switching-type output. Note that the sensor shares the power source with the external interface, and the sensor output can be connected to either a load or an external transistor circuit (shown with dotted lines). The NPN type of sensor output has a *current-sinking* characteristic. In current sinking, the output transistor of the activated sensor provides a low resistance path to the *black* lead or ground, and so the sensor output looks like a short to ground or zero volts. Under these conditions, current would flow from the external interface device into the sensor output. In contrast, the PNP output circuit has a *current-sourcing* characteristic. When active, the sensor output transistor provides a low resistance path to the *red* lead, or power supply source, so that the sensor output is approximately equal to the supply voltage. Under these conditions the sensor is supplying current to the external interface device. The last type of output in Figure 5–16 is an ac switching type. In this type of output, the sensor has two output leads (*white and black*) that act as a switch for an ac source and load in series. The interface between sensors and work-cell control devices is described in greater detail in the next chapter.

The Omron models also offer the option of normally closed (NC) operation or normally open (NO) action at the output. The wave forms in Figure 5–16 illustrate the NC and NO operation of sensors when they are close to an object (*present*) and when they are away (*absent*). When the waveform is at the *operates* level, the sensor has current flow in the output lead; however, when the waveform is at the *release* level, no current is flowing. Study the charts until the concept of NC and NO operation is clear.

Most models permit a wide range of voltages for operating power. The model E2EM by Omron (Figure 5–8), for example, allows the supply power on dc output models to range between 12 and 24 volts dc unregulated, and the ac output devices can use 24 to 240 volts ac. The output can be interfaced directly to a robot controller or a programmable logic controller, or be connected to a sensor controller provided by the sensor manufacturer. The sensor controller provides relay contact outputs for control of other work-cell machines, plus an inverting switch to change between NC and NO operation. A sensor amplifier, Figure 5–17, is used on some sensors, like the E2K-C in Figure 5–9, to supply dc power and provide the output switching. This unit provides a relay contact (SPDT) and an NPN (current sinking) output.

The application of proximity sensors is straightforward.

1. Select the sensor model geometry that corresponds to the material characteristics of the parts to be sensed, and select an output circuit (current sourcing, current sinking, or ac) that is compatiable with the input of the controller receiving the sensor signal.
2. Mount the sensor so that the object to be detected passes within the sensing distance of the device.
3. Apply power in the range specified for the sensor from an external source.
4. Wire the sensor to the contoller and adjust the sensor for target material and distance.

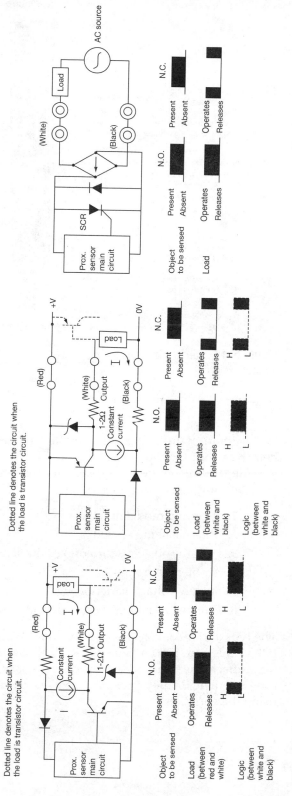

Figure 5-16 Sensor Output Circuits.

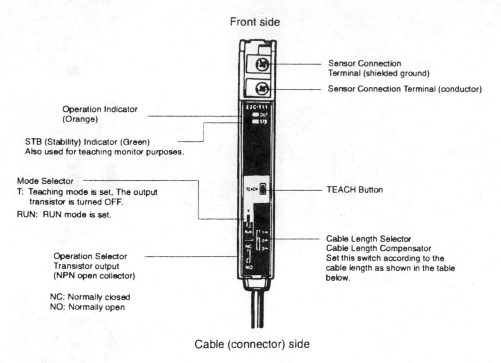

Figure 5–17 Sensor Amplifier.
(Courtesy of Omron Electronics, LLC)

Although this operation is simple, numerous design decisions must be made before each of the four steps is taken. In addition, the actual wiring diagram will vary depending on the sensor and controller selected.

Sensing distance is affected by many variables, including the following six:

1. Type and model of sensor
2. Material being sensed
3. Path the object uses to trigger the sensor
4. Ambient temperature variation
5. Supply voltage variation
6. Proximity of other objects and other sensors

The type and model of sensor chosen dictate the sensing distance; for example, the E2EM in Figure 5–8 has shielded sensing distances of 2, 4, 8, and 15 mm. The material affects the standard sensing distance as well. Table 5–1 lists the correction factors for Allen Bradley proximity sensors. Corrected sensing distances are found using the equation

$$S_u = M \times S_n$$

Table 5–1 Material Correction Factors.

	Correction Factor
400 series stainless steel	1.15
Cast iron	1.10
Mild steel (Din 1623)	1.00
Aluminum foil (0.05 mm)	0.90
300 series stainless steel	0.70
Brass MS63F38	0.40
Aluminum ALMG3F23	0.35
Copper CCUF30	0.30

where S_u is the new usable sensing distance, S_n is the nominal sensing distance, and M is the correction factor. The chart in Figure 5–18 illustrates how Omron specifies the effect of target size and material with graphed data sets. The path is also a factor because the nominal sensing distance for a given target is different if the target is moved in from the side versus an approach from the end. The ambient temperature and supply voltage also affect sensing distance. For example, Allen Bradley has a $+/-10$ tolerance on each parameter so that the variation could be 20 percent. Omron has a $+/-10$ to 25 percent tolerance on sensing distance due to temperature change, with the 10 percent value more typical. The following examples illustrate these calculations.

Example 5–2
Determine the maximum tolerance present on an Omron E2E sensor (Figure 5–18) if the target is brass and is 25 mm on one side. Most E2E have a $+/-10$ percent tolerance.

E2E-X14MD☐-N

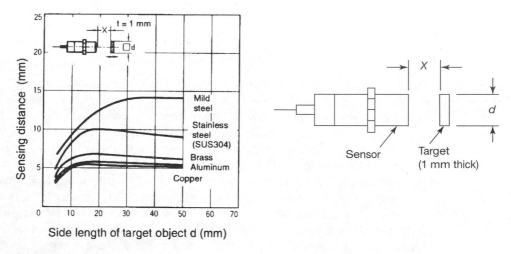

Figure 5–18 Sensing Distance Versus Material and Object Size.
(Courtesy of Omron Electronics, LLC)

Solution

a) Find the new sensing distance for a brass target of 25 mm on a side using Figure 5–18.

$$S_{new} = 7 \text{ mm}$$

b) Find tolerance for temperature change.

$$\Delta S = +/-7 \text{ mm} \times 0.1$$
$$= +/-0.7 \text{ mm}$$

c) Subtract tolerance to get safe sensing distance.

$$S_{new} = 7 \text{ mm} - 0.7 \text{ mm}$$
$$= 6.3 \text{ mm (safe minimum distance)}$$

Example 5-3

Determine the maximum tolerance present on an Allen Bradley sensor that has a nominal sensing distance of 50 mm and a target of aluminum.

Solution

a) Find the new sensing distance for an aluminum target using Table 5–1.

$$S_u = 0.35 \times 50 \text{ mm}$$
$$= 17.5 \text{ mm}$$

b) Adjust for supply voltage and temperature.

$$\Delta S_u = \pm 17.5 \text{ mm} \times 0.1$$
$$= \pm 1.75 \text{ mm}$$
$$S_u = 17.5 \text{ mm} - 1.75 \text{ mm}$$
$$= 15.75 \text{ mm (safe minimum distance)}$$

In addition, if another proximity sensor must be mounted close to the first, then the second sensor must either use a different operating frequency to avoid interference or be located a minimum distance away from the sensor with the same frequency.

Interfacing the power and output signals requires an equally large number of design considerations. One factor affecting the sensor selection process is the input characteristics of the machine to which the sensor is connected. For example, if a programmable logic controller or robot controller is interfaced to a sensor, then a low-voltage type of sensor output circuit (usually 0 to 5 or 0 to 28 volts dc) should be selected. If a motor starter control relay or contactor must be triggered, then a high-power-type sensor output (usually 0 to 110 volts ac) is the sensor interface of choice. The decision to connect the sensor directly to the machine control versus using a sensor controller will be affected by the need for a sensor power supply and for logical operations with two or more sensors.

Figure 5–19 Electrostatic Field of the Capacitance Sensor.

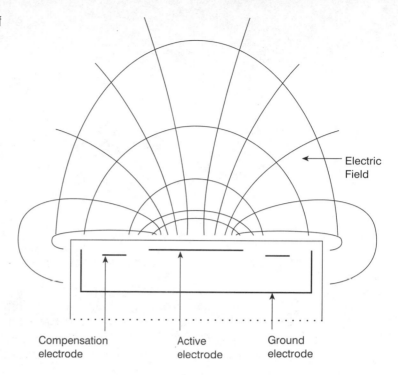

Electric Field

Compensation electrode

Active electrode

Ground electrode

Capacitive Sensors

Capacitive sensors measure the change in the capacitance as an approaching object interacts with the electric field created by the sensor. This change in capacitance causes an operational amplifier circuit inside the sensor to begin oscillating. Like the inductive sensor, this capacitance unit produces a field to sense the presence of an object. For this sensor, however, an electric field is generated by having one plate of a capacitor positioned at the sensing end of the device. Figure 5–19 illustrates how the electrostatic field is produced by the active electrode. The capacitor producing the field is part of an operational amplifier oscillator circuit in the sensor like the circuit in Figure 5–13. When no parts are within range of the sensor, called the *no target* state, the oscillator is at rest (not oscillating), and the sensor output is not active. When a part with a dielectric constant of 1.2 or greater enters the electrostatic field, the capacitance increases, and at some capacitance threshold, oscillations start and the sensor output becomes active.

Two types of capacitance sensors are used: *dielectric* and *conductive*. The dielectric type, illustrated in Figure 5–20, has both capacitance electrodes integrated into the oscillator circuit. When a target is in the field, the capacitance is raised and oscillations indicate the presence of an object. This type can be used to detect all types of materials as long as the dielectric constant is slightly greater than air, which is 1. 0. The second type, conductive, is configured with only one of the electrodes integrated into the oscillator in the sensor. The second electrode is

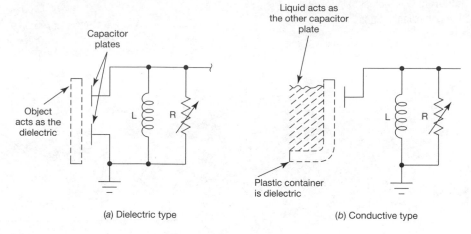

Figure 5–20 Dielectric and Conductive Type Capacitance Sensors.

introduced when a conductive material enters the field and raises the capacitance to start the oscillation. Figure 5–20 illustrates how each of these types work. Such sensors are excellent for *looking through* an insulation material, such as rubber, glass, or paper, to detect the presence of a conductive material like liquid or metal.

The capacitive sensor has the following performance characteristics:

- Oscillator frequencies in the 100 kHz to 1 MHz range.
- Detected materials include conductors, insulators, plastics, glass, ceramics, oils and greases, water, and all materials with a high moisture content.
- Detection of stationary and moving objects.
- Preferred target is a flat, smooth object.
- Operation at low levels of energy so it does not create radio interference or generate heat in the target.

The ability to detect water implies that moisture on the sensor would present a severe problem because the sensor could easily accumulate moisture in some applications. In those cases, a sensor with a compensating electrode is used. The electrode compensates for the moisture on the surface and does not affect the sensitivity at the sensing distance. The E2K sensors in Figure 5–9 are capacitive type devices.

Photoelectric Sensors

Photoelectric sensors detect the presence of an object or part when the part either breaks a light beam or reflects a beam of light to a receiver. The different types, pictured in Figure 5–21, are defined as follows:

- *Separate or through beam type:* The sensor system includes two devices: a light source to produce a beam of light and a receiving device to sense the presence of the light beam.

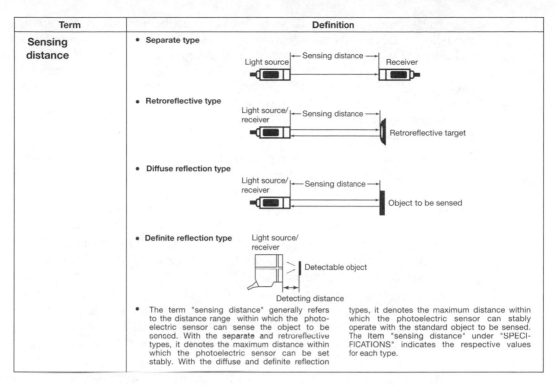

Term	Definition
Sensing distance	• **Separate type** Light source ←— Sensing distance —→ Receiver • **Retroreflective type** Light source/receiver ←— Sensing distance —→ Retroreflective target • **Diffuse reflection type** Light source/receiver ←— Sensing distance —→ Object to be sensed • **Definite reflection type** Light source/receiver — Detectable object — Detecting distance • The term "sensing distance" generally refers to the distance range within which the photoelectric sensor can sense the object to be sensed. With the separate and retroreflective types, it denotes the maximum distance within which the photoelectric sensor can be set stably. With the diffuse and definite reflection types, it denotes the maximum distance within which the photoelectric sensor can stably operate with the standard object to be sensed. The item "sensing distance" under "SPECIFICATIONS" indicates the respective values for each type.

Figure 5–21   Four Types of Photoelectric Sensors.

(Courtesy of Omron Electronics, LLC)

■ *Retroreflective type:* The sensor system includes two separate devices: a sensor with both a light source and receiver in the same case, and a retroreflective target, which is a highly reflective surface. The light beam leaves the sensor, bounces off the target, then returns to the receiver. The sensor is triggered when a part breaks either the outbound or returning beam.

■ *Diffuse reflective type:* The sensor has a light source and receiver built into the same case. Stable operation assumes that the part to be detected will return sufficient diffused light to trigger the receiver when the part is in range, and that the background equipment will not return enough light to trigger the receiver when the part is out of range. The light beam is diffused back to the receiver by using the natural reflective characteristics of the part's surface.

■ *Definite reflective type:* This sensor works like a combination of the retroreflective and diffuse reflective types. The light source and receiver are both located in the same enclosure, and the beam uses the part's surface to reflect light back to the receiver. The light reflected must be a definitive beam because diffused light alone will not be sufficient to activate the receiver. The angle of the light leaving the sensor is adjusted to improve sensitivity and establish a specific distance at detection.

These four basic types illustrate the concepts behind detecting parts using photoelectric devices. However, there are several variations of the retroreflective and diffused types. Table 5–2, from Allen Bradley (AB), lists these variations. The *definite reflective* type described earlier is called a *sharp cutoff diffused* in the AB list. The table also provides a description of typical applications along with the strengths and weaknesses of each sensor type.

Several operating parameters describe the operation of the sensors; they are defined as follows:

- *Sensing distance:* For separate and retroreflective types, the maximum distance between the light source and the receiver, or light source and target, that produces stable operation.
- *Operating distance:* For the reflective type, the distance from the sensor to the part that causes the output of the sensor to change to the *on* state.
- *Resetting distance:* For the reflective type, the distance from the sensor to the part that causes the output to change from *on* to *off* as the part is withdrawn from the sensor.
- *Differential distance:* A measure of the hysteresis present in the system. The hysteresis is the difference between the resetting distance and the operating distance for the type of reflective sensor used.
- *Optical axis:* The axis passing through the sensor along which the light beam is generated or received.
- *Direction angle:* For separate and retroreflective-type sensors, the maximum angle through which the optical axis can move and still provide stable operation.
- *Dark on operation:* For all types of sensors, it means that the output will be active, or *on,* when no light is received by the receiver.
- *Light on operation:* For all types of sensors, it means that the output will be active, or *on,* when light is received by the receiver.
- *Response time:* The time required for the output to change states after the part breaks the beam or reflects the light to the receiver. Response time can be either operate time (time to turn the output *on*) or reset time (time to turn the output *off*).

The selection and application of photoelectric sensors requires an understanding of the *physical, electrical,* and *optical properties* along with the *operational characteristics* of the different types of sensors. Some of the various types of sensors available from Omron are illustrated in Figures 5–22 and 5–23.

The photoelectric sensors displayed in the figures illustrate the wide range of devices available from Omron and numerous other sensor manufacturers. Two types are shown, completely self-contained sensors with light source and amplifier in the same case and separate amplifier types with the amplifier separated from the light source. As the *features* heading indicates, each model has distinguishing characteristics designed for specific applications. The maximum sensing distance for a photoelectric sensor is usually fixed; however, most sensor amplifiers have sensitivity adjustments so that there is some fine tuning available. The distance

Table 5-2 Photoelectric Sensors.

Sensing Mode	Applications	Strengths	Weaknesses
Transmitted beam	General purpose sensing	• High margin for contaminated environments • Longest sensing distances	• More expensive, separate light source and receiver required; more costly wiring
Retroreflective	General purpose sensing	• Moderate sensing distances • Less expensive than transmitted beam	• Shorter range than transmitted beam • Less margin than transmitted beam • May detect reflections from shiny objects
Polarized retroreflective	General purpose sensing of shiny objects	• Ignores first surface reflections	• Shorter sensing range than standard retroreflective
Standard diffuse	Applications where both sides of the target cannot be accessed	• Access to both sides of the target not required	• Can be difficult to apply if the background behind the target is too reflective
Sharp cutoff diffuse (definite reflective)	Short range detection of targets	• Access to both sides of the target not required • Optics system provides some protection against sensing of reflective backgrounds	• Only useful for very short ranges
Background suppression diffuse	General purpose sensing	• Access to both sides of the target not required • Ignores backgrounds beyond rated sensing distance regardless of reflectivity	• More expensive than other types of diffuse sensors • Limited maximum sensing distance
Fixed focus diffuse	• Detection of small targets • Detection of targets at a specific distance from sensor • Detection of color marks	• Accurate detection of small targets in a specific location	• Very short range • Not suitable for general purpose sensing
Wide angle diffuse	• Detection of large, irregularly shaped targets • Detection of very fine objects over a broad area	• Very good at ignoring background reflections	• Very short range
Fiberoptics	• Allows photoelectric sensing in areas where a sensor cannot be mounted due to size or environment considerations	• Glass fiberoptic cables available for high ambient temperature applications • Very high shock and vibration durability • Plastic fiberoptic cables can be used in areas where continuous movement is required	• More expensive than self-contained lensed sensors

(Courtesy of Rockwell Automation)

229

	 (a)	 (b)	 (c)	 (d)
Amplifier Type	Built in DC/AC	Built in DC/AC	Separate	Separate
Model	E3JM	E3G Long Dist.	E3C Probes	E3C Amplifier
Features	Built in multi-function timer module provides selectable ON-delay, OFF-delay and one-shot output with 0.1 to 5 sec. range. Switch selectable Light-ON/Dark-ON operation. Polarized retroreflective types accurately detect shiny objects.	Auto teach function with zone capability. Relay or transistor outputs. Light-ON/Dark-ON operation, NPN/PNP output are switch selectable.	Bendable stainless steel probes retain shapes for detection in hard-to-reach areas. Ideal for detection in small parts assembly and inspection. Cylindrical or rectangular model available. Wide selection of amplifiers with or without timing functions.	Amplifier provides fine sensitivity adjument, alarm output for unstable switching conditions such as sensor misalignment or dust, and a 40 ms OFF-delay for programmable controller input AC and DC outputs, built-in timing functions including ON-delay, OFF-delay and one-shot with 0.1 to 10 second timing range.
Through Beam Distance	10 m		50 mm	
Retroreflective Distance	Polarized 0 to 4 m	10 m		
Diffused Reflective Distance	0 to 700 mm	2 m	2, 5, and 10 mm	
AC Control Output	Relay 3 Amp	SPDT Relay 3 Amp	See E3C amplifiers	SPDT Relay AC
DC Control Output	NPN or PNP	NPN/PNP switch selected	See E3C amplifiers	NPN and PNP
Light Source	Infrared LED Polarized LED on retro	Infrared LED diffuse – Red LED polarized retro	Pulse modulated IR LED or red LED	2 ms or 41 ms max ON/OFF switch selected
Frequency Response	30 ms relay 5 ms transistor		See E3C amplifier	

Figure 5–22 Photoelectric Sensors.
(Courtesy of Omron Electronics, LLC)

	 (a)	 (b)	 (c)	 (d)
Amplifier Type	Built in DC	Built in DC	Built in	Built in DC
Model	E3S-GS	E3C-V	E3S-R	E3L
Features	Fast response time, ideal for packaging applications. 1 cm groove type detects marks on transparent film. Pre-aligned emitter and receiver simplifies installation. Sensitivity adjuster and stability indicator.	Detect extremely small objects including 200 micron copper wire and IC chip parts, through narrow gaps in equipment. Inspect for color, texture or part orientation. Remote sensitivity adjustment, timing functions, and an alarm output for deteriorating sensing conditions.	Detects clear glass and plastic bottles without false signals. Fast response time ideal for high-speed packaging equipment. Sensitivity adjuster and stability indicator allow fine-tuning of sensor to application.	Laser beam detects very small objects up to 2m away with high accuracy. Apertures allow detection of objects from 0.1 to 0.5 mm in dia. Sensitivity adjuster and operating stability indicator allow fine-tuning. Alarm output signals deteriorating detection conditions.
Through Beam Distance	10 mm mark detection – 30 mm general			2 to 10 m
Retroreflective Distance			300 mm and 1 m	
Diffused Reflective Distance		10 +/−2 mm to 70 +/−10 mm		200 to 500 mm
AC Control Output		See E3C amplifiers		
DC Control Output	NPN and PNP SPST open collector	See E3C amplifiers	NPN and PNP open collector	NPN and PNP open collector and NPN current
Light Source	Pulse modulated green and red LED	Pulse modulated green and red LED	Red or infrared LED	Pulse modulated laser diode in red or infrared
Frequency Response	2 ms ON/OFF		1 ms ON/OFF	1 ms ON/OFF or 3 ms ON/OFF

Figure 5–23 Photoelectric Sensors.

(Courtesy of Omron Electronics, LLC)

values for the sensors in Figures 5–22 and 5–23 specify different distances for the three types: through beam, retroreflective, and diffused reflective. Note that some models can be used for operation in all three modes, while others support only one mode of operation. For example, the E3JM has a configuration for all three sensing modes, through beam, retroreflective, and diffused reflective; however, the E3S-R only supports the retroreflective mode. Two-wire or three-wire outputs are provided with normally open (NO) or normally closed (NC) contacts options for the two-wire operation. Some two-wire outputs provide relay contacts for switching currents in the 1–3 ampere ranges. The output options for the three-wire devices are NPN (current sinking) and PNP (current sourcing) transistors, and open collector configurations. The sophistication and miniaturization in electronics permits many different output options to be incorporated into the relatively small amplifier packages. An example of two outputs available on the E3C-JP4P and -JC4P sensor are illustrated in Figure 5–24. Note that transistor A in the PNP (current sourcing) circuit acts as a switch to connect the load to the 12 to 24 volt dc power. If A is a closed switch, then the transistor acts as a *source* of current to turn on the load that is connected from the transistor to ground. Transistor B in the NPN (current sinking) circuit also acts as switch to connect one side of the load to ground or 0 volts. The other side of the load is connected to 12 to 24 volt dc power, so when B is a closed switch the current flows into the sensor from the load (sinking or taking the current) to ground. Each of these circuits are also called *open collector* circuits because in each case the collector of the transistor is connected to the output without any other internal component connection in the collector. In open collector outputs, the load completes the collector circuit to either power or ground. The output circuits for proximity and photoelectric sensors are similar, so review the discussion of the three-wire outputs provided after the proximity sensor section. The frequency response for the different models is specified as well. The light sources used for photoelectric sensors include LEDs and lasers. The frequency of radiation runs from infrared through the visible range of the spectrum with red and green the most frequently used colors. The sources are either continuous or modulated depending on the model selected. The ambient operating temperature range for most models is −13 to 158°F.

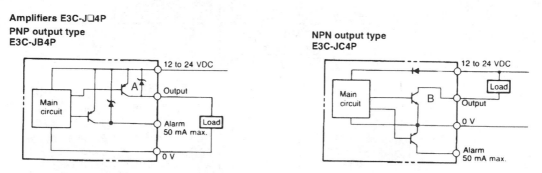

Figure 5–24   Current Sinking, Current Sourcing, and Open Collector Output Circuits on the E3C Sensor.
(Courtesy Omron Electronics, LLC)

The primary differences among the four types of photoelectric sensors are in their range, that is, the distance to the detected part, and the response time. The separate-type sensor system permits the light source and receiver to be separated by up to 30 meters for some models, so that detection can be anywhere along the 30-meter length. The retroreflective type has separation distances between the sensor and target from 1 to 5 meters. The diffused-light type has a detection distance that is less than 1 meter and typically in the 5- to 50-centimeter range. Last, the definite reflective type has the shortest detection distance—typically from 5 to 25 centimeters. Although the range is small for the definite reflective type, the ability to operate with high ambient light conditions makes this type of sensor superior to the diffuse reflective type under those operating conditions. The response times vary from 40 microseconds to 60 milliseconds.

The sensor unit consists of one or two packages, depending on the model and manufacturer, with mounting brackets designed for easy alignment of the sensors. Figure 5–25 shows the internal arrangement of parts and components of an Omron model E3B sensor. The packages are designed for operation under varying environmental conditions, and models can be found that meet world standards and those of the National Electrical Manufacturers Association (NEMA). The sensors include solid-state sensing electronics, which drive the light source in the transmitter, and the light-sensitive element in the receiver. The electronics require either dc or ac voltage for power (typically 12 to 24 volts regulated dc or 24 to 240 volts ac). The sensor output includes an electrical signal and light-emitting diode to indicate the state of the sensor. The electrical signal can drive a sensor controller that provides relay contacts for use in switching higher current loads or can drive smaller loads directly.

Figure 5–25 Sensor Construction.
(Courtesy of Omron Electronics, LLC)

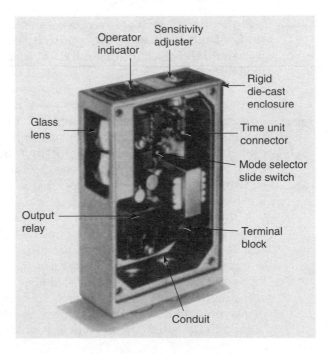

The Omron models, for example, have direct output drive capability that ranges from 80 milliamperes on some models to 3 amperes on units with relays.

Figure 5–26 shows several examples of sensors being used to check for the presence of parts. These could very easily be incorporated into a robot work cell. For example, Figure 5–21A, C, G, J, and K illustrate how a sensor could be used to check for the presence of various types of parts a robot might be moving in a material-handling application or using in an assembly application. The application in Figure 5–21B involves the tasks of unstacking thin plates. The signal from the sensor could be used by the robot controller to stop the downward motion of the arm and activate a vacuum gripper. In a robot drilling application, the drill length must be checked frequently to be sure a drill rod has not been broken. The application in Figure 5–21D shows how this is accomplished with a photoelectric sensor.

Application of photoelectric sensors involves the same steps described at the end of the section on proximity sensors. Review those steps again with the photoelectric sensors in mind. The factors affecting sensor selection include the following seven:

1. Sensing distance required by the application
2. Sensor mounting requirement
3. Work-cell area available
4. Size, shape, and surface reflectivity of the part to be sensed
5. Response time required
6. Environmental conditions in the work cell, especially background light present
7. Interface requirements

Selecting the sensor that best satisfies the work-cell conditions requires a thorough study of the application. The primary elements of the study are the factors just listed, and the results of the study are criteria that can be used to select the sensor model for the job.

The sensitivity of photoelectric sensors can be adjusted so that the sensing of transparent, translucent, and opaque materials is possible with the same model sensor. The procedure for adjustment of sensor sensitivity varies according to models and manufacturers. On some models the indicator LED (light-emitting diode) changes colors as the sensitivity is adjusted; on others, the LED changes from *on* to *off* to indicate the sensing condition. On all models with a sensitivity adjustment, the sensor can be operated in a stable mode with a variety of surface finishes on the part being sensed.

Interfacing the sensor to supply power and controller inputs requires the designer first to consider the sensor model selected. If regulated dc power between 12 and 24 volts is available, then an additional dc power supply is not required. Some sensors have a universal ac/dc power supply. When a sensor controller is not used, the sensor output must be interfaced directly to robot controller input or the input of a programmable logic controller. This requires an output circuit in the sensor that is compatible to the input circuit in the robot or programmable logic controller. Compatibility issues include current sinking and sourcing plus maximum and

- When the sensor is susceptible to the reflection from background object surface

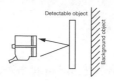

(A)

Typical examples
(1) Sensing of thin objects on the conveyor line.
(2) Sensing of objects in the presence of a background object with high reflection factor such as rollers, metallic plates, etc.
(3) Sensing of the residual quantity in a hopper or a parts feeder.

- Sensing of level or height

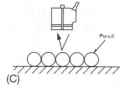

(B)

Typical examples
(1) Sensing the height of stacked plywood, tiles, etc. from above.
(2) Monitoring and control of the liquid level from above.
(3) Determination of the heights of objects on a conveyer line.
(4) Sensing a slack of sheets from above.

- Sensing of objects traveling in contiguous succession

(C)

Typical examples
(1) One-by-one sensing of pencils or metallic bars traveling successively or in contiguous succession.
(2) Similarly, one-by-one lateral sensing of bottles or cans traveling in contiguous succession.

- Sensing of small, slender or fine objects

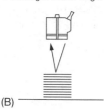

(D)

Typical examples
(1) Sensing of broken drill bits.
(2) Sensing of small parts such as electronic components.
(3) Sensing of the presence or absence of bottlecaps.
(4) Sensing of fine mesh.

- Sensing of small holes, narrow openings, or unevenness

Typical examples
(1) Sensing of holes in flat board.
(2) Sensing of protrusions.

(E)

- Sensing of objects utilizing their difference in luster

Typical examples
(1) Identifying the face or back of tiles.
(2) Identifying the face or back of lids.

(F)

- Sensing of transparent objects

Typical examples
(1) Sensing of transparent or translucent objects.
(2) Sensing of transparent glasses, film, or plastic plates.
(3) Sensing of the liquid level.

(G)

- Sensing of objects through a transparent cover

Typical examples
(1) Sensing of the contents in a transparent case.
(2) Sensing of the position of meter pointer.

(H)

- Sensing of the edge of object

Typical examples
(1) Positioning control for plywood.
(2) Positioning control for various other products.

(I)

- For sensing of presence of parts in parts feeder

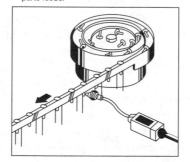

(J)

- For sensing of presence of resistors on conveyer line

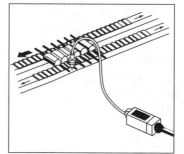

(K)

Figure 5–26 Example Applications of Photoelectric Sensors.
(Courtesy of Omron Electronics, LLC)

minimum output voltages and currents. (Interface compatibility between sensors and work-cell controllers is addressed in detail in Chapter 7.)

5-4 SENSOR SELECTION CHECKLIST

Sensors are the primary source of work-cell data for the robot and automation system, so it is critical that the correct sensor is selected. The sensor selection checklist in Table 5–3 lists questions that must be answered for the following three categories: target information, environmental information, and electrical information. The answers to the questions help to determine if a limit switch, proximity, capacitive, or photoelectric sensor is required. In addition, the type of sensor in the chosen category is identified through the data.

5-5 SMART SENSOR SYSTEMS

The computer/data network has become the backbone of automation systems. The work-cell integration described in later chapters indicates that multiple networks are used to link production and automation control machines. For example, networks used in automation include industry standard IEEE 802.3 Ethernet and IEEE 802.5 token ring, plus numerous proprietary vendor networks link together vendor specific device; for example, the Allen Bradley data highway is used to network PLCs and other Rockwell automation systems. This growth in network technology has had a great impact on the sensor with the development of field-bus technology. Using the field-bus standard, automation control vendors have developed an array of control devices including sensors that are implemented as a node on a data network. Sensors offered in this mode have imbedded microcontrollers that allow the device to exchange the work cell, environmental, and sensor operational information with other network devices using only a single network cable.

DeviceNet Network

Many vendors offer products that support field-bus technology, and one is Allen Bradley's *DeviceNet*. DeviceNet, based on Controller Area Network (CAN) technology, is a low-level network that provides connections between simple industrial devices such as sensors and actuators. In addition, DeviceNet can link high-level devices like PLC controllers and computers. The benefits of this technology are apparent from a study of one of the devices. For example, a DeviceNet motor starter has all the functionality of a conventional contactor, but it also includes the ability to report the load current and contractor temperature. Numerous devices are available in DeviceNet format including switches, pushbuttons, photoelectric sensors, limit switches, motor starters, motor controllers, and proximity sensors. A single DeviceNet network can be up to 328 feet (100 meters) long and have up to 64 nodes or devices attached. A DeviceNet network is illustrated in Figure 5–27, which shows the range of devices that can be linked by this type of technology. Study the figure to understand the range of devices supported by DeviceNet Technology.

Table 5–3 Sensor Selection Checklist.

Target Information

Parameter	Data Needed
Material	Is the target material ferrous, nonferrous, or nonmetallic?
Mass	How much does it weigh?
Size	How large is the target in square inches or cubic inches?
Shape	Is the target flat, cylindrical, cubic, or spherical? Is it compact, long, or narrow?
Surface	Is the surface opaque or transparent? If it is opaque, is it colored, shiny, dull, smooth, textured, porous, or nonporous?
Motion	What is the velocity in feet per second? Is the direction of movement perpendicular or oblique to the line of sight? If using photoelectric sensors, can the reflector or receiver be mounted behind the target?
Distance	How far away is the object to be detected? How close to the part or machine can the sensor be located? (This parameter alone can determine whether a proximity or photoelectric sensor should be used.)
Presentation	What is the rate of presentation to the sensor in units per minute? Are targets oriented randomly or regularly? Are parts separated or do they overlap? What is behind the target?
Precision	Is precise range needed? This is usually true in machine tool applications.

Environmental Information

Parameter	Data Needed
Enclosure	Is a NEMA qualified case required? Is the material aluminum or plastic?
Temperature	What is the typical annual range in °F? Is it very cold or very hot?
Relative humidity	What is the typical annual range in percent? Is it very dry or very wet?
Ambient lighting	What are the sources of light? Are they constant or variable? Is there any direct sunlight?
Electrical noise	What are the typical and peak intensities? A welding line would represent a worst-case example.
Mechanical vibration	What is the typical amplitude range? What is the maximum physical shock possible?
Air quality	Is the air clean or dirty? Is the contaminant dust, smoke, oil mist, or paint spray? Is the atmosphere explosive?
Mounting surface	Is the material ferrous, nonferrous, or nonmetallic? How thick is it? Is it flat, rounded, or some other shape? How much mounting space is available in square or cubic inches? Are precision clearances necessary? Is the mounting surface stationary or mobile?

Electric Circuit

Parameter	Data Needed
Power supply	120 Vac, 50/60 Hz, or 12/24/35 Vdc?
Contacts	Number and configuration?
Output form	Analog or digital?
Output load	Minimum and maximum current required?
Output voltage	Minimum and maximum voltage required?
Switching delay	Activation delay in microseconds? Release delay in microseconds?
Switching frequency	Hz?

(Reprinted with permission from SENSORS —*Your Resource for Sensing, Communications, and Control*, April 1986.)

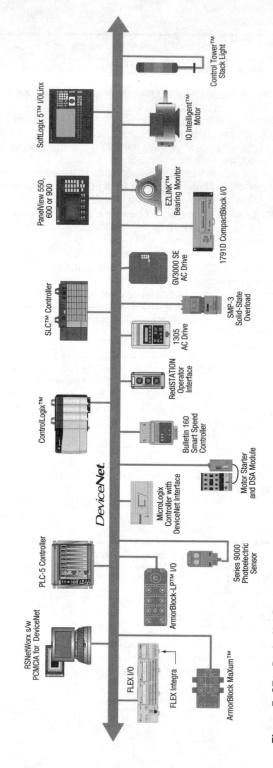

Figure 5–27 DeviceNet Devices on a Network.

(Courtesy of Rockwell Automation)

5-6 PROCESS SENSORS

In addition to the sensors required for the robot, most process or manufacturing operations need sensors to monitor parameters inherent in the process itself. Also, many of the other automated machines have sensors to alert and warn the operator about conditions that are developing within the system. A forging operation is a good example. Through warning lights, the human operator monitors the temperature of the oven that is heating the parts before forging. The level of oil and oil pressure for the press are also displayed by indicators. With a robot present, the visual indicators are of no value. If visual indicators are replaced with electronic sensors, however, then an electrical signal can be used to alert the robot or work-cell controller when corrective action is needed.

Most system or machine parameter to be monitored by the robot are discrete signals, that is, they are either an *on* or *off* signal. Analog sensors that have a variable output, 0 to 5 volts, for example, are associated with proportional control requirements. These can also be interfaced to the work cell or robot controller; however, they are usually handled by a separate control system designed to provide proportional control. A discrete signal, called an *alarm*, could be provided by the proportional controller to warn the robot that the process is off course and a correction is necessary.

5-7 TROUBLESHOOTING SENSOR SYSTEMS

The general troubleshooting discussion in Chapter 3 discussed a strategy for locating faults in electrical and electronic systems. In this section, those techniques are applied to systems using sensors; in addition, some troubleshooting tips for typical sensor problems are provided. Quickly review the Chapter 3 material before starting this section.

Sensors are never used alone because there is always some load(s) that is controlled by the sensor output. In most systems, the sensors are connected to a system controller, such as a programmable logic controller (PLC), or industrial process machine like an industrial robot or a computer numerical control machining center. As a result, the most frequent signal flow model is a linear signal flow with convergent inputs and divergent outputs. An example illustrates this best.

The die-cast operation in Figure 5–28, introduced in Chapter 2, uses numerous sensors and limit switches to ensure safe and efficient operation with a robot present. This example focuses on the process of trimming the excess metal from a complete die-cast part. Refer to the sensor system wiring and signal flow block diagram in Figure 5–29 while the operation is described. Say the robot is at a programmed point just outside of the trim press with an untrimmed part in its gripper. Before the robot program can place the part into the trim press, three sensors must give a signal to proceed. Sensor 1 (at top of the trim press) is *on* when the press is open, sensor 2 (at the rear of the press at the die level) is *on* when a part is in the trim press die, and sensor 3 (on the robot gripper) is *on* when a part is present in the gripper. So sensor 1 and sensor 3 must be *on*, and sensor 2 must be *off*

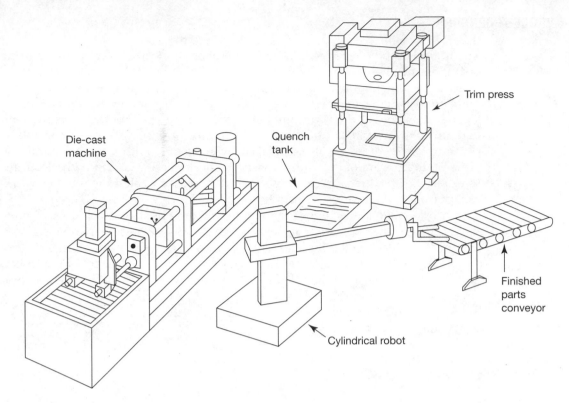

Figure 5–28   Robot Machine Tending Application.

before the signal for the robot to proceed is given. In addition to generating this signal, the PLC also turns on a panel pilot light indicating that a trim cycle has started.

A review of the signal flow block diagram indicates that a fault in numerous units could result in no motion by the robot or no trim cycle light. A fault in any block in any of the three linear paths between the sensing heads and the PLC input module would halt the robot and prevent the panel light from illuminating. An illuminated panel light but no robot motion would indicate a problem in either the PLC output module or robot input module. The problem would probably not be in the robot or PLC program if the programs worked on the previous cycle; however, if this is the first cycle of the day then a change in the program during off-shift maintenance could make a program fault a possibility.

The funneling concept suggests that initial checks use hardware indicators or switches before connectors are removed and voltage levels measured. Numerous indicators on the hardware are useful. For example, the sensor amplifiers have indicators to indicate the state of the sensor, and the PLC input and output modules have LEDs to indicate that input signals and output signals are present. If left brackets were placed at the sensors and right brackets on one or both outputs, then the LEDs on the PLC would be good middle point for the troubleshooting process.

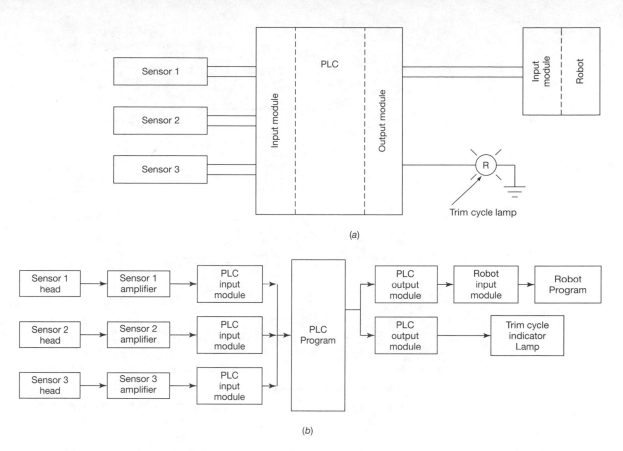

Figure 5–29   Sensor Wiring and Signal Flow Analysis (a) Wiring diagram (b) Block diagram signal flow.

Troubleshooting Tips for Proximity Sensors

The following tips may be helpful if the problem appears to be in the sensor or sensor amplifier. Sensors from different vendors have different operating characteristics, so the first requirement is to know how the sensors in the system operate. The following order of the tips does not indicate a preferred sequence. Review all the tips and consider those that apply to the current sensor problem. Verify that the sensor has power in the specified range. Checking other operating equipment connected to the same power bus is a good method for testing for power.

■ Verify that all the amplifier settings are correct. Many sensor amplifiers have a sensitivity adjustment (see Figure 5–30 for the E2C sensor amplifier); make sure the protective seal is still in place. Verify that all switch settings are correct.

■ Use the *operation indicator* on the sensor or sensor amplifier (Figure 5–30) to determine if the sensor electronics recognize that a part is present. An *on* condition for operation indicator usually indicates that the output transistor or relay is operating. On Omron devices with the output set to the normally

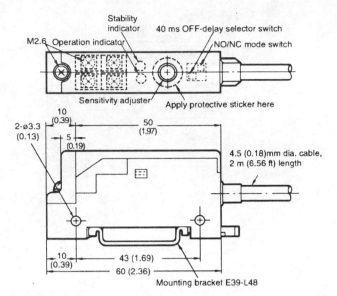

Figure 5–30 Proximity Sensor Amplifier.
(Courtesy of Omron Electronics, LLC)

open (NO) operation, the operation indicator is *on* when an object is sensed. The opposite is true for a NC setting. A good practice is to move the part toward the sensor along the same path used in the process and determine how close the required sensing distance is to maximum value.

■ Use the *stability indicator* (Figure 5–30) to indicate how close the desired sensing distance is to the maximum. If the sensing distance falls into the area where the stability indicator is off, then unstable operation will result. The function of the operation and stability indicators is illustrated in Figure 5–31 for the Omron E2C-T sensors.

■ Verify that a foreign object is not creating a problem on one of the small sensing heads.

■ Verify that the velocity of the parts past the sensor does not exceed the frequency response of the unit.

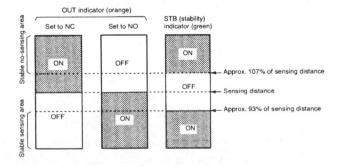

Figure 5–31 Indicator Operation on Sensor Amplifiers.
(Courtesy of Omron Electronics, LLC)

- Verify that the sensing distance was not reduced due to a change in the ambient temperature or supply voltage.
- The design process addresses a number of issues (see Sensor Selection Checklist Table 5–3) associated with placing sensors in the automated system. If the sensor appears to be fault free, then verify that some part or process parameter has not changed that would cause the sensor not to detect the part. Part size, distance, material, part speed, temperature, and supply voltage all affect proximity operation.

Troubleshooting Tips for Photoelectric Sensors

The following tips may be helpful if the problem appears to be in the sensor or sensor amplifier. The first rule is to know the operation of the sensor in the system.

- Verify that the sensor has power in the specified range. Verifying power to other operating equipment connected to the same power bus is a good method for testing for power.
- Verify that all the amplifier settings are correct. Many sensor amplifiers have a sensitivity adjustment (see Figure 5–32 for the E3C sensor amplifier);

Figure 5–32 Photoelectric Sensor Amplifier.

(Courtesy of Omron Electronics, LLC)

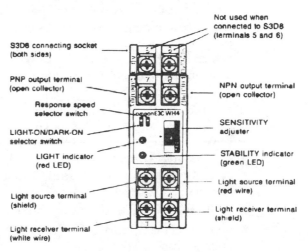

Figure 5–33 Photoelectric Sensor
Timing Diagram E3C.
(Courtesy of Omron Electronics, LLC)

E3C-WH4F Timing Chart

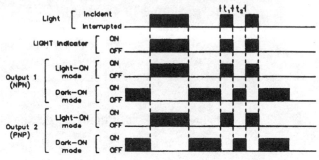

NOTE: t₁ and t₂ must exceed selected response time (1 or 2 ms) before solid-state
output states will change.

make sure the protective seal is still in place. Verify that all switch settings are correct.

■ Use the *operation indicator* on the sensor or sensor amplifier (Figure 5–32) to determine if the sensor electronics recognize that a part is present. An *on* condition for operation indicator usually indicates that the output transistor or relay is operaing. On Omron devices with the output set to *light-on* mode, the operation indicator is *on* when light is striking the sensor. The opposite is true for a *light-off* mode setting. The timing diagram for a typical photoelectric sensor is illustrated in Figure 5–33.

■ Use the *stability indicator* (Figure 5–32) to indicate how close the desired sensing distance is to the maximum for diffuse sensor models. If the sensing distance falls into the area where the stability indicator is off, then unstable operation will result. The operation of the operation and stability indicators is illustrated in Figure 5–34 for the Omron E3G sensors.

■ Verify that the lenses are clean and free of foreign objects.

■ Verify that the velocity of the parts past the sensor does not exceed the frequency response of the unit.

■ Verify that the sensing distance was not reduced due to a change in the ambient temperature or supply voltage.

■ The design process addresses a number of issues (see Sensor Selection Checklist Table 5–3) associated with placing sensors in the automated system. If the sensor appears to be fault free, then verify that some part or process parameter has not changed that would cause the sensor not to detect the part. Changes in the part surface luster, texture, color, and angle of presentation to the sensor affect the operation. In addition, air quality, ambient light, part speed, distance, temperature, and supply voltage can all affect sensor operation.

Figure 5–34 Operation of Operation and Stability Indicators for Photoelectric Sensors. (Courtesy of Omron Electronics, LLC)

■ ADJUSTMENTS

Indicators

The following illustration indicates the operation levels of the E3G.

Set the E3G so that it will work within the stable operation range.

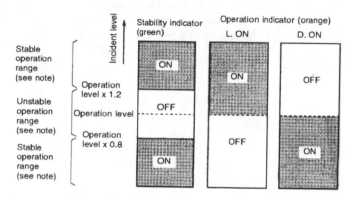

Note: If the operation level is set to the stable operation range, the E3G will operate with the highest reliability and without being influenced by temperature change, voltage fluctuation, dust, or setting change. If the operation level cannot be set to the stable operation range, pay close attention to environmental changes while operating the E3G.

5-8 SUMMARY

Sensors make the robot part of the environment in which it exists. They give the robot information about the work cell that is vital to normal operation. Sensors are used to (1) protect worker and robot from harm, (2) monitor the production system and work-cell operation, (3) analyze product quality, and (4) provide part identification and orientation. All sensors are grouped into either a contact or noncontact category, and each type can have either a discrete or an analog output signal. Contact sensors include limit switches and artificial skin, and the most commonly used noncontact sensors include proximity sensors and photoelectric devices. Limit switches are available with tripping mechanisms, including plungers and levers. Contacts are rated as either pilot duty or electronic duty, based on the level of current that is controlled. A large selection of devices supports the many different operational characteristics present in automation. The concept of artificial skin focuses on the need for a sensor that can replicate the sense of touch used by human operators. Sensors in this area provide either simple touch sensing or full tactile sensing; the latter is difficult to achieve at this time.

Proximity sensors detect the presence of a part when the part comes within a specified range of the device. The sensors are available in a variety of package shapes and sizes. The sensing technology uses either a magnetic field (metallic part) or electrostatic capacitance (both metallic and nonmetallic parts) to detect the presence of a part. Output circuits dc and ac, in normally closed and normally open configurations, are available. The dc output in either an NPN or a PNP transistor configuration is used for smaller power requirements, and the ac output is used for larger loads.

Photoelectric sensors detect the presence of a part when a transmitted light beam is broken or reflected to a receiver. The four types commonly used include separate, retroreflective, diffuse reflective, and definite reflective. The operational characteristics for the four types distinguish each category; however, the primary difference is range or distance to the detected part. As is true of proximity sensors, a variety of package shapes and output configurations are available.

Field-bus technology permits devices with imbedded microcontrollers to be connected in the work cell with only network cables. The DeviceNet from Allen Bradley is an example of this type of technology.

Process sensors measure parameters that are part of the production process. In most applications the sensors have analog output signals to provide the exact value of the measured parameter.

The troubleshooting concepts introduced in Chapter 3 are applied to automation sensors. In addition, troubleshooting tips for proximity and photoelectric sensors are described.

QUESTIONS

1. What is the basic function of sensors in an automated work cell both with and without a robot present?
2. What are the six basic reasons sensors are used in a work cell?
3. What are the two categories into which all sensors can be grouped?
4. What kinds of sensors are included in the contact sensor group?
5. What is the definition of a limit switch?
6. What are the four basic parts of every limit switch?
7. How does pilot duty differ from electronic duty in the electrical characteristics of limit switches?
8. Using a data sheet for a general-purpose limit switch, determine the switch part number and lever part number required for rotary operation using fork rollers on the same side, and double-pole–double-throw operation for electronic duty.
9. What is the function of the lamp on a limit switch?
10. What is the distinction between tactile sensing and simple touch sensing?
11. Describe how tactile sensors operate.
12. What three steps are required for intelligent acquisition of parts?
13. What sensors are included in the noncontact sensor group?
14. Describe the four package configurations now available with proximity sensors.

15. Describe the difference between metallic detection and nonmetallic detection proximity sensors.
16. What three output circuits are available on proximity sensors?
17. Describe the four different types of photoelectric sensors.
18. What is the primary difference among the four types of photoelectric sensors?
19. What type of photoelectric sensor has the greatest range?
20. What type of photoelectric sensor is most accurate when used for range finding?
21. What factors affect photoelectric sensor selection?
22. Select a photoelectric sensor and sensor controller for the following applications. Identify the part numbers of the sensor and controller required.

    (a) A sensor to operate at a distance of 8 centimeters with high background light. The output should be *on* when the object is sensed.
    (b) A sensor to operate over a distance of 8 feet to detect when a part breaks the beam. The output should be active when the beam is broken.
    (c) A sensor to count the number of resistors in a parts feeder by using the resistor leads to reflect a beam. The output should be *off* when the lead is present.
    (d) A sensor to count revolutions of a shaft using marks on a transparent disk mounted to the shaft. The output should be active when the mark is not present.

PROBLEMS

1. Design a dog to trip a D4A-1102N limit switch with a D4A-A20 lever for the non-overtravel operation of a parts carrier with a velocity of 1.03 feet per second.
2. Design a dog to trip a D4A-2505N limit switch with a D4A-B06 lever for the non-overtravel operation of a parts carrier with a velocity of 22 inches per second.
3. What are the electronic characteristics for the limit switches used in Problems 1 and 2?
4. Determine the minimum sensing distance on an Omron E3E-X14MD sensor (Figure 5–18) if the target is stainless steel and has a one-side dimension of 50 millimeters.
5. Use Table 5–1 to determine the maximum sensing distance on an Allen Bradley sensor that has a nominal sensing distance of 25 millimeters when the target is 300 series stainless steel.
6. Study all the application descriptions in Figure 5–26 and recommend the best type of sensor from the proximity, capacitive, and photoelectric devices covered in the chapter. Justify your choice in each case.
7. The signal from an E3L photoelectric sensor is used to count the bottles moving on an assembly line by passing the beam through the neck of each bottle.

The bottles are 1 inch in diameter at the neck and 2.5 inches in diameter at the base. If the response time in the data sheet (Figure 5–23) is the turn on and turn off time, calculate the maximum rate in bottles per minute that can be detected by the sensor.

8. The encoder wheel in Figure 2–36 uses an E3S-GS sensor to count increments as the wheel is turned by the robot actuator. The open slots are 0.3 inch long by 0.25 inch high, and the centers are located 2.5 inches from the center of the wheel. If the response time in the data sheet (Figure 5–23) is the turn on and turn off time, calculate the maximum angular velocity in rpm for the wheel for a 1 millisecond minimum output pulse.

9. How many additional slots can be added to the encoder in Problem 8 without reducing the angular velocity?

10. This troubleshooting problem will use Figure 5–29 (b) and system conditions of part in gripper, trim press in open position, and trim press empty.

   (a) Determine the location of the left and right brackets if the robot does not move but the trim cycle light is on. What would be the first check?

   (b) Determine the location of the left and right brackets if the trim cycle light is off but the robot has moved to put the part into the press. What would be the first check?

   (c) Determine the location of the left and right brackets if the trim cycle light is off and the robot does not move. If the failed unit is the sensor 2 amplifier, indicate the order of checks to find the bad unit and show how the brackets would move after each check.

11. Use the Internet site URLs for sensor manufacturers in Appendix B to locate a specifications sheet for a proximity and a photoelectric sensor. Then write a paper that completely describes the operation of each type.

12. Using the sensors located on the Web in the previous problem, develop a troubleshooting flow chart for each to determine if the sensor is operating correctly.

PROJECTS AND CASE STUDY PROBLEMS

1. Draw a classification tree that includes all the sensors described in this chapter.

2. Design a decision tree to select the best sensor for a given manufacturing problem. Use questions at each branch of the tree that require either a yes or no response.

3. Write a Visual Basic computer program that will execute the decision tree from Project 2.

4. Develop a list of sensors for the injection molding production cell developed in Case Study Problem 3 in Chapter 2.

5. Select a sensor requirement for the kitchenware bowl production cell design started in Case Study Problem 5 in Chapter 2, and complete the sensor selection checklist in Table 5–3 for the application.

6. Develop a list of sensors for the production cell design started in Case Study Problem 5 in Chapter 2. Include the model number and type for each.

7. Develop a list of sensors to support the production cell design started in Case Study Problem 3 in Chapter 3. Include the model number and type for each.

8. Develop a list of sensors to support the West-Electric slug production cell design started in Case Study Problem 8 in Chapter 2. Include the model number and type for each.

9. Develop a list of sensors to support the West-Electric slug lubrication cell design started in Case Study Problem 9 in Chapter 2. Include the model number and type for each.

10. Read the West-Electric case study where the work cell sensors are selected (Sections 12-9 and 12-10 in Chapter 12). Complete the questions and problems at the end of Chapter 12 that relate to the sections covered.

Work-Cell Support Systems

CHAPTER GOALS AND OBJECTIVES

The goal of Chapter 6 is to provide a broad overview of the primary support systems frequently used with robot automation. The topics include vision systems, material handling, automatic storage and retrieval (ASRS), part feeding, inspection, and automatic tracking. At the completion of this chapter you should be able to:

- Describe the five tasks that vision systems perform in automated manufacturing.
- Name the component parts for vision systems and describe image measurement, analysis, and recognition.
- Describe the three lighting techniques—front, back, and structured—used for vision work.
- Name the two functions of material handling and describe the three types of automated transfer systems: continuous, intermittent, and asynchronous.
- Describe the function and operation of an ASRS.
- Describe the function and operation of four types of parts feeders: gravity, tape, waffle tray, and vibratory.
- Define the expression *quality at the source* and describe why the concept is critical in manufacturing automation and in a CIM implementation.
- Describe the two most popular bar code technologies, interleave two of five and code 39, and the operation of radio frequency tags.

6-1 INTRODUCTION

Some combinations of the sensors described in the previous chapter are found in work cells that use robot automation. In addition to basic contact and noncontact sensors, many of the cells have various other support systems present. A description of all the supporting equipment used in robot cells is beyond the scope of this text, but we will discuss some of the systems that are used frequently. The systems associated with robot automation include *vision, material handling, automatic storage* and retrieval systems (ASRS), part feeding, inspection, automatic tracking, and safety. The first six support system areas are covered in this chapter, and the sixth, safety, is included in Chapter 10.

6-2 MACHINE VISION SYSTEMS

Vision systems are being used increasingly with robot automation to perform the following tasks:

- *Part identification:* Commercially available vision systems store data for different parts in active memory and use the data to distinguish between parts as they enter the work cell. The system can learn the characteristics of different parts and identify each part from its two-dimensional silhouette.
- *Part location:* Vision technology allows the user to locate randomly placed parts on an X-Y grid. The vision system measures the X and Y distances from the center of the camera coordinate system to the center of the randomly placed part.
- *Part orientation:* Every part must be gripped in a specified manner by the end-of-arm tooling. The vision system supplies the orientation information and data that are used to drive the gripper into the correct orientation for part pickup. Many part orientation parameters, both measured and calculated from measured data, are provided by the vision system for use in automated part handling.
- *Part inspection:* Vision systems are used to check parts for dimensional accuracy (for example, the diameter of a part) and geometrical integrity (for example, the number of holes). The parts are measured by the camera, and the dimensions are calculated; at the same time, the vision system checks the parts for any missing features or changes in the part geometry.
- *Range finding:* In some applications the system uses two or more cameras to measure the X, Y, Z location of the part. This technique is also used to measure and calculate the cross-sectional area of parts.

The use of vision to enhance the operation of an automated work cell has moved from the research laboratory to the factory floor. More than fifteen manufacturers provide equipment that gives robots the eyes they need to perform complex manufacturing tasks. Most of the equipment is for part identification, location, and orientation information necessary for automatic handling of parts plus inspection.

Vision Standards

The standards established by the AIA for the RIA include the following six ANSI standards:

1. AIA A15.08/1: Standard AIA Analog Camera Connectors.
2. AIA A15.08/2: Standard AIA Digital Camera Connectors.
3. AIA A15.08/3: Monochrome Digital Interface Specification.
4. AIA A15.08/4: RGB Digital Interface Specification.
5. AIA A15.05/5: Monochrome Analog Interface Specification.
6. AIA A15.05/6: RGB Analog Interface Specification.

Vision System Components

The block diagrams in Figures 6–1 through 6–3 illustrate three architectures used to implement vision technology. In Figure 6–1, a self-contained system is illustrated. In this type, the vision system does not use resources from any other

Figure 6–1 Block Diagram for Stand-Alone Vision System.

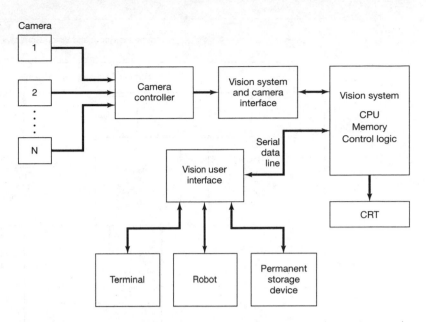

work-cell hardware. Note that vision information is passed to the work-cell robot through a serial interface. The system in Figure 6–2 has the vision system integrated into the work-cell programmable logic controller (PLC), in this case an Allen Bradley PLC-5. This type of system has the vision data and PLC control data

Figure 6–2 Machine Vision System with Programmable Logic Controller (PLC).

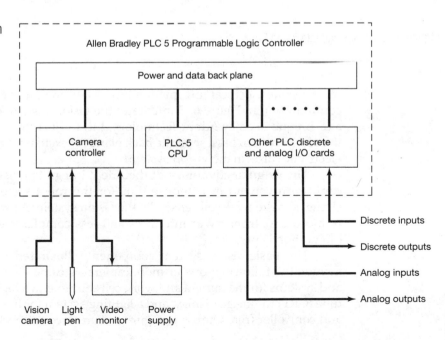

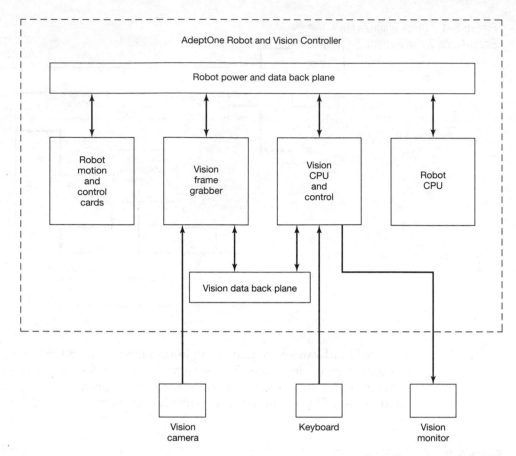

Figure 6–3 AdeptOne Integrated Robot and Vision System.

on the same electrical information bus, called the *back plane,* inside the PLC. The configuration in Figure 6–3 integrates the vision system with the robot controller, a technique used by Adept Technology, Inc. for its robot vision systems. Note that the vision system has a separate back plane for exchanging vision data between the electronic cards in the vision system.

The singular advantage of the Adept integrated vision system is the significant improvement in response time for the robot system when it responds to changes in the work cell sensed by the vision system. A secondary advantage is the reduced cost from fewer interfaces and less complex electronics required when integration is present.

The basic vision system components illustrated in all the configurations include the following: one or more cameras, a camera controller, interface circuits and systems for the camera and work-cell equipment, a high-resolution cathode ray tube (CRT) for image display, and a lighting system for the parts. The vision camera and controller from Omron in Figure 6–4 are used to check labeling on boxes.

Figure 6–4 Vision Camera Checking Box Labels.

The lighting system illuminates the critical features of the part so that they can be captured by the vision system cameras. The vision camera and lens system measure the level of light coming from the part under study; they capture a black and white representation of the part. A charge-coupled device (CCD) camera is used most frequently, but in some cases a vidicon tube camera is used. These two types of cameras are very different in operation.

The vidicon tube used in vision applications produces an analog output signal much like the cameras used in the broadcast television industry. The vidicon is a vacuum tube with a flat circular end designed to receive the image of the part from the lens. An internal electron beam is scanned across the image of the part at the end of the tube, and the level of light in the image produces a proportional analog voltage for every point that the beam strikes. A vidicon camera is not as linear or as stable as a CCD device and can have a 10 percent error due to image distortion.

The CCD camera uses a solid-state array of light-sensitive cells deposited on an integrated circuit substrate. Each cell is a small light-sensitive transistor or diode whose output is a function of the intensity of light striking its surface. The CCD systems come in two basic configurations: *linear* arrays and *imaging* arrays. Figure 6–5 shows how each of the types is organized. The linear arrays measure a single line, whereas the imaging arrays measure a complete two-dimensional image and cost more than the linear type. CCDs are accurate, rugged, and have good linearity.

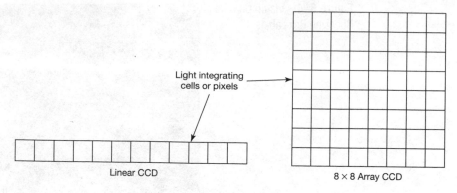

Figure 6–5 Charge-Coupled Devices (CCDs).

Image Measurement

The basic unit of measurement in a vision system is the *gray scale,* the basic parameter measured is *light intensity,* and the basic measurement element is the *pixel.* The vision system lens focuses light from the part onto the light-sensitive surface in the camera. The light-sensitive surface is divided into small regions or picture cells. Each of these regions or picture cells is called a *pixel.* The resolution of the vision system is directly proportional to the number of pixels on the light-sensitive surface. For example, a CCD with a 256×256 array will have a higher resolution than a CCD with a 128×128 array of pixels. The higher resolution results from more pixels per unit area in the 256×256 device. Higher resolution means greater accuracy when a system is used to measure part dimensions. Measurement is best understood with an example. If a 1-inch square produces an image that fills the 8×8 CCD array in Figure 6–5, then the smallest variation that could be detected is $\frac{1}{8}$ inch. If a 32×32 array with the same dimensions was substituted for the array in Figure 6–5, then the resolution of the measurement would be $\pm\frac{1}{32}$ inch. Each pixel on the sensor's surface is excited by the light focused on it by the camera lens. With no light present, the pixel is turned *off,* but when light reaches a *saturation level,* the pixel is full *on.* Between those two extremes there are shades of gray that cause the pixel to be excited to a partially *on* condition. The number of excitation states between *off* and *on* is called the gray scale. Current systems have gray scales ranging from 4 to 256. The greater the number in the gray scale, the better the system is at locating fuzzy edges and shadows of parts. The gray scale of a pixel is the numerical representation of brightness for one small spot on the part.

Image Analysis

The first step in image analysis is to find the part in the camera view. That means the system must locate all the areas in the image that correspond to the part being viewed by the camera. Most current vision systems are two-dimensional, which means that the outline of the object is the only feature of interest

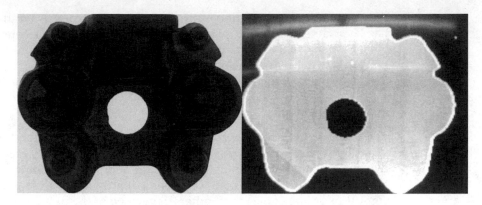

Figure 6–6 Vision System Representation of Production Casting.

to the system. The two-dimensional features of a part are enhanced by placing the part on a backlighted surface or by placing the object on a surface with contrasting color and using top lighting. The left-hand side of Figure 6–6 shows a part on a backlighted surface, and the right-hand side of Figure 6–6 shows the corresponding image generated by the vision system. The two most commonly used systems to find objects from camera data are *edge detection* and *clustering*. Both techniques attempt to locate the boundaries of objects or regions in the image so that their location, size, shape, and orientation can be computed as clues for recognition. Edge detection is based on the fact that there is a sharp difference in brightness between the object and its background. This can be seen in the first part of Figure 6–6. Areas of high contrast are found by searching through the array of pixel data for jumps in the gray level. To locate the change in gray level, a mathematical operator examines a small neighborhood of adjacent pixels and computes an *edge probability* value. If the value is above a set threshold, then a notation is stored in the vision controller memory and part data indicating a probable part edge. It is easier for the system to find fuzzy edges and to ignore false edges due to electrical noise if the size of the neighborhood or pixel group used to compute the edge probability has many pixels. However, increasing the size of the neighborhood increases the computation time for the system, so a trade-off is required.

The second method used to find objects from an array of pixel data is called *clustering* or *region growing*. The system basically tries to find adjacent pixels that have similar properties. This technique uses a *discrimination function* to determine if a pixel is part of a region being developed. The discrimination function computes the desired properties of a pixel and compares them with the properties of other pixels in the region or against a threshold value. In the simplest case, the gray level of each pixel could be checked against a threshold value to determine if the pixel is in the part image or in the background area. Many vision systems locate parts by using every pixel whose gray level value is different from the gray level value for the background. This process starts with a *seed* pixel within the part image and then grows in every direction until a boundary is reached. After the area of the part has been determined, the pixel gray scale data values are frequently changed to binary values; that is, every pixel that is part of the object is stored as a 1, and every pixel that is not part of the object is stored as a 0. Transferring these data to a CRT produces the part silhouette shown in the right-hand side of Figure 6–6.

Image Recognition

The second problem facing the vision system after the edges are detected and stored is the *recognition* or *identification* of the current image. Three frequently used two-dimensional recognition strategies are template matching, edge and region statistics, and statistical matching using the Stanford Research Institute (SRI) algorithm or one of the many others developed in the past 20 years.

In the *template matching* technique, templates of the parts to be recognized are stored in the vision system memory. The images recorded by the vision camera are compared with the templates stored in memory to determine if a matching part is present. Figure 6–7 shows the template of a part and the problems that this type of system must overcome. Any change in the object's scale or a rotation of the view makes a match with the template more difficult. The Allen Bradley vision system illustrated in Figure 6–2 uses the template matching process. In the system's teaching process, regions in the captured image are identified and recorded with the correct contents or parts present. In operation, an image of the region is captured and compared with the image recorded during the teaching process. If the image values for the regions are the same, the captured image matches the stored template, and the part is identified.

The *edge and region statistics* technique defines significant features for the parts studied, then develops a method of evaluating these features from the pixel data. Some of the more common features used in this method are the following:

- *Center of area:* A unique point from which all other points on the object are referenced
- *Major axis:* The major axis of an equivalent or best-fit ellipse
- *Minor axis:* The minor axis of an equivalent or best-fit ellipse
- *Number of holes:* The number of holes in the object's interior

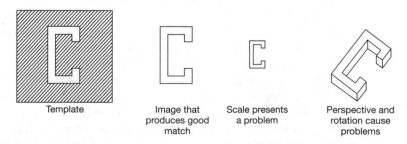

Figure 6–7 Template Matching.

- *Angular relationships:* The angular relationship of all major features to one another
- *Perimeter squared divided by the area:* A value unique to any given shape that does not change as the image of the object changes scale
- *Surface texture:* An identifying characteristic for which gray level images are used

The statistical data for parts to be identified are stored in memory, and when the camera produces an image of an unknown part, the system calculates a set of feature values from the pixel data. The feature values for the unknown part are compared statistically with the feature values stored in memory. If a close match is found, then the part is recognized and identified by the system. The SRI technique is similar to the method just described. The one exception is that an algorithm is used to apply the features identified for each part.

Vision will be an important tool in future automated work cells, but a significant amount of work remains to be done before it can perform the basic operations now done by human operators. The areas requiring additional work include three-dimensional camera systems, ranging techniques, tracking, and handling missing or extra features in taught objects.

6-3 LIGHTING FOR MACHINE VISION

Although lighting is an important consideration in every vision application, it is frequently overlooked. In many cases a greater research effort on the lighting and optics problem would result in a less sophisticated and lower cost vision system.

Selection of the Light Source

The light source must provide the vision system with the best possible images of the part under production conditions. The analysis of captured images requires high contrast between part features and the background of the captured data. To achieve the high contrast, lighting systems must minimize the effects of natural light, shop lighting, and radiation from process sources that could disrupt the vision system operation. Selection of a lighting source for a vision application is driven by three factors: (1) the type of features that must be captured by the vision system, (2) the need for the part to be either moving or stationary when the image is made, and (3) the degree of visibility of the environment in which the image must be captured. The three lighting techniques used to satisfy these factors are described next.

Lighting Techniques

The three lighting techniques used in vision applications are *front lighting, backlighting,* and *structured lighting.* In front lighting (Figure 6–8a) the camera and light source are on the same side of the part. This technique, the most common in vision applications, is used when the surface features of the part are important image

Figure 6–8 Lighting for Machine Vision.

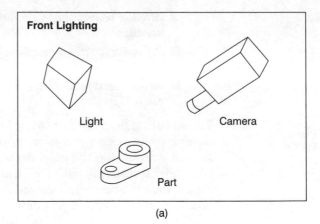

Front Lighting

Light

Camera

Part

(a)

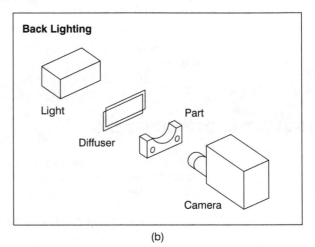

Back Lighting

Light

Diffuser

Part

Camera

(b)

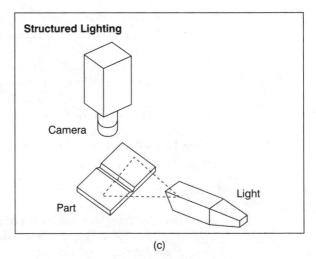

Structured Lighting

Camera

Part

Light

(c)

characteristics. (Review Figure 3–6 and the associated case study and note the front lighting used.) Continuous front lighting is used when the part is stationary or is moving slowly, whereas strobing is used to capture images of parts moving rapidly past the camera. The strobe system bathes the part in a high-intensity flash to freeze the fast-moving part and capture an image with a synchronized vision camera.

The backlighting technique (Figure 6–8b) provides an image with the greatest contrast; however, only the silhouette of the part is captured by the vision system. The part in Figure 6–6 is backlighted, and the silhouette is clearly visible. The high contrast provided by backlighting makes the identification of part edges less complex. This lighting technique is limited to vision inspection applications for measuring a silhouetted feature or determining the presence or absence of a feature. Again, the backlight source is continuous if the parts are stationary, and a strobe is used if the parts are moving when the image is captured.

The third technique, structured lighting, is illustrated in Figure 6–8c. Similar to front lighting, in structured lighting the light source and camera are on the same side of the part. However, a structured light source controls the shape and form of the projected beam to aid in the capture of part features. The light source in Figure 6–8c, for example, produces a narrow slit of light that is projected onto the part at an angle relative to the camera. Light beams are controlled with apertures, lenses, fiber optics, and coherent light sources such as lasers. Structured light is used in two application areas: (1) the capture of a specific part feature from a complex shape when only the feature of interest is lighted, and (2) the extraction of three-dimensional information from a part using only a two-dimension vision system. The second application is illustrated in Figure 6–8c. Note that the angular shape on the surface of the part is visible to the camera because the narrow band of light is shifted by the V slot in the part due to the angle of the light source.

Illumination Sources

The illumination energy source chosen for the vision application is equally important for a successful project. The major types of sources include incandescent bulbs, fluorescent tubes, xenon flash tubes, lasers, infrared light-emitting diodes, and X-ray tubes.

Each of the sources has unique properties such as the location of the source on the electromagnetic radiation spectrum, useful life, heat generation, level of illumination energy, coherent structure, and compatibility with human operators in the work cell. Significant characteristics of the more frequently used sources can be described:

- *Incandescent bulbs:* This most commonly used light source includes devices ranging from standard household bulbs with reflectors to high-power quartz halogen lamps. Fiber-optic bundles are often used to pipe the light onto specific locations on the part. Two major considerations are bulb life and removal of the heat generated by the bulb.
- *Fluorescent tubes:* The reduced infrared energy (heat) produced by this type of illumination, plus the extended life of the tube, make it more efficient than the incandescent bulb. The natural diffused light produced by fluorescent tubes is preferred for vision applications with highly reflective parts.

- *Xenon flash tubes:* When used as a strobe light source, the xenon flash tube is an important light source in vision applications. Xenon strobe light is used when the vision system must capture the image of a moving part. The 5 to 200 microsecond flash of light illuminates the part at one point in its motion. Since CCD vision cameras are temporary storage devices, the image of the part created by the flash is stored by the camera and the part features are scanned into the vision system memory. In addition, xenon tubes produce light with frequencies from a broad part of the spectrum and have high-intensity levels.
- *Lasers:* The laser is an important source because it produces coherent light. Coherent light does not disperse as it travels from the source to the target. As a result, the diameter of the laser beam at the target is very close to the beam size leaving the laser. This condition makes the laser a good source for structured light applications.

The selection and design of the light source is a critical part of the vision application. A successful vision application must put an equal emphasis on part lighting and selection of image-capture hardware and software.

6-4 MATERIAL HANDLING

Raw material and parts must be delivered to the automated work cell, and the finished parts must be removed. Material-handling systems are responsible for this transfer activity. The transfer mechanism used to move parts between work cells and workstations has two basic functions: (1) move the part in the most appropriate manner between production machines, and (2) orient and position the part with sufficient accuracy at the machine to maximize productivity and maintain quality standards.

Automated Transfer Systems

The transfer mechanisms used to achieve these two functions are grouped into three categories: *continuous transfer, intermittent transfer,* and *asynchronous transfer.*

Continuous Transfer. In a continuous transfer operation, the parts or material move through the production sequence at a constant speed. The work does not stop at a workstation so that a production operation can be performed. If the workstation includes a robot, then the robot must perform the work while the part is moving through the cell. The best example of continuous transfer in manufacturing is the assembly of cars on an automotive assembly line. The cars move at a constant rate through 95 percent of the assembly process, with the automated assembly machine and manual operations performed on a moving vehicle. The picture in Figure 1–4 shows a robot installing front windshield glass on cars as they move through the assembly process. In this case, the robot is capable of moving the glass at a rate synchronized with the material-handling system carrying the car.

Three types of mechanisms are used to achieve continuous motion in manufacturing:

1. *Overhead monorail:* This implementation usually consists of a continuous series of interconnected hooks attached to free-turning rollers that ride in overhead tracks. Parts are hung from the hooks to transport material between workstations or through a production operation. This type of system is used in Figure 6–9 to move door panel assemblies for cars to the assembly area. The overhead system pictured uses the self-powered carriers in Figure 6–10, where the motors are visible on the top of each carrier. These automatic monorail systems consist of independently controlled carriers with on-board logic for automatic control of speed and operation of system queues. Typically the payload is 1000 pounds per trolley wheel with higher capacities available using multiple trolleys and load bar arrangements.

Figure 6–9   Single Trolley Overhead Rail Automatic System.
(Courtesy of ACCO Systems, Inc.)

Figure 6–10 Motorized Carriers for Overhead Trolley Systems.

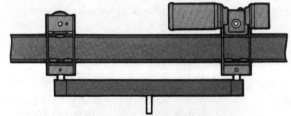

Twin trolley on horizontal track

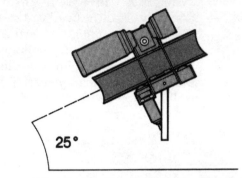

25°

Single trolley on a gradient

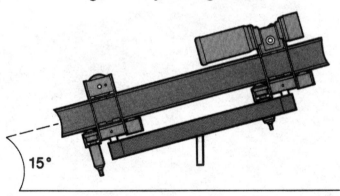

15°

Twin trolley on a gradient

2. *Monorail tow system:* This system is similar to the overhead monorail except that wheeled carts are attached to the hooks, and the overhead system pulls the carts from one destination to another.

3. *In-floor tow system or inverted monorail:* In an operation similar to the monorail tow, the towing mechanism is a continuous chain riding in a channel placed below the surface of the floor. The chain has hooks placed at regular intervals that pull a wheeled carrier through production. A current floor technique is pictured in Figure 6–11. The rails are mounted

Figure 6–11 Floor-based Tow System.
(Courtesy of ACCO Systems, Inc.)

on the floor and the carriers, with parts on board, ride on the rails. In the figure, completed car bodies are queued on an inverted monorail system with several interconnected routes.

Intermittent Transfer.   The intermittent, or synchronized, transfer system has the following characteristics:

- Workstations are fixed in place.
- The motion of the transfer device is intermittent or discontinuous, so that the parts cycle between being in motion and being stationary.
- The motion of the parts is synchronized so that all parts either move or remain motionless during the same time frame.

Asynchronous Transfer.   The term *asynchronous* means *not synchronized;* therefore, an asynchronous transfer system is one in which each part moves independently of other parts. Most of the standard powered roller belt conveyers used in production are in this group. The asynchronous transfer system is often called a *power-and-free system* to indicate that parts can be either free from the transfer mechanism or powered by the transfer device. The robot gripper in Figure 4–21 is poised above an asynchronous conveyer system. The two black strips in the figure

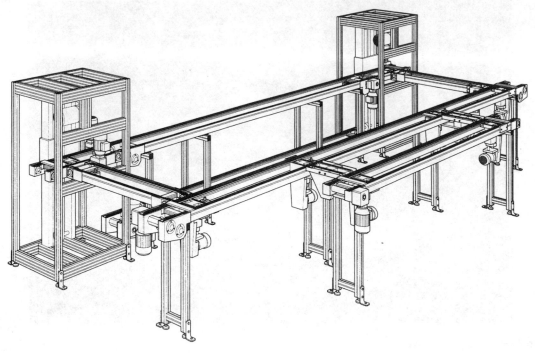

Figure 6–12   Power and Free Belt Conveyers.
(Courtesy of Menziken Automation)

form a flexible belt conveyer on which metal pallets ride. The pallets are contained by the metal edges on the outside of the belts. Part of a pallet is visible on the top right of the figure. Pallets are moved by the friction between the belt and the metal pallet base. When the parts on the pallet are needed, the pallet is stopped by a plunger, which rises up between the belts (see Figure 4–21). The belts continue to move, and the pallet slides on the belt surfaces as the robot unloads the parts. Review Figure 3–6 and note that the telephone assembly system also uses a power-and-free type of pallet conveyer system. Figure 6–12 shows how the power-and-free belt conveyer pictured in Figure 4–21 might be configured. Note that in the back of Figure 6–12 the conveyer is an *over and under* configuration with elevators at each end of the part travel. In the foreground, the configuration is rectangular. Other configurations available are *carousel* (circular or oval), *inline* (one conveyer and no return), and *side-by-side* (two conveyers next to each other).

The asynchronous transfer system performs more than 60 percent of all material-handling functions in automated production systems.

Automatic Storage and Retrieval Systems

An automatic storage and retrieval system (ASRS) is an automated version of the standard warehouse. The system pictured in Figure 6–13 illustrates this concept.

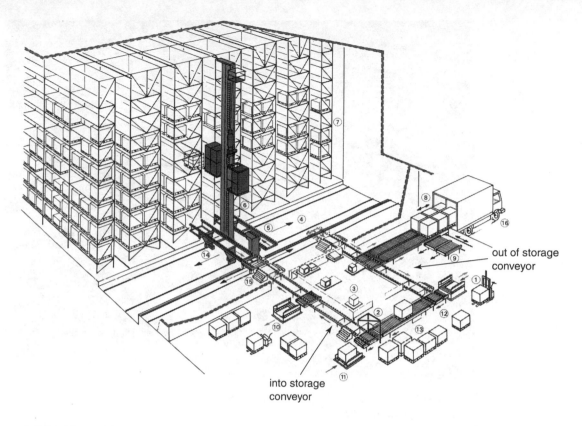

out of storage conveyor

into storage conveyor

1. Pallet pick-up and deposit
2. Pallet checker
3. Identification point
4. Storage level
5. Transfer car
6. S/R machine

7. High-bay warehouse
8. Automatic truck loading/unloading
9. Queuing for truck loading
10. Reject pallet ejection
11. Pallet delivery

12. Roller lifting table
13. Roller conveyor
14. Chain conveyor
15. Shuttle car
16. Truck with chain conveyor

Figure 6–13 Automatic Storage and Retrieval System.
(Courtesy of Rapistan Systems)

Conveyer systems bring material to the system for storage and transport material away from the system for production or shipping. A large rectangular geometry robot, called a *storage and retrieval machine (S/R)*, moves horizontally to the opening between each stack. The S/R then moves between stacks and up to the level where the material is stored. Stored material travels on the left conveyer, and material leaving the ASRS is transported on the right conveyer. ASRSs come in a wide variety of sizes. The system pictured in Figure 6–13 is a large system; however, some systems are no larger than a vertical file cabinet. The ASRSs are configured vertically (Figure 6–13), horizontally, and as a carousel. Many systems are totally automated, with all movement and material handling under computer control. In other

Figure 6–14 ASRS with AGV Transport System.
(Courtesy of Rapistan Systems)

applications, they are combinations of automatic and manual control, and in some an operator rides in the S/R machine.

Another type of ASRS is pictured in Figure 6–14, where automotive bumpers are stored on pallets. An automatic guided vehicle (AGV), in the bottom right of the picture, delivers the pallets to the production area.

6-5 PART FEEDING

Recall that the second function of material handling is to orient and position parts with sufficient accuracy to maximize productivity and maintain quality standards. In most automated cells, this task is the function of part orienters and feeders. The number of different types of part feeders is as varied as the types of parts used in products; however, several devices are standard in most automated cells.

Gravity Feeders

In the configurations of feeders in Figure 6–15, gravity provides the force to move parts to the pickup point. The configurations in the figure are just a sample of hundreds of styles used to feed single or multiple parts to a manufacturing system

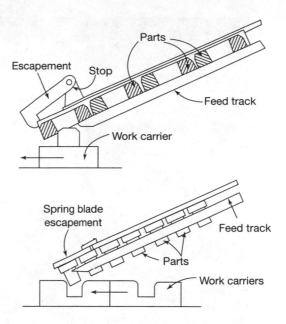

Figure 6–15 Gravity Feeders.

for automated handling. The advantage of gravity feeders is the relatively simple design, and the primary disadvantage is that they do not hold a large supply of parts.

Magazine Feeders

The magazine feeder is another type of gravity-based part feeding mechanism. The name is derived from the shell magazines used in automatic pistols and rifles. Two views of a magazine feeder for round plastic parts is pictured in Figure 6–16. The parts are stacked in a vertical tube with a pneumatic cylinder ejection mechanism (Figure 6–16a) to push out one part at a time. The exiting part slides out and the ejection mechanism prevents the parts in the tube from dropping (Figure 6–16b) until the ejection mechanism is completely retracted.

Tape Feeders

Tape feeders have components trapped between two layers of tape with equal spacing and a fixed orientation. The tape feeders are clearly visible in Figure 3–6; note the rolls of components at the rear of the feeder. The feeder moves the part to a fixed point and removes the top layer of tape so that the component is ready for automated handling. Tape feeders are used most frequently with small components that are not suitable for a vibratory type feeder.

Waffle-Tray Feeders

A waffle-tray feeder is usually a plastic tray with equally spaced compartments, each containing a single component for retrieval by automatic part-handling

(a)

(b)

Figure 6–16 Magazine Feeder for Plastic Parts.

equipment, such as robots. An example of a parts tray is visible on the right side of Figure 3–6 under the robot arm.

Vibratory Feeders

Two types of vibratory feeders are common: linear track and bowl. A vibratory bowl feeder is pictured in Figure 6–17 as a part of an automatic parts feeder. In each case the parts holder is isolated from the feeder base by rubber pads, and the holder is vibrated at a variable rate to move the parts. Track vibrations are along the length of the track, and the vibration of the bowl is rotary about the centerline of the bowl. Three bowl feeders are visible in Figure 3–6; the different bowl sizes are dictated by the size of the component part. Bowl feeders have the parts placed randomly at the bottom of the bowl. The vibrating motion on bowl feeders causes

Figure 6–17 Vibratory Bowl Feeder.

(*Source:* **COMPUTER INTEGRATED MANUFACTURING,** 2nd ed. by Rehg and Kraebber©, p. 381. Reprinted by permission of Pearson Education, Inc., Upper Saddle River, NJ)

parts to move up the circular ramps around the inside of the bowl (see Figure 6–17). When the parts are near the top of the bowl, tabs on the ramp push parts that are not in the correct orientation off the ramp, and the rejects drop to the bottom of the bowl. Bowl feeders are also visible in Figures 1–7 and 2–33.

6-6 INSPECTION

The use of CIM-based work cells demands that quality be verified at each manufacturing step; as a result, the term *quality at the source* was coined. Quality takes many forms, because quality is everyone's job. Quality must be designed into the product and into the system used to make the product. Every person and every automated machine has the responsibility to guarantee that a quality product is produced. To achieve this goal, the production standard is a product free of defects, and the product must be checked frequently to verify that it is defect-free.

Several different devices are available to verify that no defects are present in a product. The devices include manual inspection gages and measuring instruments and automated inspection systems. Automated systems range from manual measuring instruments with digital outputs to coordinate measuring machines (CMM). The CMMs are programmed to measure part parameters with little or no human interaction. In many applications, the quality data are transferred to a computer in the work cell through a digital interface present on the measuring device. Because of the computer interface, product quality can be analyzed quickly and saved as part of the product database.

6-7 AUTOMATIC TRACKING

Tracking the movement of material, parts, and tools through production is a major task. The primary technique used to track the movement of parts and material is *bar coding,* with more than 50 different coding systems in use. Bar code scanning has replaced the keyboard and other data entry devices in most automated systems for recording manufacturing information and data.

A bar code is a symbol composed of parallel bars and spaces with varying widths. The symbol is used as a graphical code to represent a sequence of alphanumeric characters, including punctuation. For example, the three bar codes pictured in Figure 6–18 represent the numbers written below the codes. The codes can also represent combinations of letters and numbers. The advantages offered by the various coding systems include robustness of character sets (many different kinds of characters, for example, numeric, alphabetic, Greek, can be recognized), ease of printing and reading, and the ability to represent large numbers with a small code. The two most popular codes in industry are called the *interleave two of five* and *code 39.*

The interleave two of five is a numeric code, so only numbers can be represented by the code. The code was designed with two key features: (1) the code has a high density so that a large amount of information can be encoded in a short

Figure 6–18 Sample Bar Codes.

(*Source:* COMPUTER INTEGRATED MANUFACTURING, 2nd ed. by Rehg and Kraebber©, p. 454. (Reprinted by permission of Pearson Education, Inc., Upper Saddle River, NJ)

space, and (2) the code can be printed on most types of printers and read with most scanners because the code has a wide tolerance limit for both devices. This code is a continuous code, meaning both bars and spaces are coded. The odd-numbered digits are represented by the bars and the even-numbered digits are represented by the spaces. The specification requires that the code represent an even number of digits. To satisfy this requirement, a zero is placed in front of all numbers with an odd number of digits before converting to the interleave code. The interleave two of five is effective for coding information on corrugated boxes. In this application, bars must be wide to permit automatic scanning while the boxes are moving on conveyers; however, space on the boxes is usually limited. The code is used widely in the automotive industry and also in warehousing and heavy industrial applications.

Code 39, also called the 3 of 9 code, is the most widely used and accepted industrial bar code symbology. The code can represent the twenty-six letters of the alphabet, the ten digits, and seven additional characters. The code, which can vary in length, has unique start and stop bits. All characters are self-checking, and the intercharacter space does not contain information. Major advantages of this code are the high level of data security provided by the self-checking feature and the same wide tolerance for printing and scanning. For example, with a good laser bar code printer and properly selected bar code scanner, the error rate is less than one improper character substitution in 3 million scanned characters. If high-quality printers and scanners are used, the value drops to one substitution error in 70 million scanned characters.

The scanners used to read bar codes have three major components: (1) a light source (laser, light-emitting diode, or incandescent lamp) to illuminate the code, (2) a photodetector to sample the light reflected from the code, and (3) a microcomputer to convert the photodetector output into the series of letters and numbers represented by the code. Scanners fall into three categories: *handheld contact type, handheld noncontact type,* and *fixed-station systems.* The scanners connect to either portable data storage devices or computers, or they transmit the data over radio frequency links to a host computer.

In many tracking applications the environment does not permit bar codes to be applied or read. For example, the cutting tools used on CNC machines are often covered with cutting fluid; as a result, a bar code attached to the tool would not work. A common tracking technique for items that cannot use bar codes is the *radio frequency tag.* Tags are passive electronic circuits that transmit a code when subjected to radio frequency energy. The passive electronic circuit, shaped like a small computer chip, is held in a slot in the tool base by epoxy. Each circuit transmits a unique code when the tracking device focuses radio frequency energy at the tool. The tracking device receives the tool code and records the number. Radio frequency tag hardware from Omron Electronics is pictured in Figure 6–19.

Another tracking technique, frequently used on production pallets, uses a binary code to track objects. The pallets have a sequence of holes placed somewhere on the surface. At preselected points along the production process, the holes are aligned with proximity sensors or limit switches, with a sensor or switch over each hole. To give a pallet a unique code, some combinations of holes are filled with pins. When the pallet passes under the sensors or switches, the holes with pins trip the sensor or switch.

Tracking techniques are also used to direct the manufacturing process for a product. In an application at Allen Bradley, for example, a bar code placed on the

Figure 6–19   Radio Frequency Tracking Hardware.
(Courtesy of Omron Electronics, LLC)

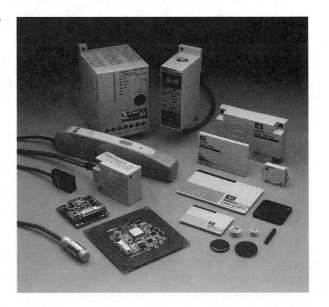

base of a blank motor contactor tells the production system what type of contactor to build. At every step in the totally automated process, machines add the correct parts to the contactor based on the information read from the bar code. Tracking of material, tools, and products is critical in manual and automated systems.

6-8 SUMMARY

Vision is used to identify parts, determine part position and orientation, inspect parts, and find ranges for three-dimensional location problems. Vision systems are available in three architectures: stand-alone systems, systems based on a programmable logic controller, and integrated systems. In each case, the basic system components include a lighting system, one or more cameras, camera controller, interfaces to external equipment, video image monitors, and programming devices. Most vision systems use the charged-couple device camera; when larger images of surfaces are needed, a vidicon tube is employed.

The basic unit of image measurement is the gray scale, the basic parameter is light intensity, and the basic measurement element is called a pixel. Pixels are arranged in either a linear or rectangular array, and the output of each pixel varies between a fully *on* condition and *off* state as a result of light from the image. The image of the part captured by the camera must be analyzed. The first step in analysis is identification of the part edges and features using either edge detection or clustering. Identification of the captured image involves three techniques: template matching, edge and region statistics, or statistical matching using an algorithm.

Lighting for machine vision is a critical element because the cost of the vision system and the success of the application are directly affected by the techniques used to light the part. Three factors affect the selection of the lighting source: (1) the type of features that must be captured by the vision system, (2) the need for the part to be either moving or stationary when the image is made, and (3) the degree of visibility of the environment in which the image must be captured. The light source is applied to the application by means of three techniques: front lighting, backlighting, and structured lighting. Front lighting is used most frequently. If objects are in motion during the image capture time, then a strobe light is appropriate. The illumination sources in vision applications include incandescent bulbs, fluorescent tubes, xenon flash tubes, lasers, infrared light-emitting diodes, and X-ray tubes.

Material-handling systems are responsible for two basic functions: moving parts in the best manner and orienting parts for easy handling. The transfer mechanisms used to achieve these two goals include continuous transfer, intermittent transfer, and asynchronous transfer. In a continuous transfer operation, the parts or material move through the production sequence at a constant speed. The types of mechanisms used to achieve continuous motion in manufacturing include overhead monorail, monorail tow systems, and in-floor tow systems. The second type of material movement system is called intermittent, or synchronized, transfer. This transfer system has fixed workstations, and all the parts on the transfer device stop in unison at the work cell. The third type of material movement system is called asynchronous transfer. An asynchronous transfer system is often called a power-and-free system to indicate that parts can be either free from the transfer mechanism or powered by the transfer device.

In the second function of material handling, orientation and positioning, various part orienters and feeders are used. The standard devices include gravity feeders, tape feeders, waffle-tray feeders, and vibratory feeders.

Quality takes many forms, because quality is everybody's job. Every person and every automated machine has the responsibility of guaranteeing that a quality product is produced. The devices that check for a quality product include manual inspection gages and measuring instruments and automated inspection systems called coordinate measuring machines.

Tracking the movement of material, parts, and tools through production is a major task. The primary technique for tracking the movement of parts and material is bar codes. A bar code is a symbol composed of parallel bars and spaces with varying widths. The symbol is a graphical code that represents a sequence of alphanumeric characters plus punctuation. The two most popular codes in industry are called the interleave two of five and code 39. The scanners that read bar codes can be grouped in three categories: handheld contact type, handheld non-contact type, and fixed-station systems. A common tracking technique for items that cannot use bar codes is radio frequency tags. The tags are passive electronic circuits that transmit a code when subjected to radio frequency energy. Another tracking technique, common on production pallets, utilizes a binary code pattern generated by mechanical switches or electronic sensors.

QUESTIONS

1. Describe five application areas for vision systems.
2. What are the basic components of a vision system?
3. What two types of cameras are currently used?
4. Describe the operation of a solid-state CCD camera and define the terms *pixel* and *gray scale*.
5. Describe edge detection and clustering.
6. Describe the three commonly used techniques for two-dimensional recognition with vision systems.
7. Describe the three factors used to select lighting for vision applications.
8. Describe the three lighting techniques used in vision applications.
9. What conditions in the application require the use of strobe lighting?
10. What are the advantages and disadvantages of the six lighting sources?
11. What are the two basic functions of material handling?
12. Describe the three basic types of material-handling systems.
13. Describe the four standard part feeders used most often in automated cells.
14. What is a CMM and how is it used?

15. Describe the three techniques most frequently used for automatic tracking of production parts and material.
16. Compare and contrast the interleaved two of five and code 39 bar codes.
17. List and describe the major components of bar code scanners.

PROJECTS AND CASE STUDY PROBLEMS

1. Apply the concepts associated with work-cell support systems presented in the chapter to one or more of the automation work-cell designs listed below. Specify part feeders, conveyers, storage systems, and any other support device using the resources in Appendix B or other sources. Use the design data from previous chapters as necessary. The design component of the case studies varied in complexity and design detail, so it may be necessary to make some assumptions as you work on the support system design. Document the design and support all assumptions.

   (a) The die casting production cell started in Case Study Problem 3 in Chapter 2
   (b) The kitchenware bowl production cell started in Case Study Problem 5 in Chapter 2
   (c) The West-Electric slug production cell design started in Case Study Problem 8 in Chapter 2
   (d) The West-Electric slug lubrication cell design started in Case Study Problem 9 in Chapter 2
   (e) The kitchen products production cell design started in Case Study Problem 3 in Chapter 3

2. Read the West-Electric case study (Section 12–10 in Chapter 12) where the support systems are discussed. Complete the questions and problems at the end of Chapter 12 that relate to the cell support systems.

Robot and System Integration

CHAPTER GOALS AND OBJECTIVES

The goal of Chapter 7 is to look at solutions to the integration problems created by the many different hardware and software products that are used to build an automated work cell or flexible manufacturing system. At the completion of this chapter you should be able to:

- Describe the architecture and provide a system overview for the hardware and software found in a typical automated work cell and CIM system.
- Name the component parts for a programmable logic controller (PLC) and a computer numerical control (CNC) machine and describe their basic operation.
- Describe the three types of controller architectures used to control nonservo robots.
- Discuss the component integration issues present when a PLC is used to control a multi-axes pneumatic robot in a work cell with displays, limit switches, sensors, and network connections to other smart systems.
- Describe the controller architectures used to control servo robots.
- Define a simple sensor interface and list six guidelines to minimize potential interface problems.
- Describe the function of the complex sensor interface using an automated work cell example.
- Describe the function and operation of the enterprise data interface.
- Describe a typical architecture for an automated work cell that includes a cell controller, product tracking, servo and/or nonservo robots, switches, sensors, PLCs, and process machines.

7-1 INTRODUCTION

The CIM wheel introduced in Chapter 1 (Figure 1–5) identified the major areas present in manufacturing enterprises and indicated the interdependence and operational integration that must be present. The integration of information among the elements in the CIM wheel is illustrated by the enterprise network architecture in Figure 7–1. Note that information flow in the three major elements— product and process definition, manufacturing planning and control, and factory automation—is supported by local area networks (LANs). Network communication between the major areas occurs over the enterprise network backbone.

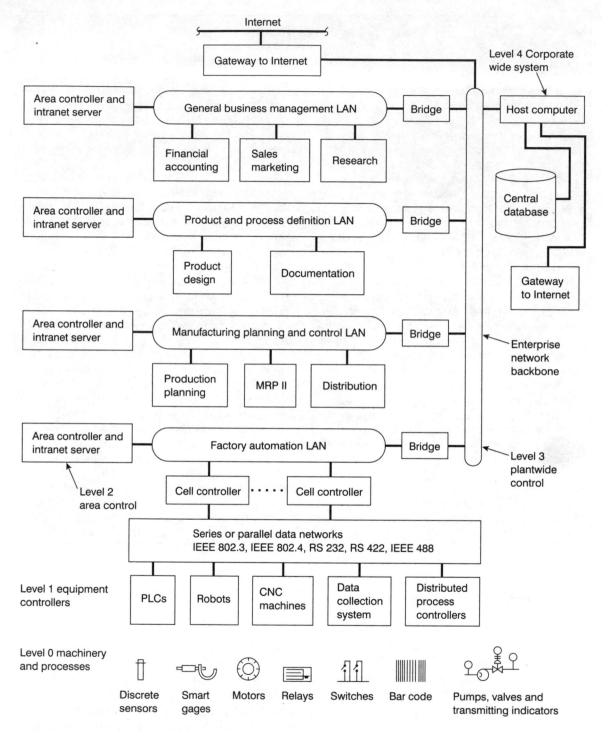

Figure 7–1 CIM Computer Architecture.

The CIM architecture for an enterprise is built around a common central database, or single image of the product, and production data that support the needs of every user. The generation of product data begins with the design of the product. The part specifications and drawings and lists of material for the product are often the first additions to the product database. The design data are used by many other departments; for example, purchasing uses the list of material, called the bill of materials, to purchase parts that are not manufactured in-house. The production planning area uses the part drawings to plan the manufacturing sequence. In the past, the design information was given to other departments through paper copies and interoffice mail. It was difficult to keep all of the distributed documentation current because design updates occur throughout the development of a product. As a result, sometimes parts were ordered that were no longer necessary, or planning was done on parts that had changed significantly. The CIM architecture avoids these problems by keeping a single image of the product data available to all departments. For example, the bill of materials is entered into the manufacturing planning and control software system from the design and updated as changes occur. Therefore, the current parts list, stored in the central database (Figure 7–1), is available to anyone on the enterprise network.

The architecture also provides a core of common services for the central data: communication and distribution of data, management of the data, and presentation of the data. The data take three distinct forms: (1) alphanumeric characters, called *text* data; (2) graphics data, called *vector* data; and (3) bit-mapped scanned pictures, called *image* data. Each data type presents a unique integration problem for the three core services just described. The degree to which the CIM architecture performs the three core services on all data types is a measure of the quality of the CIM implementation. Therefore, the general CIM architecture includes the physical system and the enterprise data. The physical system is divided into hardware and software. The hardware includes the computers, input devices, output devices, physical networks, and network interfaces. The software, as the name implies, is the software used throughout the system.

The following sections describe the hardware and software used to integrate the shop floor with the enterprise database and to control the operation of computer-driven automation in production cells and systems.

7-2 SYSTEM OVERVIEW

The primary focus of this text is the integration of robots into automation cells and systems. Nevertheless, an understanding of the larger enterprise-wide information system is essential for an understanding of the detailed operation of the automation cells. The next two sections provide a general overview of the hardware and software that support the enterprise network illustrated in Figure 7–1.

Hardware Overview

The hardware is distributed across the enterprise and, as Figure 7–1 illustrates, it includes host, area, cell, and production machine computers; computer peripheral devices; different types of networks; and network interface devices. Users often have microcomputer workstations networked to the system at the area level. The most frequently used network hardware is *ethernet* and *token ring,* with the former found in most shop floor applications. If both networks are used in the CIM system, then a gateway computer is required to interface the two networks. All of the physical network is part of a company intranet that usually has a gateway to the Internet and World Wide Web.

Host computers are usually mainframe machines with large data storage capability. The area-level computers are frequently midsized machines, like the IBM AS400 or a minicomputer like a Sun workstation that run local user applications and act as data concentrators. In some smaller installations, microcomputers are used as area controllers. The cell controllers are either industrial microcomputers like the industrial PC or a smaller Sun workstation. The computers used for production machine control vary widely. Included in the group are microcomputers, microprocessors, microcontrollers, and programmable logic controllers.

Software Overview

The software distributed across the enterprise in a CIM implementation takes many forms. Each computer from production machine controller up to the host has an *operating system.* Examples of operating systems at workstation or cell-control level include DOS, UNIX, and Windows NT. UNIX and Windows NT support multiple sessions and threading so that the user can have two or more applications executing and present on the monitor simultaneously. All the computers in the enterprise that share the network must have *network software* running. The network software permits communication between active computer nodes in the enterprise system. For example, a cell-control application program could request a different robot program file from the host.

All the computers in the CIM network have one or more application programs running under the operating system. Using network software, the applications can access data from other computers in the system. The applications are as varied as the departments in which they are used. The front office can be using word processing, desk-top publishing, and electronic mail, whereas the product design department can be using CAD, finite-element analysis, word processing, and electronic mail. The marketing department can retrieve product drawings from the central database that were developed on the CAD computers in the design department. All departments would use an Internet browser to retrieve Web data and view corporate intranet information and data.

7-3 WORK-CELL ARCHITECTURE

The architecture of automated cells includes layers of computer control. The area controller is at the top, and smart, or microprocessor-controlled, devices are at the bottom. Often the architecture includes programmable logic controllers for sequential control of the cell and some production machines, robot controllers to drive the servo robots, and computer numerical control (CNC) for process machines. Commonly used cell control devices are described in the next sections; then an example of the architecture used in a machining and assembly cell is presented.

Cell Controllers

The cell controllers are either industrial computers that use standard microprocessor chips or minicomputers with proprietary processors. The industrial computers using the latest Intel type processors are designed for operation in harsh production environments found in manufacturing. Minicomputers from Sun are the most frequently used machines for area and cell control applications. The system configuration for cell controllers includes internal hard drive and floppy drive for data and program storage, internal memory (RAM), keyboards, pointing devices, monitors, and network cards.

Cell Control Software Structure

The operating system software used for cell controllers in most applications is Windows NT or UNIX. These operating systems can run multiple application programs concurrently. This multitasking and multithreading feature permits the cell controller to have multiple application programs running at the same time. For example, the cell controller can execute a program to collect quality data from a smart gage while another program downloads a program to a CNC production machine to cut the next part. Work-cell controller applications that demand this level of flexibility require multitasking operating systems.

The cell controller interfaces with other intelligent machines and devices at different levels in the control architecture using standard serial data interfaces and networks. For example, a cell controller is often attached to the factory LAN (Figure 7–1) and interfaced with devices in the cell. Figure 7–1 indicates that the interface to work-cell devices uses a variety of serial and parallel data networks: IEEE 802.3 (ethernet), IEEE 802.4 (token ring), RS 232, RS 422, and IEEE 488 general-purpose instrumentation bus (GPIB). A typical software configuration in the cell controller to manage these many interfaces is illustrated in Figure 7–2. Study the figure until you are familiar with the terms in each box.

The operating system (OS) is usually one of the two listed at the bottom of the figure. The application program interface (API) software resides between the OS

Figure 7–2 Work-Cell Controller Software Layers.

(*Source:* COMPUTER INTEGRATED MANUFACTURING, 2nd ed. by Rehg and Kraebber©, p. 432. Reprinted by permission of Pearson Education, Inc., Upper Saddle River, NJ)

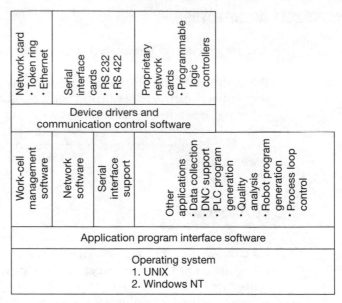

and the applications. An API, such as Microsoft's Windows API, handles some of the operational overhead common to all applications; in addition, an API helps manage some of the system's resources. The use of API software permits the developers to spend more time programming the specific application and less time programming routines required by the operating system. The applications can include a wide variety of software (see Figure 7–2).

The cell controller frequently has different interface cards to support network standards like ethernet and token ring. The device drivers and communications control software reside between the interface cards and the application software (see Figure 7–2). In some applications, a proprietary interface card is installed to permit direct communication between a work-cell device, like a programmable logic controller or robot, and the cell controller.

Work-Cell Management Software

The cell management software used in cell controllers manages all of the cell activity. The primary role of the cell controller is communication and information processing. For applications higher in the architecture, the cell controller is a concentrator of information generated in the cell and a link between production machines and applications running on computers in the design and the production control areas. On the downstream side, the cell controller is a distributor of data, information, and program files to the computer-controlled devices of the work cell. The cell controller interfaces with production machines and transfers data files as small as a single bit and as large as files containing megabytes of data. Other cell control responsibilities include management of program libraries for devices in the cell, support for engineering change control, and production tracking. The information and communication tasks frequently consume 80 to 90 percent of the resources of the processor in the cell controller.

Applications typically loaded into the cell controller include these:

- Production monitoring
- Process monitoring
- Equipment monitoring
- Program distribution
- Alert and alarm management
- Statistical quality and statistical process control
- Data and event logging
- Work dispatching and scheduling
- Tool tracking and control
- Inventory tracking and management
- Report generation on cell activity
- Problem determination
- Operator support
- Off-line programming and system checkout

7-4 PROGRAMMABLE LOGIC CONTROLLERS

The *programmable logic controller* (PLC) was developed in the early 1970s for General Motors. The early PLCs were special-purpose industrial computers designed to eliminate relay logic from sequential control applications. Current PLCs can control discrete and analog processes, attach to general-purpose and proprietary LANs, provide servo and sequential control for robots, and serve as the primary controller for cells or complete automated systems. An understanding of robot automation is not complete without an understanding of PLC operation and programming. The following sections give an overview of PLC operation, and programming is described in Chapter 8.

PLC System Components

The PLC is a computer designed for control of manufacturing processes, assembly systems, and general automation. The mechanical configuration of PLCs includes a *rack* into which *modules* are inserted. Figure 7–3 shows a processor removed from the rack. The modules and processor are inserted into electrical connectors on the back plane at the back of the rack. The back plane contains the power and signal conductors to interconnect all the modules inserted into the rack. The modules generally fall into one of the following categories: interfaces to production equipment, network and serial communications, special-purpose modules, power supplies, and a microcomputer. A description of systems using each of these modules is included in the next section. Figure 7–4 shows the rack of an Allen Bradley series SLC PLC with modules, a processor, and power

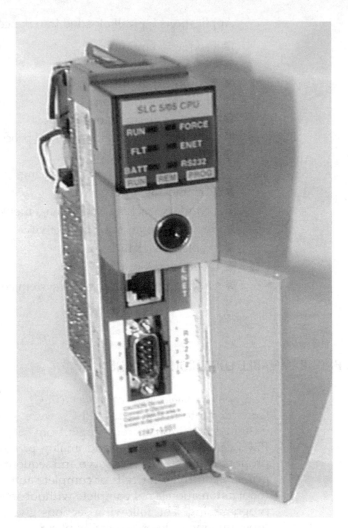

Figure 7–3 Allen Bradley SLC 505 Programmable Logic Controller.

supply installed. The power supply is on the far left, the first module is the processor module, and the balance of the modules are a combination of inputs and outputs.

Basic PLC System Operation

The operation of PLC components, for the control of automated systems, is described in the block diagram in Figure 7–5. Study the figure and note the components that are part of the PLC (inside the dashed box) and those that are external to the PLC.

The computer, called the *processor* in PLC systems, is at the heart of the PLC operation. *Input modules* receive electrical signals (Figure 7–5) from a wide variety of sensing devices (switches and sensors) and outputs from other systems

Figure 7–4 Processor and Input Module for an SLC 500 Programmable Logic Controller.

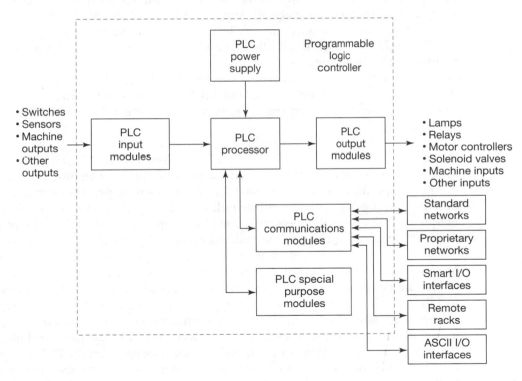

Figure 7–5 Programmable Logic Controller System Block Diagram.

(*Source:* COMPUTER INTEGRATED MANUFACTURING, 2nd ed. by Rehg and Kraebber©, p. 442. Reprinted by permission of Pearson Education, Inc., Upper Saddle River, NJ) (machine outputs and other outputs). The source labeled *other outputs* requires some clarification. In complex automated cells with more than one PLC vendor present, the output from one PLC is often connected to the input of another PLC. Under these conditions, the information processed by the first PLC is used by the second PLC to control another part of the process. In that case, the output from one PLC is the input to a second PLC. For example, often a manufacturer using one vendor's PLC will have a production machine that has a PLC from a different vendor used inside for machine control.

The processor performs arithmetic and logical operations on input data and turns *on* or *off* outputs in the *output modules* based on the control program resident in the process. The output modules are wired to system components that control the process (lamps, relays, motor controller, solenoid values, and machine inputs). The input module, processor, and output module blocks are the only ones required for most PLC automation applications. Inputs are scanned and outputs are changed based on the input conditions present and the logic programmed into the PLC processor.

The PLC *communications modules* are not used as frequently as input and output modules; however, communication is a critical part of every robot automation project because production data must be available to departments across the enterprise. The *standard networks* box in Figure 7–5 indicates that the PLC can be placed directly on the factory LAN to communicate with other cell or area controllers. The standard network supported by most vendors is ethernet, and common protocol standards are TCP/IP, manufacturing message specification (MMS), and DeviceNet. In addition, proprietary LAN software from most major vendors is also available.

The *proprietary networks* box in Figure 7–5 indicates that most PLC systems have a proprietary network available to link PLC processors together in a LAN. In most cases only PLCs from the same vendor are compatible with the network interface and protocol.

The box labeled *smart I/O interfaces* in the figure includes PLC hardware that interfaces smart devices to the PLC through a serial data link. The term *smart* implies that the external device has a microprocessor and can be programmed. For example, operator panels have switches for control of process machines and devices, and the panels have lights to indicate the condition of process equipment. Frequently the operator panels are located in a control room away from the process itself. The traditional approach in building an operator panel requires a minimum of one wire per switch and lamp, plus a number of return wires between the operator panel and PLC input and output modules. As a result, the wire bundle between the panel and the PLC can have hundreds of wires running hundreds of feet in conduit. In contrast, the smart operator interface uses the same number of switches and lamps but controls them with a microprocessor located in the operator panel. The communications interface between the operator panel and the PLC is just a single coaxial cable for the transfer of the data between the PLC and panel processor. A large number of smart external devices are available: motor controllers, process controllers, text readout devices, programmable CRT displays supporting full color and graphics, voice input and output devices, and discrete input and output devices.

In an effort to distribute the control capability across a large automation system, PLC vendors provide *remote rack* capability (Figure 7–5). The rack uses the standard I/O modules for the control of machines and processes; however, the processor module is replaced with a remote rack communications module. The remote rack is controlled by the program and processor in the main PLC rack, and communication between the two racks is supported by a single coaxial cable.

Another communications box illustrated in Figure 7–5 is the ASCII interface. This communication resource is either built into the processor module or comes as a separate module. In both cases the ASCII interface permits serial data communication using a number of standard interfaces (e.g., RS 232 and RS 422).

7-5 COMPUTER NUMERICAL CONTROL

The definition of computer numerical control (CNC) was restricted initially to just numerical control (NC) machines associated with metal cutting. Currently, CNC represents all production machines that use internal computers to control the movement of tooling with production programs. The basic robot system described in Figure 1–9 is an example of a machine that fits the definition of CNC. Consult the generic block diagram for CNC machines in Figure 7–6. Compare the robot controller in Figure 1–9 with the CNC block diagram and verify that the functions are similar.

The input to the CNC system comes from either an operator or another machine. Frequently the input for CNC machines is another computer. For example, many of the CNC machines received the program they used to cut parts through a serial connection to a computer, such as a cell controller. In other cases,

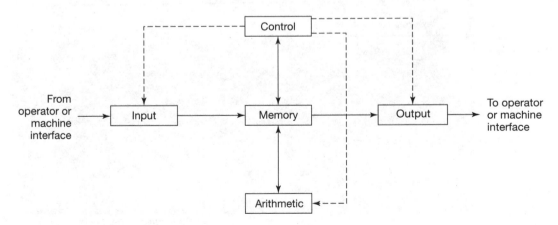

Figure 7–6 Five Major Functional Units of Computer Numerical Control.

(*Source:* COMPUTER INTEGRATED MANUFACTURING, 2nd ed. by Rehg and Kraebber©, p. 450. Reprinted by permission of Pearson Education, Inc., Upper Saddle River, NJ)

the CNC machines and the cell control computer are connected to a direct numerical control network to download programs to the CNC machines. The production machines using CNC also include robots, coordinate measuring machines, and most material processing machines.

7-6 CONTROLLER ARCHITECTURE

The controller architecture for servo-type robots is different from that used in nonservo systems. A description of each is provided in the following sections.

Nonservo Robot Controllers

The sequential movements of nonservo robots are controlled by devices external to the robot's arms. The types of controllers used by pick-and-place robots (like the robots in Figures 2–4 and 2–10) include the following:

- *Drum programmers,* such as the Seiko sequencer illustrated in Figure 7–7, are used in some applications. In this unit, as in most mechanical drum programmers, the operational sequence is set by the arrangement of tabs

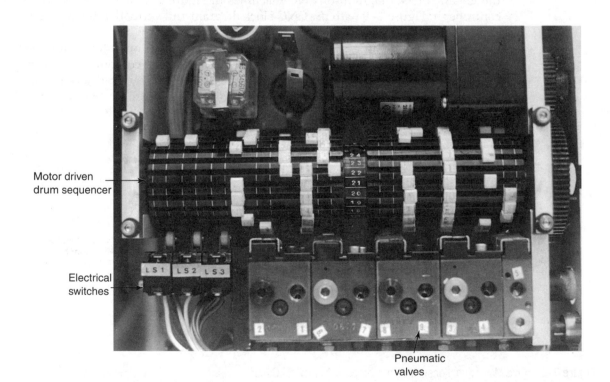

Motor driven
drum sequencer

Electrical
switches

Pneumatic
valves

Figure 7–7 Drum Programmer for the Seiko 700 Robot.

or cams on the drum surface. The tabs actuate either electrical switches of hydraulic-pneumatic valves that control the movement of each robot axis. The time for each sequence is determined by the number of tabs used and the speed of rotation of the drum.

■ *Pneumatic logic* or *air logic programmers* are also used to control the sequences of pick-and-place pneumatic robots. An air-powered logic network is built using fluidic logic elements to provide the sequential control the robot requires. The air logic system is built and programmed by connecting the fluidic elements together by small plastic air lines. The time and sequence of robot movements is determined by the fluidic elements used and the way in which they are interconnected.

■ *Programmable logic controllers* are the most frequently used devices to control the sequential movement of pick-and-place machines. Figure 7–8 illustrates the use of a PLC in a typical low-technology robot system. The figure indicates that the PLC is used for many functions in the cell and as an interface to the CIM system. The interface between a

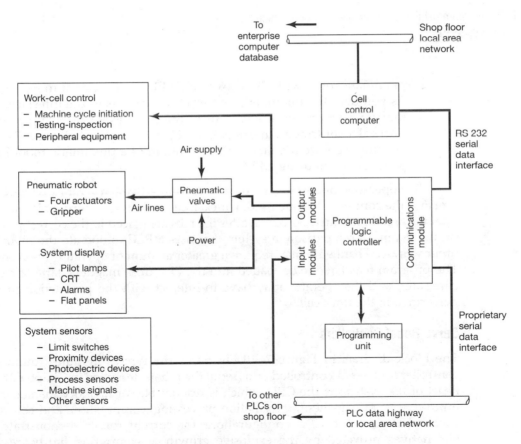

Figure 7–8 Robot System Driven by Programmable Logic Controller.

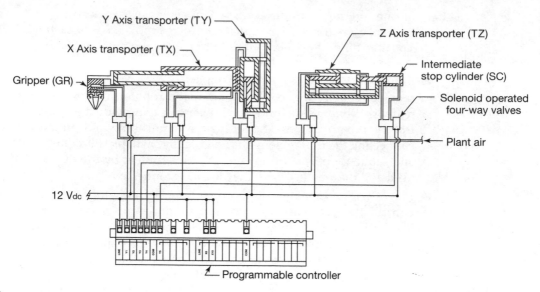

Figure 7–9 Programmable Logic Controller and Robot Interface.
(Courtesy of MACK Corporation)

pneumatic robot with three axes and a PLC is illustrated in Figure 7–9. As you trace the pneumatic and electrical circuits in the interface, note that four pneumatic valves control the three axes and gripper, and a fifth valve controls an intermediate stop actuator. Study Figures 7–8 and 7–9 until you are familiar with integration of a pneumatic robot into a production cell using a PLC.

The operation of controllers for nonservo robot systems varies with system size and the control technique used. For example, if a drum controller is used, the operator would just turn power on after the tabs are placed in the correct position on the drum. In comparison, a system that uses a PLC to operate the robot and other work-cell hardware may require a greater amount of operator involvement. The program may have to be loaded into the PLC from magnetic tape or from a computer, and the operator may have to interact with the production process performed in the work cell.

Servo Robot Controllers

The block diagram in Figure 7–10 illustrates the typical modules found in a controller for a servo-controlled arm (recall the robots described in Chapter 2). The heart of the systems is the CPU, which is responsible for memory management, input/output management, information processing, computation, and control of each robot axis. The CPU configurations for current robots demonstrate the alternatives provided by the explosive growth of computing hardware. The options adopted by some robot manufacturers include the following:

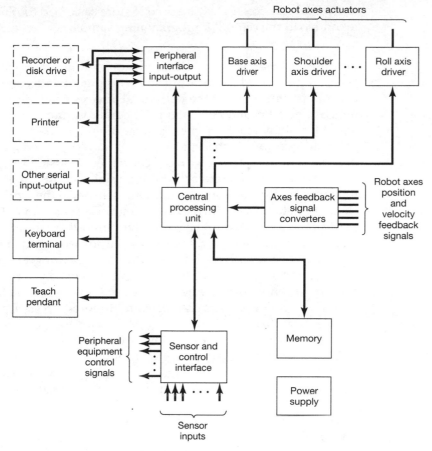

Figure 7–10 Controller Block Diagram for Servo System.

- Using off-the-shelf minicomputers, such as the Sun workstation, to configure a special-purpose computer capable of handling robot CPU duties
- Using either a 32- or a 64-bit microprocessor as the base of the CPU design
- Using a network of 32- and 64-bit microprocessors linked together by hardware and software to perform the functions of the CPU. Most of the current state-of-the-art machines follow this parallel processing practice.

The last technique described, networking, offers several advantages for robots in the high-technology group. The system operation is much faster when the duties of the CPU are divided among different parallel processors.

Memory. The memory used in most robot controllers is solid-state with battery backup for program protection during loss of primary ac power. The operating system and application program are usually stored in volatile devices (RAM), and nonvolatile devices (ROM) are used to store basic load routines to wake the system up and load the controller programming language.

Peripheral Interface. The CPU controls the interface between two different types of peripheral hardware. The first is digital data communications between the robot controller and the peripheral devices needed to operate the robot system. The most commonly used peripherals are the teach pendant and the keyboard terminal. More complex systems often include a recording device for program storage and playback, a printer for hard copy, and a general-purpose serial port for communications with cell controllers or vision systems. The second type of interface is for discrete or *on/off*-type signals. This interface in the servo-type robot controller monitors the output of sensors in the work cell to determine if a change of state has occurred. For example, a sensor to detect the presence of a part in the gripper would be monitored through an input of this interface. If the gripper acquired the part, the sensor output would change from *off* to *on,* and the controller would drive the arm to the next programmed point only when this input change was received. The output portion of the interface has the capability to control discrete devices. For example, in a punch press loading application, the robot program can activate the press to start processing a part that has just been loaded.

The discrete interface section of the robot controller is basically a built-in PLC for use in the work cell. Typical input values of 80 to 140 volts ac and 6 to 40 volts dc are available on the discrete input interface of some robot controllers. The output interfaces will switch either ac or dc currents in ranges that are compatible with the input section. Figure 7–11 shows a standard PLC monitoring work-cell sensors and passing the logical results of their status to the robot controller.

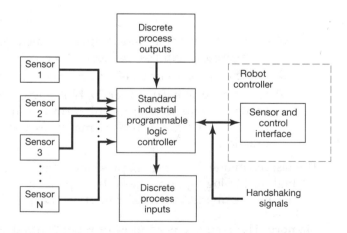

Figure 7–11   Block Diagram of Robot Controller and External Programmable Controller.

The integration of production cells into CIM systems dictates that robots and automated systems must work with other equipment in the enterprise. If this is so, then robots must be interfaced to the other hardware.

> *An interface is defined as a place at which independent systems meet and act on or communicate with each other.*

In an automated system, there are devices, like proximity sensors, pneumatic valves, PLCs, and CNC machines, that must be interfaced so that communication among the systems is possible. The numerous independent systems are grouped into three interface categories: devices supplying data, devices receiving data, and smart devices. Sensors and switches are usually in the group of devices that supply data. The sensors fall into two categories called *simple sensor interface* and *complex sensor interface*. Devices receiving data include pneumatic valves, motor starts, and other types of actuators. Smart devices include machines with embedded controllers or computers that exchange data with the enterprise data interface. All interface requirements of the automated robot cell fall into one of these categories.

Simple Sensor Interface

Currently the most well-defined interface area is the simple sensor interface. A simple sensor is basically one that has its signal originate in some peripheral hardware or device. Peripheral hardware is defined as the equipment used in the design of a work cell with an industrial robot present. An additional requirement for a simple sensor interface stipulates that the communication with the robot must be by discrete signals only (*on* or *off*). The group does not include peripheral devices such as disk drives, printers, or equipment using higher-frequency binary-coded signals with a data format. Equipment in this group does include all the discrete sensors, such as limit switches, proximity sensors, and photoelectric sensors. In addition, all discrete process signals necessary to control the cell are included. These would be signals from machine tools, welders, and material handlers, plus the signals that the PLC, cell controller, or robot originates to operate these peripherals.

The standard logic signal levels for this interface are 0 volts for the low level, and 110 volts ac or 24 volts dc for the high level. Some robot controllers use logic levels of 0 volts and 5 volts, however, and are identified as *TTL* signal devices. The input/output interface uses optical coupling between the PLC, cell controller, or robot and the peripheral to assure isolation of power and grounds between the different systems. Figure 7–12 illustrates a typical input/output module for a PLC or robot with the typical wiring required. Figure 7–13 shows the opto-isolators provided by the robot controller for input-output interfacing. Note in Figure 7–12 that the only connection between the robot and the external circuit is the light that activates the solid-state switch. The output module acts as a switch to apply power to the external load. The input module has one lead connected to the electrical common and the other lead connected to the external sensor that switches power to the input and activates the light coupling and the solid-state switch. The

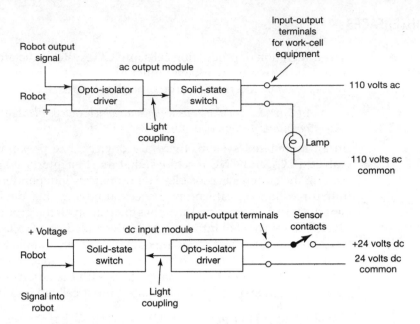

Figure 7–12 Simple Sensor Interface in Robot Controller.

opto-isolated driver is usually an LED, and the solid-state switch is a light-activated transistor for dc or a triac for ac modules.

Electrical noise from industrial machines and contact bounce on the sensor switch contacts can cause input/output interface problems in the system. Adopting the following six guidelines will eliminate many of the potential problems:

1. Eliminate ground loops and provide good grounds.
2. Shield wires when excessive electrical noise is present.
3. Use damping in the sensor contacts or delays in the PLC or robot program to avoid the multiple signals produced by bouncing contacts.
4. Use only opto-isolated input/output modules.
5. Use arc suppression circuits to reduce noise from switching inductive devices.
6. Route 110-volt ac and 24-volt dc signal lines in separate wire bundles and cable trays.

No standard connector type or size is recommended for the simple sensor interface. Most manufacturers use screw-type terminal strips for termination of signal wires.

Simple Sensor Interface Design

Simple sensor interface design is used anytime a sensor is connected to a controller or any time two controllers are interfaced as illustrated in Figure 7–11. In the figure, sensors are connected to the PLC, and the PLC and robot controller

Figure 7–13 Input/Output Modules in Robot Controller.

must communicate through *handshaking* signals. The handshaking signals usually permit the two controllers to communicate with each other using discrete signals. For example, the PLC would indicate with a change in a discrete signal that the robot could precede to the next programmed point, and the robot would use a discrete signal to inform the PLC that the move was completed. To accomplish this data exchange, an output from the PLC is connected to an input on the robot controller for data to pass from the PLC to the robot. In addition, an output on the robot is connected to an input on the PLC so that the robot can pass data back to the PLC.

The interface used at the input and output of the controllers is illustrated in Figure 7–12. The optical isolation permits different power supplies to be used without introducing noise problems from grounding issues. Isolation, however, is not the only concern when sensors and controllers are interconnected. The interface must also be consistent from a *current sourcing* and *current sinking* standpoint. These two modes of operation and illustrations of sensor output circuits for each type are provided in Chapter 5. The first interface rule is:

A current sourcing output must be interfaced with a current sinking input, and a current sinking output must be interfaced with a current sourcing input.

The interfaced output and input must have opposite modes of operation, or if current flows *out* from the output to indicate an *on* condition, then it must flow *in* the input to transfer the *on* condition. If the input expects current to flow *out* for the *on* state, then the output must able to *sink* current (current flowing *in*) to generate the *on* condition. Many proximity and photoelectric sensors have output operations that can be switched between NPN (current sinking or in) to PNP (current sourcing or out). As a result, the sensor can be matched to either a sourcing or sinking controller input condition. The interface between controllers usually doesn't offer that flexibility. PLC vendors offer input and output modules with either sinking or sourcing operation, so the PLC module is often selected to match the requirements of the robot controller since robot vendors offer fewer options.

The second rule is:

The specified current and voltage levels for the on and off state of the discrete signal must be compatible between the interfaced systems.

An example illustrates this concept best. Assume that a current sourcing output from a sensor is connected to a current sinking input on a PLC. The sinking input requires a current of 50 ma to indicate an *on* condition and less than 2 volts to indicate an *off* condition. Ideally, a sensor *off* output voltage would zero, but many solid-state sensors never get to zero in their *off* state. So you must make sure that the actual *off* output voltage is consistent with the other interfaced device. Therefore, the sensor output connected to this PLC input must be capable of supplying or sourcing greater than 50 ma for an *on* condition and have an *off* output voltage that is less than 2 volts. If the sensor has a 100 ma maximum *on* current and a 1 volt *off* output voltage, then the interface will function correctly. However, if the output is 3 volts for an *off* condition, then the PLC input will always appear to be *on*.

Considerable care must be taken in the design of the simple sensor interface to assure trouble-free data exchange between work cell devices and systems.

Complex Sensor Interface

The complex sensor interface is used with sensors that require some signal conditioning before the data are transferred to the PLC, cell controller, or robot controller. A complex sensor is defined as one that requires some type of preprocessor to perform analog-to-digital conversion, scaling, filtering, formatting, analysis, or coordinate transformation on the raw data before they are presented to a higher level controller. A complex sensor communicates with the complex sensor interface using either digital or analog signals, but the complex sensor interface communicates with the higher level controller using only digital signals.

The complex sensor system gathering the data frequently must perform computational operations on the data to do the following:

- Convert from analog values to digital values that are compatible with the higher level controller.
- Detect features or recognize patterns present in the information.
- Compare measured data with values previously stored in the controller or sensor interface.
- Transform measured values or images into x, y, z, coordinate values referenced to the work cell equipment.

Data from vision and torque/force sensors illustrate the way in which complex computation must be performed on the sensory information before it can be used by the higher level controller. For example, an image of a part captured by a vision camera is processed by the vision interface electronics to produce a coordinate value for use by a robot controller. The robot controller uses the part location coordinate values to change the programmed pickup point to the actual part location captued by the camera.

In addition to the computation requirements placed on the complex sensor interface, it must also operate at high speed. The very nature of the job to be accomplished, for example, trajectory control, requires that the closed-loop response time of the interface be very short. Decisions cannot be made and errors corrected until the data are presented to the higher level controller. Delays of more than a few milliseconds do not permit high-performance closed-loop operation.

Bidirectional data communication is another important characteristic of the complex sensor interface. Not only must the interface talk to the PLC, cell controller, or robot controller, but the PLC and controllers must communicate with the interface. In the most complex interfaces, the two-way communication includes control signals, serial or parallel data, and addressing information. For example, in vision applications, the robot controller could signal the vision system when a part needs to be analyzed, and the vision system sends the coordinates and orientation of the part to the controller.

The choice of complex sensor interface is dictated by the type of complex sensory information present. The system may require one sensor per interface if processing speed is important, or it may permit one interface to handle many sensors. At present, vision, torque/force, and remote positioning tables require the most complex interfaces. Figure 7–14 is a block diagram of a robot controller and complex sensor system. In one of the interface modules, the vision sensor requires a minicomputer with memory for storage of data and vision system programs. In another, a pressure sensor needs an analog-to-digital converter to change data in a analog representation, for example, zero to five volts, to the equivalent digital value in binary code. In a third, a digital-to-analog converter is used to drive the heating unit by converting the digital or binary data output from the controller to an analog (0 to 24 volts) output

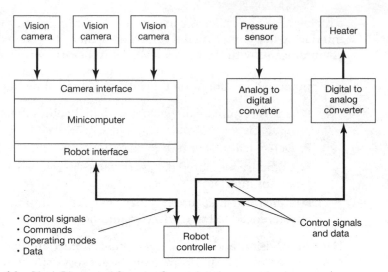

Figure 7–14 Block Diagram of Complex Sensor Interface.

voltage. As automated work cells become more complex, the demands on sensor interfacing will increase. In some current work cells, for example, more than 1000 sensors are required for control of the production line system.

Enterprise Data Interface

The enterprise data interface represents the link between the robot controller and other computers in the work cell and CIM architecture. The interface passes blocks of digital information, usually in the form of ASCII files. The interfaces available in robot controllers include *node-to-node* or *direct* communication devices and also LAN capability. The direct communication offered as a standard feature on most robot controllers uses either the RS 232 or RS 422 standard for serial connectivity. In addition, most controllers offer LAN options for IEEE 802.3 (ethernet standard), IEEE 802.5 (token ring standard), or the MAP/TOP (Manufacturing Automation Protocol/Technical Office Protocol) standard.

The RIA is working on several standards that affect the communications-information interface area. One is the American National Standard, ANSI/RIA R15.04, for Industrial Robots and Robot Systems—*Communications-Information*. It is a companion for the Manufacturing Message Specification standard ISO 9506. A second standard supported by the AIA, ANSI/AIA A15.08, has six components addressing sensor interface issues in the vision area.

7-8 AN INTEGRATED SYSTEM

The work cell in Figure 7–15 illustrates the hardware and software interfaces present in state-of-the-art production systems. Study the system carefully before reading the following description.

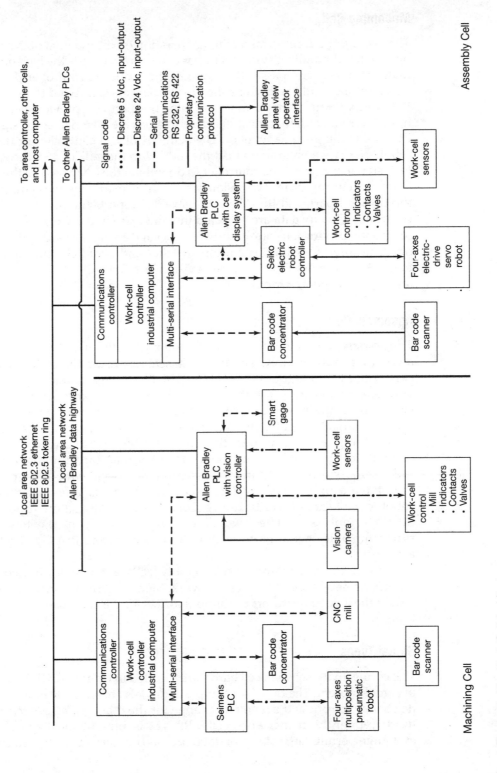

Figure 7–15 Work-Cell Architecture.

Machining Cell

The *machining cell* performs a milling operation on the material delivered to the cell by a material-handling system. The bar code scanner reads the bar code on the raw material carrier and passes the value to the work-cell controller that verifies the stock number with the current shop order. Sensors connected to the Allen Bradley PLC verify (fixture is open, mill is in home position, and previously machined part has been removed) that the mill is ready to receive material, and a signal from the PLC starts the pneumatic robot. The Seimens PLC, a controller for the pneumatic robot, executes a program to load the mill. With the mill loaded, the Allen Bradley PLC signals the work-cell controller to download the CNC code to the mill and initiates the cutting cycle. The vision system and smart gage measure the quality of the finished part, and the Allen Bradley PLC reports the results to the work-cell controller. Quality data are collected in a database in the work-cell controller and sent to the production control software package on the host at regular intervals. An operator interface with pushbutton switches and pilot light indicators, interfaced to the Allen Bradley PLC, provides a visual indication of cell activity and problems and permits operator input.

Assembly Cell

The *assembly cell* assembles the machined part with several purchased parts into a complete subassembly. The bar code scanner reads the code on all the parts carrier delivered to the cell by the material handling system. The cell controller verifies that all parts presented to the cell are mating parts for the shop order received from the production control software system. The cell-control software sends a signal to the PLC that indicates the assembly procedure should start. After verifying from sensor data that the assembly fixture is ready, the PLC sends a 5-volt dc signal to the Seiko robot controller that starts the robot program and the assembly process. During the assembly process, the robot controller and the PLC have a preestablished data exchange, called *handshaking,* that forces the robot to wait for PLC verification of correct sensor data before proceeding to the next assembly step. The *panel view* system attached to the PLC provides a programmable operator interface and dynamic graphic display of cell status and activity.

Both the machining and assembly cells are nodes on two LANs: the enterprise network and the PLC network. Many enterprise networks use the IEEE 802.3 ethernet standard on the shop floor because it has been available for some time.

Signal Types

The connectivity between cell hardware is coded in Figure 7-15 to indicate the type of signal present. The two *discrete* types of signals are either *on* or *off,* with a 24-volt dc change for one and a 5-volt dc change for the other. *Serial data communication* is used frequently in the work cells; RS 422 is present most often. Serial data exchange permits large blocks of data, such as programs, to be transferred between computer-controlled machines. Data exchanged between hardware from the same vendor frequently use a *proprietary communication protocol,* such as the Allen Bradley data highway, that is not an industry standard. This type of communication is usually a serial data exchange.

Work-Cell Controller

The work-cell controllers using Pentium or higher microprocessors are microcomputers hardened for the shop floor environment. The work-cell controller could also be a Sun workstation for support of cell data communications and control. The PCs have multiple serial parts and LAN peripheral cards installed to permit data transfers to the work-cell hardware and the enterprise computer network. The LAN card using LAN software makes the work-cell controller a node on the enterprise-wide data and communications network.

The software running on the controller includes an operating system, network software, a work-cell control program, application programs for the hardware in the cell, and all the necessary drivers. The work-cell control program is a software package that permits the cell controller to

1. Exchange data (single values or blocks of code) with the devices interfaced through the serial ports
2. Exchange data (single values or blocks) with computer nodes on the enterprise network
3. Create dynamic graphic screens for the monitor to display work-cell activity and operating conditions
4. Execute mathematical functions using data from the cell and/or evaluate cell data using logical expressions
5. Initiate cell activity

Programmable Logic Controller

The PLCs serve several functions:

1. A machine to correctly sequence all work-cell movement and activity
2. The interface to all sensors monitoring cell conditions
3. The control device for a vision system, smart gage, and work-cell display
4. The controller for a stop-to-stop pneumatic robot
5. A device to exchange data with other PLCs and work-cell machines using the LANs linking cell hardware

As the description of the cell indicates, the PLC provides for the work cell what a conductor provides for an orchestra, that is, a method for keeping every event in the correct order. The Allen Bradley PLC in the machining cell could act as the pneumatic robot controller in addition to controlling the cell activity. It is not uncommon, however, to use a separate PLC, often supplied by the robot vendor, to keep the two programming functions separate.

The system illustrated in Figure 7–15 is an example of work-cell architecture but does not imply a singular solution. The same function could be achieved by many variations in hardware connectivity. In addition, hardware and software from different vendors could be used to achieve similar results.

7-9 SUMMARY

The robot is part of the total CIM architecture and must interface with the work-cell hardware and with the enterprise-wide network. The architecture of the enterprise system includes layers of computers ranging from mainframe machines to microprocessor-based production machines. The system is controlled by software operating systems that run application software. The computers used to link the many shop floor functions include area controllers, cell controllers, programmable logic controllers, computer numerical control machines, and robot controllers.

The architecture for servo-controlled robot systems includes a CPU, memory, peripheral input/output, discrete input/output, power converter or supply, axes drivers, and feedback signal conditioning. Although the nonservo robot system also has these elements, the axes drivers are part of the discrete input/output, and the feedback network is absent. The nonservo robot controller is often manufactured by the vendor who builds the robot arm, but in some cases a standard industrial PLC is used for arm control.

The robot interface permits the controller to communicate with all the systems in the work cell. These systems range from passive switches with no onboard intelligence to complex vision systems with dedicated minicomputers. The controller interface must be compatible with this broad range of signal requirements. All robot interfaces fall into one of the following categories: simple sensor interface, complex sensor interface, and enterprise data interface. The simple sensor interface handles discrete signal sensors that do not require signal conditioning. The complex sensor interface provides signal conditioning for complex sensory signals before they are routed to a higher level controller. Finally, the enterprise data interface establishes the data connectivity between the robot and computer nodes in the enterprise network system.

QUESTIONS

1. Describe the function of the types of software present in cell controllers.
2. Define the term *multitasking* and give three examples of software that support that type of operation.
3. Compare the strengths and weaknesses of the three work-cell management software options.
4. List the components of a PLC system and describe the function of each.
5. Compare and contrast the robot controller block diagram from Chapter 1 with the block diagram for a CNC machine presented in this chapter.

6. Compare the function and operation of the controller used with servo and nonservo robots and describe the differences and similarities.

7. Compare the programming, control techniques, and operation of the servo and nonservo cell control.

8. Compare the simple and complex sensor interface.

9. What is the function of the cell controllers in Figure 7–15?

10. List the devices that would use a simple sensor interface and those that would be considered complex sensor interfaces in Figure 7–15.

11. How do each of the work cells in Figure 7–15 verify that the programs in the production machines are correct for the stock and parts delivered by the material-handling system?

12. Why are the work-cell sensors in Figure 7–15 interfaced to the PLC and not the robot controller?

13. Describe the difference between the multi-serial interface and communications controller in the work-cell controllers in Figure 7–15.

14. How do the functions of work-cell controllers and PLCs differ?

15. What are the characteristics of sensors using the complex sensor interface?

16. Name the two different types of interfaces handled by the robot controller CPU.

17. Describe one example of each type of interface named in Question 16.

PROJECTS AND CASE STUDY PROBLEMS

1. Apply the cell control concepts presented in this chapter to one or more of the automation work-cell designs listed below. Specify the work-cell control architecture and integration of the robot controller with the other cell control devices; show simple and complex sensor interfaces; and select PLC vendors, I/O modules, and other necessary devices using the resources in Appendix B or from other sources. Use the design data from previous chapters as necessary. The design components of the case studies varied in complexity and design detail, so it may be necessary to make some assumptions as you work on the support system design. Document the design and support all assumptions.

   (a) The die casting production cell started in Case Study Problem 3 in Chapter 2
   (b) The kitchenware bowl production cell started in Case Study Problem 5 in Chapter 2
   (c) The West-Electric slug production cell design started in Case Study Problem 8 in Chapter 2
   (d) The West-Electric slug lubrication cell design started in Case Study Problem 9 in Chapter 2

2. Read the West-Electric case study (Section 12-11 in Chapter 12) where cell control architecture is discussed. Complete the questions and problems at the end of Chapter 12 that relate to cell control.

Work-Cell Programming

CHAPTER GOALS AND OBJECTIVES

The general goal of this chapter is to introduce programming of automation work-cell systems with specifics and detail associated with robot and programmable logic controller (PLC) programming. At the completion of this chapter you should be able to:

- Describe the three categories of work-cell control software.
- Describe three techniques used to create a PLC program and be able to write a generic PLC program.
- Describe robot program language development and the classification process.
- Describe how robot reference frames are used in programming and how they are referenced.
- List and explain the five translation programming techniques for servo robots.
- Describe how PLCs are used to program nonservo robots.
- Describe the difference between on-line and off-line robot programming.
- Use the eight-step strategy to create a task point graph and a functional robot program for a servo robot application.

8-1 INTRODUCTION

Programming languages are the basic communication mechanisms between human beings and intelligent machines. Initially, these intelligent machines were the computers themselves, programmed to solve problems in business or scientific areas. More recently, however, computer microcontrollers have been incorporated into other industrial and office machines to increase their efficiency and capability. Microcontrollers are computers composed of just a few integrated circuit chips

This career spotlight focuses on the many different types of careers in programming. The need for programmers started much earlier than the invention of the computer in the 1940s. As the history time line in Chapter 1 indicates, examples of industrial machine programming can be found as early as the seventeenth century. The development of the computer increased the need for programming and created an enormous career field for programmers. Just as good communication skills are critical when you send messages to other team members, well-developed programming skills are necessary to indicate how you want the robot or other work-cell computer to perform. Initially most programming was centered on the creation of software related to business applications. However, with the development of the microprocessor, computers could be integrated into any machine or device that would benefit from numerical analysis or programmed control. As a result, computers have found their way into ever aspect of our daily life and into the machines that create the products we use.

This exponential growth of computer use has created a demand for programmers that has yet to be satisfied. The strong demand for programmers in the automation area mirrors that found in other computer areas. Programming careers in manufacturing automation fall into two general categories: *software development* and *program development*. Software development focuses on the development of the programming languages and application programs. Examples include program languages like Visual BASIC and application programs like Wonderware for control of automated work cells.

The second programming category, program development, focuses on using the languages and application programs to control an automated process. Businesses need programmers to develop the programming tools (i.e., languages and applications) and additional programmers to use the tools to develop programs for industrial applications. The small numbers of programmers needed to develop the tools are usually computer science graduates, who use just a few programming languages, like C or C++, to create the languages and applications. A significantly larger number of engineers and technicians are required to use the languages and application programs to develop solutions to industrial control and production problems. For example, a single team of programmers is needed to develop the software for a robot controller, but every plant that purchases a robot must have one or more engineers or technicians who can use the robot controller software to program the robot for the specific manufacturing work-cell requirements. As a result, the need for technicians and engineers skilled in programming work-cell hardware continues to expand.

The career opportunities in automation programming include the generalist who handles programming for a number of automation machines like robots, PLCs, vision systems, sensors, or cell-control software. If a difficult programming problem occurs, then help from the specialist or the hardware vendor is contracted for the job. Other career paths include focusing on a single area of application programming, such as programming PLCs from one or more vendors. Often the systems development houses have individuals with this type of programming depth. Other areas of specialty include CNC machine programming using one of the standard application programs like MasterCAM; programming a robot vendor's machine as a field service representative for the company; specializing in one aspect of PLCs such as network communications or vision; doing application program development for PLCs, work-cell control software, or robot software for specific industries like automotive, chemical, or food processing.

If work in manufacturing automation and program development appear interesting, then you may want to investigate an application programming career in industrial automation.

usually mounted to a single small printed circuit board. Despite this rapid integration of computers into machines of all types, the function of the programming language remains unchanged. Communication continues to be the primary function, but the ever-increasing number of communication sources requires an ever-increasing variety of data and information.

The first robots and automated cells were driven by drum-type sequencers (Figure 7–7) and provided little programming flexibility. Present-day automation is controlled by powerful digital computers, many with multiple processors, which permit a high level of user and machine communication. These computers are responsible for the following system functions:

- *Manipulation:* The control of the motion of all robot joints and work-cell actuators. This includes position, velocity, and path control of the robot arm during all programmed motion and sequencing the motion of drive devices such as motors and pneumatic actuators.
- *Sensing:* The gathering of information from the physical work cell. This includes the collection of sensory information and the control of peripheral equipment.
- *Intelligence:* The ability to use information gathered from the work cell to modify system operation or to select various preprogrammed paths and routines.
- *Data processing:* The capability to use databases and to communicate with other intelligent machines over networks, including the capability to keep records, exchange programs, generate reports, and control activity in the work cell.

A manufacturing system with these four characteristics would have production flexibility not found in many current small-batch systems. If these characteristics are to be realized in small-batch systems, however, the users must be able to specify how a system must operate to produce the product or to perform the task required. Consequently, the need for a suitable programming language that satisfies the system requirements is apparent. This chapter analyzes the current cell and machine control languages used by robots and other cell control devices.

8-2 WORK-CELL CONTROLLER PROGRAMMING

Work-cell software runs on the work-cell control computer and is responsible for the coordination of work-cell activity and data collection. Work-cell software falls into one of three categories: in-house developed systems, application enablers, and open system interconnect (OSI) such as the *manufacturing message specification* (MMS) ISO standard 9506. The first two categories refer to proprietary software systems, because they are either written for a specific application or developed around a third-party software management shell, called an *application enabler*. The third category is a standard information exchange software language supported by many of the vendors supplying automation equipment. Application enabler software, however, is used in most automation work cells to program the flow of information throughout the work cell.

Let us consider an example application to clarify the three programming techniques. Review the operation of the machining cell in Figure 7–15 and note that the cell controller must pass, or download, a CNC program to the mill when the material for a new or different part is identified by the bar code data on the material carrier.

Software Developed In-house

As the name implies, in-house software is written by the end user using a programming language such as C or Visual BASIC. Programs developed in-house are written to address the specific control needs of the cell, and the work-cell control program is well integrated with other software developed in-house for other enterprise areas. For example, the in-house-developed program in the cell controller (Figure 7–15) polls, or electronically interrogates, the bar code concentrator to determine if a new material code was read. If a new code is present, the software in the controller reads the new code, finds the name of the correct CNC program for the part present from a data table, and downloads the CNC code to the mill. This data exchange assumes a compatible electrical interface between the devices and that communication software is available in the controller to talk to the concentrator and mill. In-house program development provides the best opportunity for tight integration of information and data with other enterprise software solutions. In addition, machines that need custom communication drivers present no problems because the expertise is present to do the programming. Nevertheless, these advantages are offset by major disadvantages in development time and cost and the inability to change the software easily when the cell hardware or configuration changes.

Enabler Software

Common impediments to the development of a fully implemented CIM work cell are the exorbitant cost, time, complexity, and inflexibility of custom-programmed CIM solutions. The introduction of enabler software shifts the development of application programs from software engineers to manufacturing engineers. This shift occurs because enabler software provides a set of software productivity tools for the development of control programs for the CIM cells. Products such as *Plantworks* from IBM, *Industrial Precision Tool Kit* from Hewlett-Packard, *CELLworks* from FASTech, *Factory Link* from U.S. Data, *Fix DMACS* by Intelution, and *Wonderware* from Wonderware help reduce the difficulty in developing cell control and management applications.

These enablers have a library of driver programs that permit the cell control computer to exchange data and information with most commonly used production machines and machine controllers. In addition, they offer LAN and serial data communications support, mathematics and logic functions for internal computations, links to commonly used mainframe and microcomputer relational databases, real-time data logging, real-time generation of graphics and animation of cell processes, alarm and event supervision, statistical process control, batch recipe functions, timed events and intervals, counting functions, and the ability to write custom programming applications.

Advantages of enablers are an estimated tenfold improvement in cell control and ease of program development. The primary disadvantage is that the cell control is tied to a third-party software solution, so selecting the software that closely matches the application is critical. In addition, an enabler is written to satisfy the average process requirement; therefore, if a specific requirement is not covered, then additional programming inside the enabler is necessary.

OSI Solution

The most frequently used open system solution for cell control is the manufacturing message specification (MMS). The MMS is a standard (ISO 9506) for network communication between intelligent devices in a production environment. The standard has three parts: service specifications, protocol specifications, and robot interface and protocol specifications. The MMS standard defines a set of objects that exist within a device; for example, the MMS object could be the position of one axis of the robot arm. In addition, the MMS defines a set of communication services to access and manipulate the objects and describes how the devices will respond.

Implementing an MMS solution requires that all the devices support the MMS protocol and that they be linked by a manufacturing automation protocol (MAP) ethernet or broadband manufacturing network. The work-cell system in Figure 7–15 is reconfigured as an MMS solution in Figure 8–1. Take a few minutes and compare the two implementations.

Note in Figure 8–1 that all the computer-controlled devices support the MMS and are a node on an ethernet-driven network. There is no longer a need for the proprietary PLC network because the PLCs share data directly with every machine over the single LAN. The interface to the sensors would not change from that shown in Figure 7–15. With the MMS, the work-cell controller is no longer a data

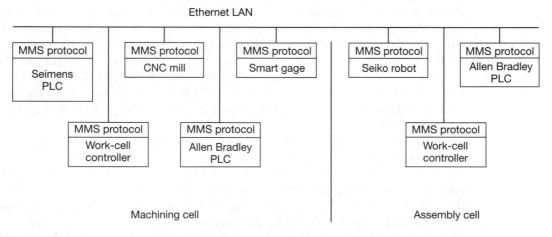

Figure 8–1 Manufacturing Message Specification.

(*Source:* COMPUTER INTEGRATED MANUFACTURING, 2nd ed. by Rehg and Kraebber©, p. 436. Reprinted by permission of Pearson Education, Inc. Upper Saddle River, NJ)

concentrator because MMS machines exchange data directly over the LAN. In the MMS configuration, each smart device provides real-time data without the need for work-cell management software to collect cell information. Many computer, robot, and PLC vendors have MMS support available as an option.

An advantage of the MMS is that it is a common communication standard, not a third-party vendor solution. Information is requested directly from the target device and delivered directly to the end user; no intermediate computer and software system are required. The primary disadvantages currently are the small user base and the limited number of equipment vendors who have agreed to support the standard.

8-3 PROGRAMMING SEQUENTIAL CELL ACTIVITY

The software described in the previous section provides programmed control of data and information exchange between the computer controller devices in the work cell. At a lower level in the cell control architecture, the action and movement of individual devices must be controlled. For example, when a sensor is activated by the placement of a part into the milling fixture in Figure 7–15, an orderly sequence of events must occur. The part clamping devices must be activated and a start-cycle signal must be sent to the mill. In most implementations the signals are discrete, with voltages ranging from 5 volts dc to 110 volts ac. In most applications, sequential control is performed by programmable logic controllers (PLCs); in some cases, however, the robot controller provides this control function.

PLC Programming

The graphic technique used to illustrate the logic in PLC programs has changed little since the introduction of the first programmable logic controller in the 1970s. The process, called *ladder logic programming,* is a variation of the *two-wire diagrams* used to document the wiring of industrial control circuits. For example, the necessary switches, sensors, and controls to sequence the operation of a pump are drawn between the two vertical lines (L1 and L2) in the two-wire diagram in Figure 8–2. S1 and S2 are switches, C is a relay coil with contacts C1 and C2, M is a

Figure 8–2 Two-Wire Diagram of Pump System.

(*Source:* COMPUTER INTEGRATED MANUFACTURING, 2nd ed. by Rehg and Kraebber©. Education, Inc., Upper Saddle River, NJ)

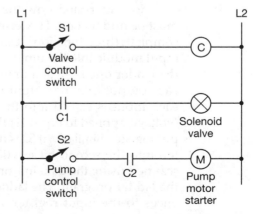

motor starter coil, and a valve is also identified. Study the circuit and determine the operational logic present.

The circuit works as follows:

1. Switch S1 is closed manually and causes the relay coil C to be energized.
2. When the relay is energized, the contacts C1 and C2 close.
3. The closing of contact C1 energizes the input valve and allows liquid to flow into the tank.
4. The closing of contact C2 causes no immediate action.
5. Switch S2 is closed manually and causes the pump motor starter to operate.

Note that the change in contact C2 in step 4 was necessary for pump operation. If the valve switch S1 is opened manually while the pump is operating, the change in contact C2 opens up the circuit that is energizing the pump motor starter and it stops. Review this operation and study the two-wire diagram in Figure 8–2 until the operation of this the circuit is clear. Also, verify that you understand the logic conditions required for the valve and pump to operate.

The PLC ladder logic program that would provide the same logical control as the circuit in Figure 8–2 is illustrated in Figure 8–3b, and the wiring for the PLC solution is illustrated in Figure 8–3a. A study of the circuit in Figure 8–2a reveals that the PLC for this solution has three elements, an input and output module and the PLC processor. Note that the switches, S1 and S2, are connected to inputs I1 and I2 respectfully. Also, the solenoid valve is connected to output O1 and the motor starter is connected to output O2. Compare the PLC ladder logic (Figure 8–3b) to the relay ladder logic solution in Figure 8–2. Note that the hardware relay C is replaced with a virtual or software relay (B3:0/1) represented by the symbol B3. The virtual relay is actually a binary register in the PLC processor module. It is the first register identified by the zero in 0/1, and is the first bit identified by the 1 in 0/1. So each of the 16 bits (0 through 15) of the binary registers (0 through 255) can be used as an internal control bit or virtual relay. Note also that the contacts used in rungs 2 and 3 are referenced to the same binary register location. So if the output B3:0/1 is true (*on or active*), then the contacts with the same label change states.

To understand how the PLC solution operates, the concept of a PLC *scan* must be understood. The term *scan* refers to the time required for the processor to complete three functions, transferring the input states (on or off voltages) of the input module into the input register as zeros and ones, checking the input logic in the ladder one rung at a time starting with the first, and transferring the state of each output into the output register. During the input module scan, the state of each input is copied into the binary input register. For example, if S1 is closed (voltage applied to input I1) then a 1 is written into the first bit location in the input register. Similarly, if S2 is open (no voltage applied to input I2), then a 0 is written into the second bit location in the input register. Next the ladder rungs are scanned using the data in from the input register to evaluate all of the contacts in the ladder program. The ladder solution in Figure 8–3b has two rungs with references to the input register, rung 1 and rung 3. The condition of the contacts

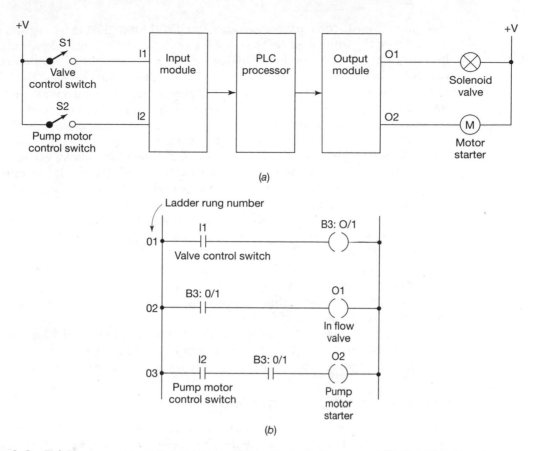

Figure 8–3 PLC Program and Input/Output Interface: (a) Ladder Logic Program and (b) PLC Wiring Interface.

(*Source:* COMPUTER INTEGRATED MANUFACTURING, 2nd ed. by Rehg and Kraebber©. Reprinted by permission of Pearson Education, Inc., Upper Saddle River, NJ)

(drawn normally open (NO) or normally closed (NC)) when the ladder was programmed is the contact condition when the referenced input register value is a 0 (no input voltage present). If there is a 1 in the referenced input register location (input voltage present), then the contact referencing that input condition is changed to the state opposite of the condition drawn during programming. This can be confusing, so let's apply the results from that previous example. When the ladder rungs in Figure 8–3b were programmed, the contacts referencing input locations I1 and I2 were drawn in a NO representation. During the input scan I1 had voltage present and I2 did not, so input register location I1 is equal to a 1, and I2 is equal to a 0.

The ladder logic is scanned starting with rung 1 in Figure 8–3b. The I1 contacts in rung 1 have change from NO to NC because the I1 input register location has a 1 present. Since the input logic in rung 1 is true (I1 contact is now closed) the output is set to a true state or the B3:0/1 bit is turned on or set to a 1. Next the input logic contact in rung 2 is evaluated. The NO contacts labeled B3:0/1 change to the NC configuration because the B3:0/1 bit in the binary register is true or a 1. Since the input logic on the rung is true, the rung is true and the output O1 becomes true. Finally the input logic or contacts on rung 3 are evaluated. The contacts labeled I2 remain unchanged (NO representation) because the I2 input register has a 0 in that bit location (no input voltage). The contacts labeled B3:0/1 have changed from the NO to the NC representation for the same reason that the contacts in rung 2 changed. The contacts on rung 3 are arranged in a logical AND connection so both must be true for the rung to be true. Since I2 is false (NO representation) the rung is false and output O2 is therefore false or off as well.

With the scan for the logic completed, the condition of all the outputs is placed into the output register. Since O1 was true, a 1 is written into the 1st bit of he output register. Similarly, a 0 is written into the 2nd bit of the output register because rung 3 evaluated false so O2 was false as well. Since O1 is true, the output module turns on the solenoid valve, but the pump motor starter is not turned on since O2 was false.

The scan time on PLCs is dependent on the size of the ladder logic program, but for most applications scans occur every few milliseconds. If on the next scan the S2 switch was closed, then at the completion of the scan the pump motor starter would also be turned on.

In addition to the AND logic illustrated with the pump motor control example, PLCs also support OR logic (parallel contacts) and combinations of AND and OR logic. Other output functions frequently used include *timers* and *counters*.

Other Sequential Programming Options

Although ladder logic offers a familiar programming environment for industrial applications, the process has some disadvantages. For example, the unstructured nature of the process results in many different program solutions for the same control problem. Also, large ladder logic programs are difficult to troubleshoot because the contacts of a virtual or software relay could appear anywhere on the ladder. The lack of structure has been addressed at the standards level. In 1979 the International Electrotechnical Commission (IEC) formed a committee to establish a common standard, IEC 1131 for programmable logic controllers. As a result of the committee work, the standard that was developed allows programming in any one of the five following languages.

1. *Ladder logic:* the traditional ladder logic programming used initially with PLCs and still the most frequently used language in North America
2. *Instruction list:* a programming language based on assembler languages native to all microprocessors and the language commonly used in Europe
3. *Structured text:* which looks a lot like the C programming language
4. *Sequential function chart (SFC):* a structured language based on the French GRAFCET language that is especially useful for control of sequential process like those used for nonservo type robot applications
5. *Function block diagrams:* a language that looks a little like electrical schematics with ladder logic elements included

Two of the standards, one and four, have an application in robotics so deserve additional discussion.

Since the first standard, ladder logic, is used most often, a procedure for program development of ladders is appropriate. The technique is often called the *empirical design process* and uses the following seven-step process.

1. Clearly state and define the control problem.
2. Identify all of the control outputs and the control requirements for each.
3. For each control output identify all combinations of input conditions that are required to produce an active output state.
4. Using the input conditions identified in Step 3, create ladder rungs for each control output.
5. Scan the program from the first rung to the last to verify that the outputs produce the desired system operation when the input states change.
6. Add additional rungs or input logic to correct problems identified in Step 5.
7. Document the ladder logic by identifying all inputs and outputs and by labeling the program and all the program rungs.

When the empirical approach is used, the solution just evolves as one approach fails and fixes to the ladder are added. As a result, you rarely find that two programmers will reach the same solution when trying to solve the same problem. A much more structured approach for the sequential type of problems found in many automated cells is that in PLC language standard five four, *sequential function charts* (SFC). The PLC vendor can provide a supported SFC language or the programmer can implement the SFC type structure using conventional ladder logic. In the SFC approach, the machine sequence is broken into a sequence of actions or states. A transition condition is identified that moves the machine from one sequential state to the next. The ladder logic drives the machine through the desired states using the transition condition to trigger the next action.

8-4 ROBOT LANGUAGE DEVELOPMENT

Initially, robot languages were designed using two techniques. The first approach focused on developing a language that satisfied the control needs of the robot arm. The new control language is expanded to include language structures, for example, conditional branching and input/output interfacing. The T3 language developed by Cincinnati Milacron in the early 1970s for its industrial robot family was an example of this type. This approach produced good control of the robot's manipulation and tool path, but the language produced did not fit the structure normally associated with computer languages and did not support the data processing function well.

The second technique started with an existing general-purpose computer language such as BASIC or FORTRAN and added the robot control commands. VAL, a language developed in the 1970s for the early Unimation PUMA robots, used BASIC as a base and added motion control commands. This technique produced a robot language that was well defined, operational, and easy for programmers to use.

However, starting with a fully functional computer language forced some design compromises that made motion control commands less efficient.

Currently, robot language design starts with both the language control structure and the robot arm control needs. As a result, the current development process produces a new general-purpose robot language capable of supporting all four system functions: manipulation, sensing, intelligence, and data processing. DARL II from Seiko and V from Adept are just two examples of modern robot languages using this approach.

Few standards currently exist for robot control languages. There is no interchangeability of programs among manufacturers—and in some cases, there is only limited interchangeability of programs between models from the same manufacturer.

8-5 LANGUAGE CLASSIFICATION

One way to classify the many languages used by robot manufacturers is according to the level at which the programmer must interact with the system during the programming process. For example, if the programmer must handle all the details associated with arm movement by specifying the joint angles for each move, then it is a low-level operating system. Compare that with a high-level language that permits the user to specify the motion required in statements such as "Pick up the part." Using this criterion, we can group robot programming into four loosely formulated levels. Of course, overlaps between levels exist and some languages appear to straddle two levels, but the classification process is still valuable. Figure 8–4 shows the four basic levels into which all robot languages are grouped.

Joint-Control Languages

Languages at this level concentrate on the physical control of robot motion in terms of joints or axes. The program commands must include the required angular

Figure 8–4 Programming Language Levels.

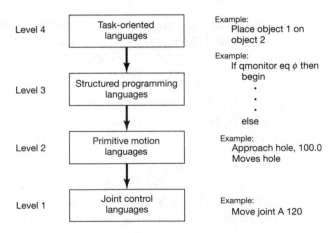

		Example:
Level 4	Task-oriented languages	Place object 1 on object 2
Level 3	Structured programming languages	Example: If qmonitor eq φ then begin ⋮ else
Level 2	Primitive motion languages	Example: Approach hole, 100.0 Moves hole
Level 1	Joint control languages	Example: Move joint A 120

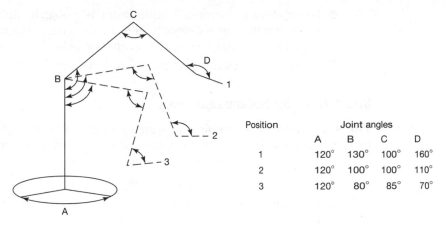

Position	Joint angles			
	A	B	C	D
1	120°	130°	100°	160°
2	120°	100°	100°	110°
3	120°	80°	85°	70°

Figure 8–5   Joint-space Programming.

change of rotational joints or the extension lengths of linear actuators. The language usually does not support system or work-cell commands, such as INPUT or OUTPUT, which can be incorporated into the programs of higher-level robot languages for control of external devices.

In this level of robot programming, the user must program in joint space. The term *joint space* means that all the programmed points in the work envelope of the robot are expressed as a series of axis positions for all the axes on the arm. Figure 8–5 illustrates a robot arm that has three programmed points. The table included in the figure shows the joint angles required for the robot axes for each programmed point.

This level of robot language is used on most *stop-to-stop* pneumatic robots controlled with PLCs. The language used for stop-to-stop robots (e.g., the robot in Figure 2–10) is the general-purpose language of the PLC.

Primitive Motion Languages

Point-to-point primitive motion languages are now usually confined to older robot programming languages, or they may be an optional programming mode for a more sophisticated robot language. Although the languages included in this group vary widely, they all exhibit the following characteristics:

- ■ A program point is generated by moving the robot to a desired point and depressing a program switch. A sequence of points is saved in this manner, thereby producing a complete program.
- ■ Program editing capability is provided.
- ■ Teaching motion of the robot is controlled by a teach pendant, terminal, or joystick.
- ■ The programmed and teaching motion can occur in the Cartesian, cylindrical, or hand coordinate modes.

- Interfacing to work-cell equipment is possible. Robot controllers can interact with external signals by using the external signals for control or by signaling external events. Work-cell control by the robot system is possible.
- The language permits simple subroutines and branching.

Structured Programming Languages

The structured programming languages offer a major improvement over the primitive motion level and have become the standard for the major vendors of robots. This level is composed of languages that have the following characteristics:

- A structured control format is present.
- Extensive use of coordinate transformations and reference frames is permitted.
- Complex data structures are supported.
- Improved sensor commands and parallel processing above the previous language level are included.
- System variables (called state variables), whose value is a function of the state or position of the system, are permitted.
- The format encourages extensive use of branching and subroutines defined by the user.
- Communication capability with local area networks is available.
- *Off-line* programming is supported.

Task-Oriented Languages

The primary function of a task-oriented language is to conceal from the user the commands and program structure that normally must be written by the programmer. The user must be concerned only with solving the manufacturing problem. Languages at this level have the following characteristics:

- Programming in natural language is permitted. A natural language command might be "Put bracket A on top of bracket B."
- A plan generation feature allows replanning of robot motion to avoid undesirable situations.
- A world modeling system permits the robot to keep track of objects. This feature provides the systems needed to locate and identify objects, to determine the pickup point and orientation of objects, and to move objects relative to one another. In addition, the system can store and use the new relationship between two joined objects.
- The inclusion of collision avoidance permits accident-free motion.
- Teaching can be accomplished by showing the robot an example of a solution.

Currently, no languages at this level are operational, but significant efforts are underway at university and industrial research laboratories to produce an opera-

Origin	Level 2	Level 3	Level 4
ABB		RAPID	
Adept		V	
		V+	
Cincinnati Milacron	T3		
GMFanuc		KARL	
IBM		AML	AUTOPASS
		AML/E	
Kawasaki		AS	
McDonnell Douglas		MCL	
Panasonic		PARL-1	
Rhino	RoboTalk		
Sankyo		Sankyo language	
Seiko		DARL II	
Unimation	VAL	VAL II	

Figure 8–6 Programming Languages by Level.

tional language in this category. A review of industrial applications indicates that level 3 structured programming languages are now the most frequently used in robot work cells. Figure 8–6 provides a partial list of languages and the corresponding level of their operation.

8-6 ROBOT PROGRAM FUNDAMENTALS

A robot program is a set of instructions that causes the robot system to perform the desired task. The program has two basic parts: the programmed points and the program statements.

Translation or Programmed Points

The first part is a set of points in space, called *translation points* or *position points*, to which the robot will move when the program is executed. These points are also called programmed points. The number and location of programmed points for servo and nonservo robots are quite different. Any location in the work envelope of a servo robot can be programmed as a position or translation point. For example, the Seiko robot in Figure 8–7 has a cylindrical work envelope and three examples of translation points (T5, T10, and T20) are identified in Figure 8–8. However, an almost unlimited number of points could be programmed, with the total number limited only by the resolution of the servo system driving the arm.

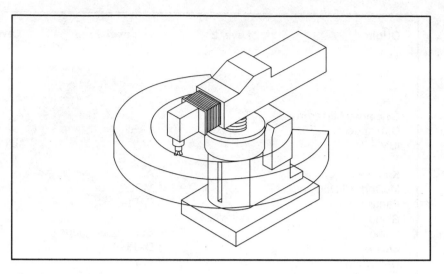

Figure 8–7   Seiko Robot Using DARL Programming Language.
(Courtesy of Seiko Instruments USA Inc.)

In contrast, the nonservo robot has a limited number of positions or programmed points, which are determined by the number of fixed and programmable stops. The nonservo robot system in Figure 8–9, for example, is a three-axes arm with one programmable intermediate stop and a wrist-roll axis. The standard arm, without the intermediate stop actuator, would have 8 programmable

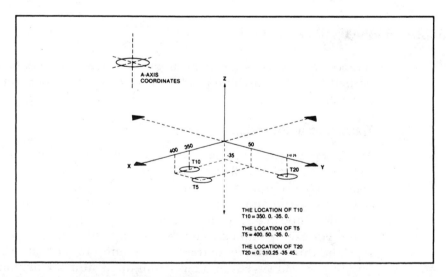

Figure 8–8   Programming Coordinate System for Seiko Robot.
(Courtesy of Seiko Instruments USA Inc.)

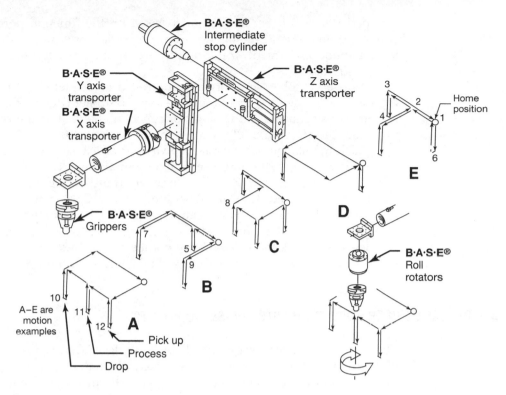

Figure 8–9 Nonservo Position Points.
(Courtesy of MACK Corporation)

positions; with the intermediate stop added, the total number of positions is 12. Find the 12 positions for the robot arm in Figure 8–9. A study of Figure 8–9 indicates that each of the locations has been numbered from 1 through 12. If the robot moves as indicated in example A, it would move through the following numbered locations: home (1)–9–12–9–8–11–8–7–10–7–3–home (1). Note that the parentheses ( ) on the first location indicate a HOME position. Try to identify the position locations for some of the other sample moves.

Programmed Statements

The second part of a robot program consists of program statements developed using the many commands in the robot programming language. This second part of the program determines the order of motion, defines the logic and conditions for decision, gathers information on the operation of the cell, and analyzes and prepares data for communication with other machines. Most servo robot languages have a robust set of command statements available to the programmer for control of the arm. The statements are similar to the structured commands used in computer languages such as Qbasic and Pascal. The statements are generally grouped into a number of categories based on the function performed by the command. In the PARL-I language from Panasonic, for example, the control functions include movement, input/output, external signal, communications, time, and program.

Nonservo robots are normally controlled by sequencing the pneumatic or hydraulic values with a programmable logic controller (PLC). Therefore, the programming language is ladder logic or function charts, and the command statements are dictated by the PLC chosen for robot arm control.

Each translation point contains a value for each axis in the robot coordinate system. For example, the four-axes Seiko robot and work envelope in Figure 8–7 requires four values (X, Y, Z, and A) for every translation point. Three translation points (T5, T10, and T20) are illustrated in the Cartesian coordinate system pictured in Figure 8–8. For example, the T5 programmed point is located at X = 400 mm, Y = 50 mm, and Z = −35 mm, and the A rotation of the gripper is 0 degrees. Check the values for the other points. In stop-to-stop or nonservo robots, the translation points are defined by the fixed or hard stops.

8-7 TRANSLATION OR POSITION POINTS FOR SERVO ROBOTS

The translation points, programmed points, are locations inside the work envelope of the robot and represent positions through which the tool center point will pass. The translations are always defined in relation to the origin of a Cartesian coordinate system located in a frame of reference. In the programmed points in the example illustrated in Figure 8–8, the first three values in each data point (see bottom right of figure) represent the distance in millimeters from the origin of the Seiko reference frame to the translation point; the last value is the rotation of the gripper in degrees. As a result, the reference frame system used in programming is a natural place to begin our detailed discussion of translation or programmed points.

Reference Frames

Reference frames are sets of Cartesian coordinates used to define locations with respect to the robot, the gripper, the tool center point, the workpiece, and the universe (world reference frame). The robot system in Figure 8–10 has all of the reference frames displayed. Note that all the frames are referenced to the world reference frame. Coordinate values can be translated between frames using mathematical equations called *transformations*, using transformation equations. The robot controller can calculate the necessary joint angles that will align the tool center reference frame with the part or workpiece reference frame.

The program stored in the controller performs the calculations using the transformation equations written for the type of robot arm used. Although the transformation equations are completely transparent to the robot user, the programming languages use the reference frame as a programming tool. For example, the DARL language from Seiko has a command, **DEF FR** (define frame), that permits an entire sequence of programmed points to be changed by changing

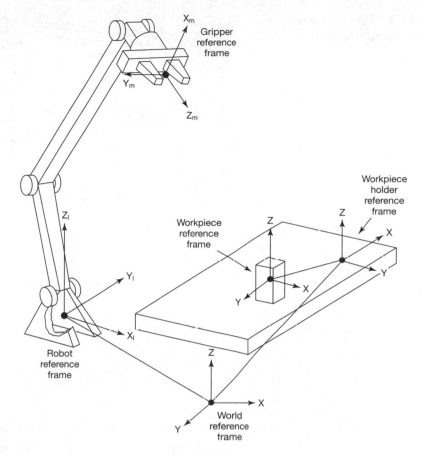

Figure 8–10 Reference Frames in Work Cell.

the reference frame. A move command in DARL, such as **MOVE T10**, can produce different motions since the point identifier T10 can have different coordinate values in each reference frame. Reference frames are also used in work-cell simulation systems to determine the optimum location of all the equipment in a production work cell that uses a robot. The Application Note on robot system simulation in this chapter describes a system developed by the McDonnell Douglas Corporation. Although a complete description of reference frames and transformation equations is not appropriate for this text, a basic understanding is essential for the robot technician, the applications programmer, and the work-cell system designer.

The position or angle of every servo-driven robot joint is measured by sensors and the information is fed back to the controller. Based on the position information, the location of the gripper with respect to the robot frame is calculated by the robot controller. For example, the X, Y, and Z values for the gripper in the robot reference frame in Figure 8–10 are calculated using equations

Figure 8–11 Reference Frames in Work Cell with Parts Aligned.

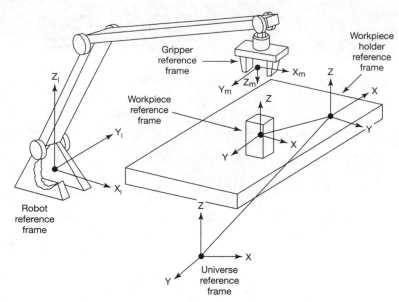

that define the relationship between the gripper reference frame and the robot reference frame. The equations have variables for each joint angle of the robot arm and constants such as the length of each arm element and the distance from the tool plate to the tool center point. As the joint angles change, the displacement of the gripper from the X, Y, and Z axes of the robot is calculated. Similarly, the equations that relate the robot reference frame to the universe reference frame determine the position and orientation of the gripper in the universe reference frame. Once the relationship between the reference frames is defined by transformation equations, the displacement of the part and the gripper from the universe reference frame is determined by the controller. Thus, for the robot to grasp the part, the controller needs only to change the joint angles until the gripper reference frame and part reference frame have the same displacement and orientation from the universe reference frame. Under these conditions (Figure 8–11), the reference frame of the gripper and the reference frame of the part are aligned. This alignment would not have been possible without the equations that describe the relationship between the reference frames in the system.

Reference frames for the robot and work-cell environment are required for some programming operations. For example, the initial reference frame for the Seiko robot is called *Frame 0* or the *HOME* frame. Nine additional reference frames can be defined, and the relationships between the frames can be established. The reference frames identified in Figure 8–10 (gripper, workpiece, workpiece holder, and universe) could be assigned to the nine available in the Seiko system.

Programming Servo Robot Translation Points

Five methods are used to define translation points: (1) teaching of translations by robot positioning, (2) teaching of translations by coordinate input, (3) creation of

APPLICATION NOTE
Robot System Simulation

Complex manufacturing work cells represent a considerable capital investment, and the electrical and mechanical requirements of all the systems in the cell must be considered for an effective design. The addition of a robot into a complex manufacturing work cell creates many new and unique problems. For example, the location of the robot in the cell is a function of the robot's geometry, payload capacity, end-of-arm tooling, and the speed and location of other cell hardware. Positioning the robot is a three-dimensional problem because the robot has three-dimensional movement. Work-cell simulation systems solve this problem by graphically displaying the robot arm

and work-cell hardware on a graphics computer monitor with full dynamic motion capability. The advantages of robot work-cell simulation include the ability to (1) select the most efficient geometry and robot size for the given work-cell application, (2) simulate the actual size and motion of a wide range of currently available industrial robot arms, (3) construct a scaled three-dimensional computer model of the work cell, (4) test the model for interference and reach over the full range of robot arm movement, (5) run the work cell and robot over a range of production speeds to determine optimum cycle times, and (6) generate the robot program required for the application and for the arm selected.

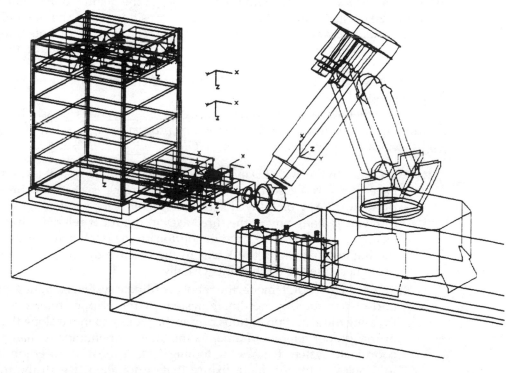

Output from Screen of Work-Cell Simulation System.

The figure in this Application Note shows a printout of a simulation for an ABB robot palletizing parts in bins. Note the X, Y, Z reference frames for each part in the work cell. The following simulation software is frequently used to program and simulate robot and work-cell environments: WORKSPACE, Robcad, IGRIP, and Grasp. These tools, called computer-aided software engineering (CASE) software, enhance and shorten the work-cell design and robot programming process. The features supported by these packages are three-dimensional CAD editor, three-dimensional solid shading, off-line programming language translators, intertask communications, cycle time calculations, collision detection, reach analysis, integrated robot program editor, robot calibration, work-cell calibration, AutoCAD .dxf file import capability, solid modeling with constructive solid geometry features, dimensioning, torque analysis and motion simulation, user definable kinematics, modeling of parallelograms, auxiliary robot axes, linking of user-defined programs, automatic inverse kinematic solution generator, computer-aided learning (CAL), and support for a microcomputer platform.

new translations by algebraic expressions, (4) creation of a new translation by means of an external interrupt, and (5) downloading new translations through the RS-232C port.

Method 1—Teaching Points by Robot Positioning. This type of programming is available at all language levels and remains the most frequently used technique for all types of robots. With the controller in the manual or monitor mode, the tooling is visually moved into the correct position using the manual teach pendant; then the translation point is programmed. On some robots, programming occurs by just pressing a program button on the teach pendant. On other systems, for example, the Seiko, a command (**HERE T <0–399>**) is entered from the keyboard. The Seiko allows 400 translation points (0 to 399) to be programmed.

Most robots have several different arm movement modes to assist in visually positioning the tooling when this method of programming is selected. The VAL language, for example, has four modes: *WORLD, TOOL, JOINT,* and *FREE.*

- *WORLD:* In the WORLD mode, the robot coordinate system is fixed to the base of the robot so the robot reference frame is used. In this mode, pressing the X, Y, or Z direction buttons on the teach pendant causes linear movement along any of the three axes in the robot reference frame. For example, pressing the +X motion button on the teach pendant causes the tooling to move along the X axis in the positive direction. This mode is used most often to teach points visually.

- *TOOL:* In the TOOL mode, the robot coordinate system is aligned with the end-of-arm tooling coordinate system or the gripper reference frame. Pressing the Z motion buttons causes the tooling to move along the Z axis of the gripper reference frame. As the tooling orientation is changed, the coordinate system follows the tooling. This is ideal for insertion of pins into holes or for loading a fixture that is not aligned with the robot or universe reference frame.

- *JOINT:* In the JOINT mode, the movement of the tooling is not associated with any coordinate system. Each robot joint can be moved independently and is used for coarse moves in the work cell.
- *FREE:* In the FREE mode, the joints can be physically moved by manually overcoming the small holding torque present on each joint axes motor. This mode is used to move the arm physically away from a problem situation or after a crash of the arm with a work-cell object.

The four modes just described support visual programming by means of the teach pendant and are used only in method 1 for teaching translation points. Most robot languages have some equivalent to the first three modes.

Method 2—Teaching Points by Coordinate Input. Teaching by coordinate input requires that the position and orientation of the tool center point be entered as either a program statement or through the controller keyboard. The Seiko robot in Figures 8–7 and 8–8, for example, would require four values (X, Y, Z, and A), whereas the jointed-spherical arm in Figure 1–10 would require the six axes variables shown.

The following program statements demonstrate coordinate input programming on the Seiko robot:

    T25 = 300. 350. −10.5 15.
    MOVE T25

The first line creates the translation point T25 with values for X, Y, Z, and A; the second line directs the robot to move to the point just programmed. All level 3 languages permit this type of position programming.

Position points can also be programmed by direct input from the keyboard. On the Seiko robot the controller is put into the *Immediate Execution* mode, called *Monitor,* and the following command is entered from the keyboard:

    DO T25 = 300. 350. −10.5 15.

The new translation point is created, and the robot moves to the new position. Level 2 and 3 languages support this type of procedure.

Method 3—Creating Points by Algebraic Expression. New translation points can be created using previously defined translations in algebraic expressions. Translations can be added or subtracted and modified by an integer value. The following examples for the Seiko robot illustrate this operation:

Adding Translation Points

    T10 = 20. 30. 5. 0.
    T20 = 30. 20. −5. 50.
    T30 = T10 + T20

The third line adds the corresponding coordinate values so the new translation point has a value of 50. 50. 0. 50.

Subtracting Translation Points

T10 = 20. 30. 5. 0.

T20 = 30. 20. −5. 50.

MOVE = T10 − T20

The third line subtracts the corresponding coordinate values and moves the robot to the new location at −10. 10. 10. −50.

Multiplying a Translation Point by an Integer

T10 = 20. 30. 5. 0.

T20 = 30. 20. −5. 50.

Y = 2

MOVE = T10*(−1) + T20*Y

(*Note:* *is a symbol for multiplication)

The fourth line multiplies point T10 by a negative one and point T20 by a variable Y that is equal to 2. It then adds these two values to obtain the MOVE value. Adds the two modified translations (−20. −30. −5. 0.; 60. 40. −10. 100.) and moves to the new location (40. 10. −15. 100). Level 3 languages permit this type of position point modification.

Method 4—Creating Points by Use of External Interrupt. An external signal can be used to interrupt the current move command and trigger the creation of a new translation point. The new point is the position of the tool center point at the time of the interrupt. The DARL language from the Seiko robot is used to illustrate this command.

SEARCH + 1E3 T10 THEN HERE T20

ELSE STOP

In this example the statement helps the robot find the top of a stack of a random number of plates (Figure 4–7). When this command is executed, the robot is moving a vacuum gripper and proximity sensor toward the top of a stack of plates (point T10 is at the bottom of the stack) that must be loaded into a machine. As the tooling moves toward T10, the controller continually scans the input 1E3, which is connected to the proximity sensor. The robot will continue moving vertically down until the top plate in the stack is sensed and input 1E3 goes *true*. The interrupt generated by 1E3 causes the *THEN* part of the statement to be executed and the location of the top plate becomes translation point T20. If translation point T10 is reached and no interrupt was generated, then no stack is present and the *ELSE* part of the statement is executed and the robot *stops*.

Creation of position points through interrupts is provided in most level 3 languages.

Method 5—Downloading Points Through the RS-232C Port. This method permits translation or position points to be specified from an external source such as a host computer, cell controller, work-cell simulator, off-line programming station, or vision system. The data are transmitted in ASCII format over a serial communications protocol. Many level 3 languages support translation point data interchange over a serial link.

The creation of the translation points through which the tool center point will pass during the execution of the program is one part in the development of a manufacturing program for a robot.

Programming Nonservo Robot Translation Points

The translation or programmed points for nonservo robots are determined by the fixed stops built into the system. The motion of the stop-to-stop robot is controlled with a sequencer such as the drum controller in Figure 7–7 or by a programmable logic controller. The robot in Figure 8–9 has five example programmed sequences (A through E) illustrated. Note that the HOME position is identified by the small circle in the path. The programmed motion is confined to the twelve positions defined by the end stops and the intermediate stop feature. The PLC interface for the robot is drawn in Figure 7–9.

8-8 PROGRAM STATEMENTS FOR SERVO ROBOTS

The second part of the robot program to satisfy the production requirements is a list of *program statements* that will move the arm and tooling through the translation points described earlier. The steps in developing a robot program are

1. Establish a basic program structure.
2. Analyze the manufacturing process in which the robot will work.
3. Divide the robot action into tasks and subtasks.
4. Draw a *task point graph* that describes the desired motion and Identify the translation points on the task point graph.
5. Assign values for all system variables to control the motion.
6. Write and enter the command statements.
7. Create or teach the translation points.
8. Test and debug the program.

Several of these steps require clarification.

Basic Program Structure—Step 1

The first step in developing a robot program is the establishment of the basic program structure (Figure 8–12). All programs start at the robot's *HOME* position and move out to a start point in the cycle called *CYCLE START*. The translation point for CYCLE START should be located as close as possible to the main motion path of the tooling because the robot returns to this point after every manufacturing cycle.

The final point in the program, called *END OF CYCLE,* provides a branching command to direct the program execution back to the CYCLE START point.

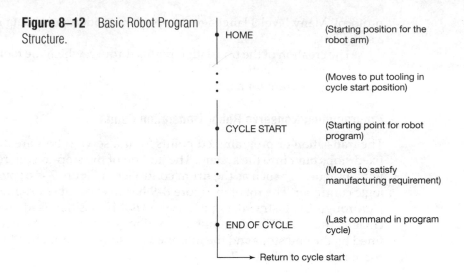

Figure 8–12 Basic Robot Program Structure.

HOME — (Starting position for the robot arm)

(Moves to put tooling in cycle start position)

CYCLE START — (Starting point for robot program)

(Moves to satisfy manufacturing requirement)

END OF CYCLE — (Last command in program cycle)

Return to cycle start

The program points and commands that solve the manufacturing problem are located between the CYCLE START and END OF CYCLE points. The operations are divided into tasks and subtasks, with the tasks embedded into the main body of the program and the subtasks included as subroutines called from the main body.

Process Analysis—Step 2

The development of the program starts with an analysis of the production problem that will be solved by the robot. The programmer must have a complete knowledge of the manufacturing process in which the robot functions. From this knowledge of the operation of the process, the programmer can identify the required motion and commands; divide the motion into tasks and subtasks; and establish the values of the system variables, such as tool velocity, for each translation point.

Tasks and Subtasks—Step 3

Based on a complete understanding of the production process, the programmer divides the required robot motion into tasks and subtasks. The following application of an injection molding machine illustrates this concept. A robot unloads a part from the injection molding machine, passes the part under a vision inspection system, and places good parts on an exit conveyer and bad parts in the recycling bin. Consider the following task breakdown. The major tasks include unloading the machine and submitting the part to the vision inspection system. Placing good parts on the conveyer and bad parts in the rework area would be subtasks. In this example the major tasks are in the main program; the subtasks are included as subroutines called from the main program. The following rules of thumb are used to separate tasks and subtasks:

- If alternative production actions are selected based on input from external sources, then make each alternative path a subtask in a separate subroutine. In the previous example the finished part is moved to either the exit conveyor or to the rework area, so both of these would be subtasks.
- If a particular robot motion or command sequence (e.g., turning on an alarm to call an operator) is required in two or more places in the program, then make the action a subtask, and put it in a subroutine.
- If the program requires two or more sequences of major tasks, then put each major task sequence in separate subroutines called from the main program. Assign subtasks for each major task to subroutines called from the major task subroutine.
- If part of the robot's action is likely to require frequent modification or updating, then make that portion of the program a subtask, and place it in a subroutine.

After the robot action is divided into tasks and subtasks, the structure of the total robot program is established.

Task Point Graph—Step 4

A *task point graph* (TPG), step 4 in the programming steps, is a visual tool to illustrate the program flow and arm motion required for a manufacturing problem. The TPG includes all translation point data (programmed locations), the motion variables (e.g., gripper conditions, delays, and tool point velocity) used at every point, and the logic used to make decisions during the program execution. A TPG for the robot and injection molding problem in the last section is drawn in Figure 8–13. Each program point on the TPG has several pieces of information: the name of the programmed point and a brief description, where the tooling will move next, tool speed, and other commands as required. Note that the subroutines are not shown.

System Variables—Step 5

The motion variables associated with translation points in Figure 8–13 include *velocity, tool center dimensions, Cartesian coordinate values, language functions,* and *commands*. These variables are usually included on the task point graph as part of the program development.

Write and Enter the Program—Step 6

With the translation points identified and named and the task point graph developed, the final step is to write the robot program code using the command structure and syntax for the controller. An example of this process for the task point graph in Figure 8–13 is provided later in this chapter.

Teach the Translation Points—Step 7

The translation points, step 5 in the programming steps, are taught or created using one of the five methods described earlier. Method 1 is an *on-line* programming

Figure 8–13 Task Point Graph.

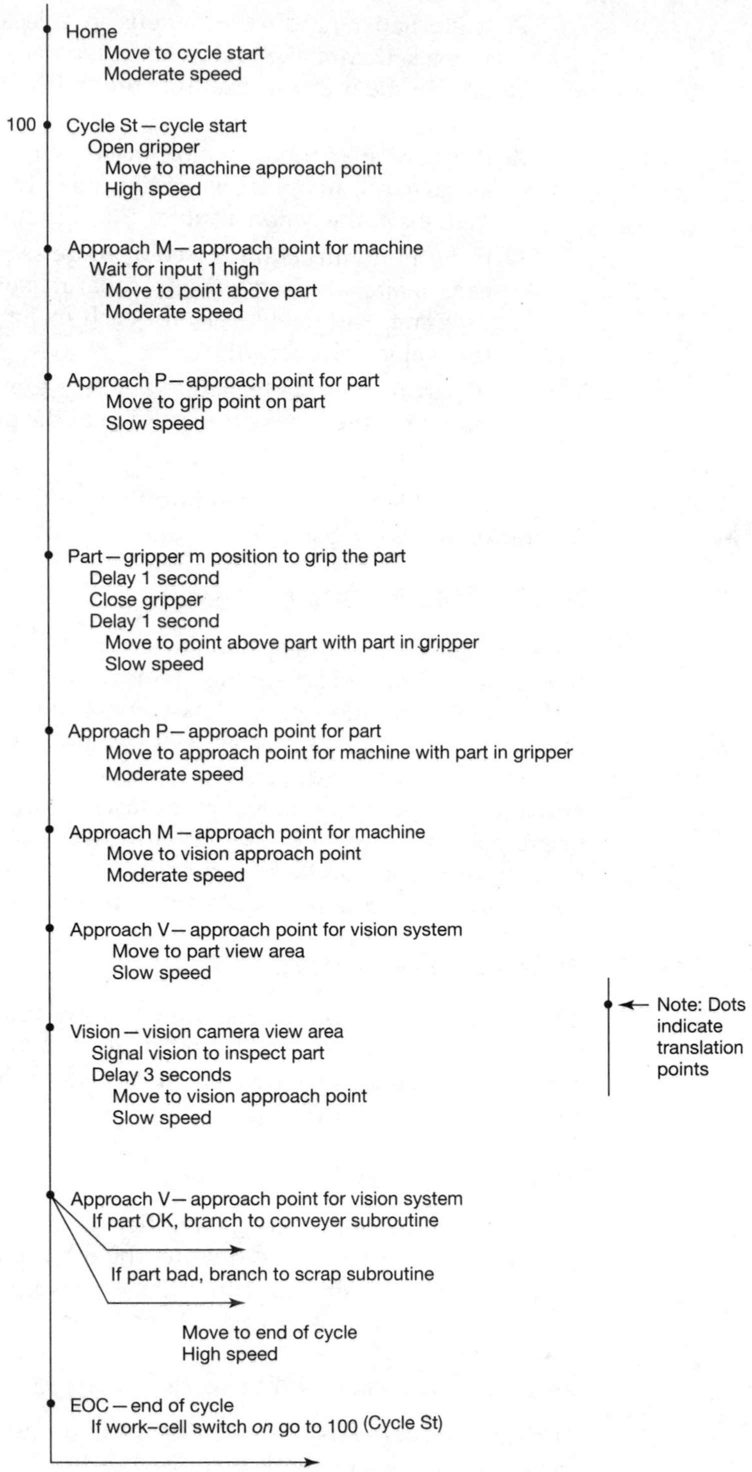

Home
  Move to cycle start
  Moderate speed

100 ● Cycle St — cycle start
    Open gripper
    Move to machine approach point
    High speed

Approach M — approach point for machine
  Wait for input 1 high
  Move to point above part
  Moderate speed

Approach P — approach point for part
  Move to grip point on part
  Slow speed

Part — gripper m position to grip the part
  Delay 1 second
  Close gripper
  Delay 1 second
  Move to point above part with part in gripper
  Slow speed

Approach P — approach point for part
  Move to approach point for machine with part in gripper
  Moderate speed

Approach M — approach point for machine
  Move to vision approach point
  Moderate speed

Approach V — approach point for vision system
  Move to part view area
  Slow speed

Vision — vision camera view area
  Signal vision to inspect part
  Delay 3 seconds
  Move to vision approach point
  Slow speed

Approach V — approach point for vision system
  If part OK, branch to conveyer subroutine

  If part bad, branch to scrap subroutine

    Move to end of cycle
    High speed

EOC — end of cycle
  If work–cell switch *on* go to 100 (Cycle St)

● ← Note: Dots indicate translation points

technique in which visual judgment establishes the fit and alignment between the tooling and parts. Although it requires a stop in production, it is the most frequently used method for establishing the translation points in a cell. Methods 2 and 5 are *off-line* programming techniques or are calculated inside a program to establish translation points. In each case, the translation point is established using coordinate data relative to the robot reference frame. Therefore, these techniques require some system to measure accurately the position of work-cell hardware and calibrate the robot arm for positional accuracy.

The creation of the translation points is the most costly part of the programming if production must be stopped; therefore, it is important to teach as few points as possible for a given program. Using *base* or *reference* points, the remaining position points in the program can be calculated using the process described in method 3.

Test and Debug the Program—Step 8

The final step in the development of a robot program is to test the program and correct any problems. Several suggestions make this last step safer for the robot and the other work-cell hardware.

- Keep the speed between translation points at a low value during the initial run so that any unexpected moves can be aborted using the emergency stop function.
- Use a single-step function. If the controller supports that mode of operation, it permits the program to be executed one move at a time with a pause at the end of each move.
- Do not have any parts in feeders or in the machines.
- Keep one hand on the emergency stop button for all moves.

On-line and Off-line Programming

The terms *on-line* and *off-line* programming define the location where the robot program is developed. For on-line programming, the production operation is stopped and the programmer puts the robot into the programming mode. Then the programmer teaches the robot the required position, motion, and control sequences. The positions or translation points are taught by visually moving the production tooling to the exact work-cell location and entering the position into the program with a teach button on the teach pendant. In this method, the exact location of the work-cell components and the native accuracy of the robot system are not critical for good operation. The automation will work as long as the robot's repeatability is good and the location of work-cell machines and parts does not change. The major disadvantage in on-line programming is the lost production time.

Some robot languages use variable names for the translation points and permit the control structure, moves, and program logic to be developed on a word processor on a microcomputer or engineering workstation. The program, complete except for the assigning of work-cell locations to translation point variables, is downloaded to the robot controller over a serial communication channel, and production is stopped only to teach translation points for all the location variables named in the program. A significant reduction in lost production time results from this modified on-line programming technique.

The term *off-line programming* means that *all* of the programming is performed away from the robot and the production area. All translation points are calculated by the robot controller from translation point coordinate values entered into the program in the off-line mode. For this technique to work, several conditions must exist:

- The accuracy of the robot and the controller must be excellent and the same for all the robots used in the production area.
- The exact distance from the robot reference frame to all the production equipment and machine fixtures must be known.
- Work-cell simulation software or a robust robot programming language must be available to program the cell off-line.

The first condition rarely occurs in standard industrial robots. For example, when a robot is programmed off-line to go to a translation point in the work envelope, it misses the point because of the mechanical tolerances in the arm linkages and feedback mechanisms. Effective off-line programming is possible only with highly accurate placement of work-cell hardware and calibration of the robot arm. This type of robot and work cell calibration is possible with integrated systems such as WORKSPACE and ROBOTRAK from Robotic Workplace Technologies Inc. ROBOTRAK provides the calibration of the robot arm and work-cell hardware used in the WORKSPACE program. WORKSPACE is simulation software that builds a manufacturing automation simulation and integrates the work-cell equipment location data and robot arm signature captured with ROBOTRAK. The robot signature data include the zero position of each joint, the length of each link, the distance offset at each link, and compliance at each joint. These types of systems measure static position and motion paths to an accuracy of 0.2 millimeters (0.008 inches) in three dimensions. As a result, a functional off-line program is developed with WORKSPACE by integrating the location data for all work-cell devices and compensating for the variations present in the mechanical linkages in the robot.

8-9 PROGRAMMING A SERVO ROBOT

The material provided in this chapter offered some guidelines for programming servo robots. The robot languages from all of the major suppliers are different, so it is not possible to learn one standard that would apply to all models and types. However, if you know how to develop a program using the structure and syntax from one vendor, then learning to program a different robot is not so difficult. With that in mind, some of the most often used commands and syntax for the Yaskawa Performer MK3 robot are provided next. As the commands are introduced, the syntax used to program the task point graph in Figure 8–13 is also discussed.

Command Modes

The software programming language, called ACL, has two command modes, *DIRECT* and *EDIT*. In the DIRECT mode, commands are executed immediately after they are entered from the keyboard, whereas in the EDIT mode, the commands are not executed until the entire program is run.

Two modes of program operation are supported, *RUN* and *STEP*. In the RUN mode, programs are executed starting at the first program line and continuing until the program is terminated. In the STEP mode, the user can execute the program one line at a time.

Coordinate Systems

The robot controller supports three coordinate systems, *JOINT* (resolver), *XYZ* (Cartesian), and *TOOL*. Figure 8–14a illustrates the joint coordinates, Figure 8–14b

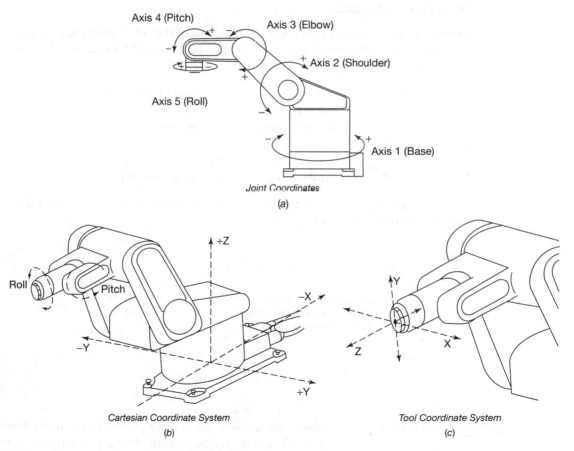

Figure 8–14 Joint, XYZ, and Tool Coordinate Systems.

shows the XYZ coordinates, and Figure 8–14c displays the tool coordinate system. These three modes are used to move the robot in the work envelope and are selected using a key on the robot teach pendant.

Data Types

The ACL language has two types of variables: *global* that can be referenced in all programs, and *private* that can be used only in the program edited at the time the variable is defined. The following examples illustrate how these data types are defined

DEFINE *X*	Defines the private variable *X*
GLOBAL *Top*	Defines the global variable *Top*
DIM A[20]	Defines an array named A with 20 private variables
DIMG App[5]	Defines an array named App with 5 global variables

Axis Control Commands

A number of commands provide for the control of the robot axes, joints, and gripper. Some of the commands are used together, such as the **MOVE** and **SPEED** commands; therefore, that is how they will be described.

MOVE W	Moves the axes to the taught point W using the current joint speed
SPEED *var*	Sets the axes speed to the value of the variable *var* where *var* is a value (percent of full speed) from 1 to 100 with default at 50. After the speed is set it remains at that value until changed in a later program step.

The following example illustrates how these commands would be used for the task point graph in Figure 8–13.

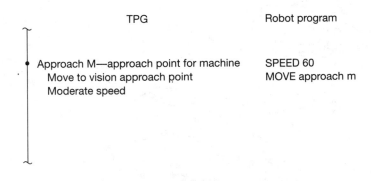

TPG   Robot program

Approach M—approach point for machine SPEED 60
Move to vision approach point MOVE approach m
Moderate speed

MOVEL Y	Moves the robot's tool center point (TCP) to the taught point Y along a linear path using the current linear speed.

SPEEDL *var*   Sets the linear speed to value of *var* where *var* is a value expressed in mm per second up to 500 mm/sec.

The following example illustrates how these commands would be used for the task point graph in Figure 8–13.

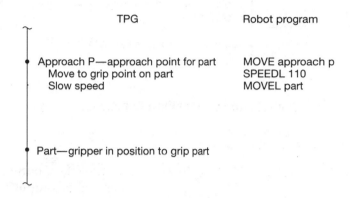

TPG	Robot program
Approach P—approach point for part	MOVE approach p
Move to grip point on part	SPEEDL 110
Slow speed	MOVEL part
Part—gripper in position to grip part	

Five other commands are frequently used as well.

OPEN	Opens the gripper
CLOSE	Closes the gripper
DELAY *var*	Suspends program execution for the time specified by the *var* value with *var* defined in hundredths of a second.
HOME	Searches for microswitch home positions for all robot axes
SET *var1* = *var2*	Assigns the value of *var2* to *var1*.

The following example illustrates how these commands would be used for the task point graph in Figure 8–13.

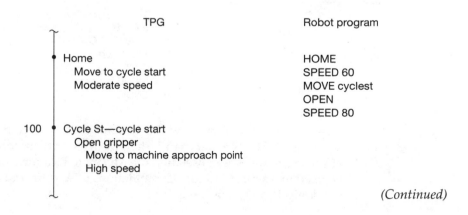

TPG	Robot program
Home	HOME
Move to cycle start	SPEED 60
Moderate speed	MOVE cyclest
	OPEN
	SPEED 80
100 Cycle St—cycle start	
Open gripper	
Move to machine approach point	
High speed	

(Continued)

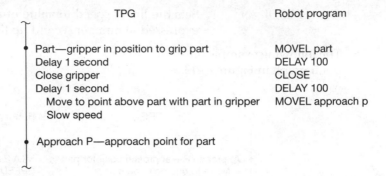

TPG	Robot program
Part—gripper in position to grip part	MOVEL part
Delay 1 second	DELAY 100
Close gripper	CLOSE
Delay 1 second	DELAY 100
Move to point above part with part in gripper	MOVEL approach p
Slow speed	
Approach P—approach point for part	

I/O Control and Program Flow Commands

The following commands have to do with the 16 discrete input and output ports on the robot controller.

SET OUT $[n] = \{0/1\}$ Sets the state of output n to either on $\{1\}$ or off $\{0\}$ where n is an output between 1 and 16.

IF IN $[n] = \{0/1\}$ A conditional **IF** command that checks to see if the variable IN[n] is equal to the value in the brackets $\{0/1\}$. Any of the standard operation signs can be used in place of the = sign. Operators include greater than (>), less than (<), equals (=), less than or equal to (<=), greater than or equal to (>=), and not equal (<>). If the condition is met and the result true, then the next sequential program line is executed. If it is false, then the execution jumps to after the **ELSE** command if it is used or to after the **ENDIF** command if no **ELSE** is present. The **ANDIF** and **ORIF** commands allow for additional **AND** and **OR** condition's to be considered.

The following example illustrates how these commands would be used in general and for the task point graph in Figure 8–13.

```
If A = B
    AND IF C > 1
MOVE top
ELSE
MOVE bottom
ENDIF
```

This code demonstrates how the **IF** command is used. If the values A and B are equal AND if C is greater than 1, then the **MOVE** to location *top* is executed . If any of those equalities are false, then the **MOVE** to location *bottom* occurs. **ORIF** would be used in a similar fashion. Combinations of **ANDIF** and **ORIF** can be used to get any logical combination required.

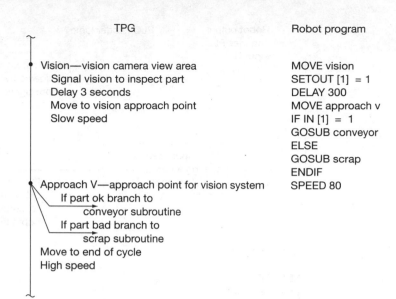

TPG	Robot program
Vision—vision camera view area	MOVE vision
Signal vision to inspect part	SETOUT [1] = 1
Delay 3 seconds	DELAY 300
Move to vision approach point	MOVE approach v
Slow speed	IF IN [1] = 1
	GOSUB conveyor
	ELSE
	GOSUB scrap
	ENDIF
Approach V—approach point for vision system	SPEED 80
If part ok branch to	
conveyor subroutine	
If part bad branch to	
scrap subroutine	
Move to end of cycle	
High speed	

The **SET OUT [1] = 1** command tells the vision system to check the part. The vision system puts a signal on robot input number one, and the **IF** command uses the IN [1] = condition to know if the part is good. (IN [1] = 1) or bad (IN [1] = 0) so the correct subroutine is run. The following command is used with the **SET OUT** command to create a handshaking exchange with a PLC.

WAIT *var1* oper *var2*   Suspends program execution until the condition *oper* is satisfied. The two variables *var1* and *var2* can be any allowed variable or system parameter, and the *oper* can be any of the standard operators defined earlier. The value of *var1* in the handshaking routine will be the **IN** [*n*] command. The interface between the robot and the PLC are shown in Figure 8–15.

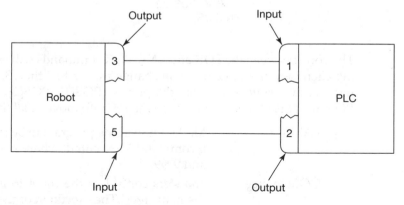

Figure 8–15   Robot and PLC Interface.

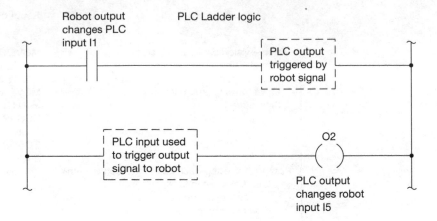

Robot output changes PLC input I1

PLC Ladder logic

PLC output triggered by robot signal

PLC input used to trigger output signal to robot

O2

PLC output changes robot input I5

Robot program

SET OUT [3] = 1
WAIT IN [5] = 1

The robot output bit 3 is wired to the PLC input bit 1, and the PLC output bit 2 is wired to the robot input bit 5. The following commands create FOR loops, permit program jumps, and branch to subprograms.

FOR *var1* = *var2* TO *var3*
Executes all of the program lines between the **FOR** command and **ENDFOR** command as many times as the *var2* and *var3* values indicate. The *var1* must always be a variable, but *var2* and *var3* can be either a variable or a constant.

Robot program

FOR L = M to N   Assume M = 1 and N = 5
  MOVE POS [L]
ENDFOR

The code between the **FOR** and **END FOR** commands will execute five times. During each loop the value of *L* will change from 1 to 2 then 3, 4, and 5. The POS [*L*] term is an array of five taught points (POS[1], POS[2], POS[3], POS[4], and POS[5]). This FOR loop will cause the robot to move to all five array positions.

LABEL *n*
Marks a line in the program to be executed after a **GOTO** *n* command is executed, where *n* is any value between 0 and 9999.

GOSUB *prog*
Transfers control of the robot to another program with the name *prog*. The execution of the main program is suspended until the subroutine is completed.

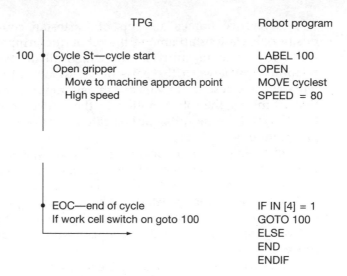

	TPG	Robot program
100	Cycle St—cycle start	LABEL 100
	Open gripper	OPEN
	Move to machine approach point	MOVE cyclest
	High speed	SPEED = 80
	EOC—end of cycle	IF IN [4] = 1
	If work cell switch on goto 100	GOTO 100
		ELSE
		END
		ENDIF

The work-cell switch is connected to robot input 4. While there are more language commands available to the programmer, the commands covered in this section are the ones most frequently used.

8-10 SUMMARY

The computers used in robot systems are responsible for four functions: machine manipulation, sensing, logical decision making, and data processing. The programming languages currently used to execute these four functions have no common format or standard form. At a higher level, work-cell control software manages the flow of information between cell devices and the enterprise. Work-cell software falls into one of three categories: in-house developed systems, application enablers, and adoption of an open system interconnect. One level below the cell control software, programmable logic controllers or robot controllers control sequential cell activity.

There are as many robot programming languages as there are robot manufacturers. The robot languages were developed by means of three techniques. The first method starts with the design of a pure manipulator control language and adds the other high-level language features as demand requires. The second adopts an existing high-level language and augments it with the manipulator commands required for robot operation. The third technique includes the design of a high-level language for robot control and support of other external data processing functions.

The many current languages can be classified into four groups based on the level at which the user must interact with the system during the programming process. The four groups are joint control languages, primitive motion languages, structured programming languages, and task-oriented languages. At present, the most frequently used languages are in the structured programming category.

Reference frames are sets of Cartesian coordinates that describe the positional relationship among the robot, the gripper, the tool center point, the workpiece, and the universe in which all of these exist. The dimensional relationships between the parts of the robot and the production parts in the work cell are described by means of mathematical equations called transformations. Linking the robot motion to the reference frames using mathematical transformations permits sophisticated control of robot motion and off-line programming.

In on-line programming, the production system must stop to allow the robot programmer access to the production robot for program generation or modification. Lost production time is reduced if some of the program development is performed off-line and downloaded to the robot before teaching the position points. True off-line programming is performed away from the production area on work-cell simulation software. Off-line programming requires a work-cell hardware and robot calibration process.

Five basic methods are available to define translation points for servo robots: (1) teaching of translations by robot positioning, (2) teaching of translations by coordinate input, (3) creation of new translations by algebraic expressions, (4) creation of a new translation by use of an external interrupt, and (5) downloading new translations through the RS-232C port. Translation points for nonservo robots are limited to the fixed stops provided by the system. Nine steps are used to develop a robot program: (1) establish a basic program structure, (2) analyze the manufacturing process in which the robot will work, (3) divide the robot action into tasks and subtasks, (4) draw a task point graph that describes the desired motion, (5) identify the translation points on the task point graph, (6) assign values for all system variables to control the motion, (7) write and enter the command statements, (8) create or teach the translation points, and (9) test and debug the program.

Programming commands and syntax for the Yaskawa Performer MK3 robot are provided. The command modes, EDIT and DIRECT, are covered along with data types and coordinate systems used in the programming process. Fifteen of the most often used robot commands are described with example code for an application included.

QUESTIONS

1. Describe the four system functions performed by the computer in high-technology robots.
2. Describe the three categories of work-cell controller software.
3. Describe the differences in programming robots controlled by PLCs and those with special-purpose controllers and languages.
4. What are the four levels of robot languages?
5. What are the advantages and disadvantages of languages at each level?
6. How do Grafcet and ladder logic programming for PLCs differ?

7. What is the function of the CYCLE START and END OF CYCLE programmed points in a robot program? Where is the optimum position for the CYCLE START point?

8. What is a TPG and what information is included in the graph?

9. Describe the two parts of a robot program for servo and nonservo robots.

10. Describe reference frames and explain how they are used in the DARL language for the Seiko robot.

11. Describe the difference between on-line and off-line programming.

12. Describe the five methods used to program a translation point for a robot program. What program development situation or work-cell condition would dictate the use of each method? Which method is used most often on robots in operation now?

13. How do the WORLD, TOOL, JOINT, and FREE operational modes differ? Identify the mode(s) that would be used for each of the following situations:

    (a) Teaching move points between fixtures in the work cell

    (b) Rotation of the tooling from one side of the work cell to the other with no change in the elevation or radial extension of the arm

    (c) Movement along the row of a pallet when the pallet coordinates are aligned with the robot coordinates

    (d) Movement to insert a part in a fixture plate when the fixture coordinates are not aligned with the robot coordinates

    (e) Separation of the robot tooling from the fixture after a system fault caused the controller to shut down in the middle of an assembly operation

14. Describe the steps used in developing a robot program.

15. How are program tasks and subtasks identified? How is each handled in the robot program?

PROBLEMS

1. Modify the task point graph in Figure 8–13 to unload two production machines instead of one.

2. Develop a ladder logic diagram for the nonservo robot motion A in Figure 8–9. Assume the system is interfaced to a PLC and assign necessary input and output module identification.

3. Develop a complete robot program for the task point graph (TPG) in Figure 8–13 using the programming commands for the Yaskawa robot.

4. In the TPG in Figure 8–13, the robot sends a signal to the vision system that indicates that the part is in position for the vision system to view it. The 3-second delay is used to give the vision system time to capture the image. Modify the TPG and robot program to allow full handshaking between the robot and vision system so that 3-second delay is not necessary. Assume vision system output bit 1 is connected to robot input bit 6.

1. Apply the cell robot programming concepts presented in this chapter to one or more of the automation work-cell designs listed below. Complete the task point graph with all robot and work-cell parameters specified. Use the design data from previous chapters as necessary. The design component of the case studies varied in complexity and design detail, so it may be necessary to make some assumptions for the programming. Document the design and support all assumptions.

   (a) The die casting production cell started in Case Study Problem 3 in Chapter 2

   (b) The kitchenware bowl production cell started in Case Study Problem 5 in Chapter 2

   (c) The West-Electric slug production cell design started in Case Study Problem 8 in Chapter 2

   (d) The West-Electric slug lubrication cell design started in Case Study Problem 9 in Chapter 2

2. Apply the cell PLC programming concepts presented in this chapter to one or more of the automation work-cell designs listed below. Complete the PLC ladder logic program using the PLC software of your choice. Identify all rungs and rung elements in the program documentation. Use the design data from previous chapters as necessary. The design component of the case studies varied in complexity and design detail, so it may be necessary to make some assumptions for the programming. Document the design and support all assumptions.

   (a) The injection molding production cell started in Case Study Problem 3 in Chapter 2

   (b) The kitchenware bowl production cell started in Case Study Problem 5 in Chapter 2

   (c) The West-Electric slug production cell design started in Case Study Problem 8 in Chapter 2

   (d) The West-Electric slug lubrication cell design started in Case Study Problem 9 in Chapter 2

3. Read the West-Electric case study (Section 12-12 in Chapter 12) where cell programming is discussed. Complete the questions and problems at the end of Chapter 12 that relate to cell programming.

Justification and Applications of Work Cells

CHAPTER GOALS AND OBJECTIVES

This chapter has two primary goals. First, the concept of justification of a manufacturing automation is introduced and a fully operational spreadsheet is introduced in the Case Study projects to demonstrate return on investment for an automation project. The second primary goal is to demonstrate how robots are used in six major application areas, using industrial case studies. At the completion of this chapter you should be able to:

- Describe how automation projects are justified using financial planning concepts including *payback period, return on investment (ROI),* and *discounted cash flow.*
- Perform the financial analysis for an automation project and calculate the payback period, ROI, and discounted cash flow.
- Describe a typical robot automation application in the following six areas: welding, material handling, machine tending, paint spraying, assembly, and the process area.

9-1 INTRODUCTION

Manufacturing is a collection of interrelated activities that include product design and documentation, material selection, planning, production, quality assurance, management, and the marketing of goods. The fundamental goal of manufacturing is to use these activities to convert raw materials into finished goods efficiently, on

When you observe a robot performing a task like welding or spray painting, it is easy to forget that it is just responding to the programmed commands covered in the last chapter. The robot cannot weld or spray paint unless a programmer who does know how to weld or spray paint programs it. Therefore, an individual who seeks a career in the robotics application development area must know more than just robots. For example, to design an automated robot welding cell, you must have some proficiency in welding in addition to your mechanical, electrical, and robotics knowledge.

The primary emphasis of this career spotlight is the importance of a broad base of technical, business, and interpersonal skills that extends beyond robotics. Having some background in economics, finance, and accounting permits you to feel more confident performing the justification analysis on the cell. In most organizations, automation projects must have a cost and payback analysis performed in order to get the budget allocation to fund the project. Some education in business principles will give you confidence in that presentation to upper management for project funding.

The same argument can be made for some knowledge in the manufacturing process areas, such as welding, spraying and coating, material handling and machining, or other types of material processing. If you are called on to automate one of these areas, then the process knowledge will be valuable. When automation projects fail to meet productivity, quality, or up-time design goals, the fault often is not in the designer's understanding of robots, but in his or her lack of the process knowledge. Welding is a perfect example. A manual MIG welder must have the correct voltage and current settings on the weld unit, hold the weld tool at the correct angle and distance above the joint, feed the weld wire at the correct rate, and move the welding tool at the correct speed. In a robotic welding application, all of these parameters must be part of the robot program. In every process area, you can't program what you don't know.

So how does industry address these issues in application development? There are a number of solutions, such as:

- A process expert or specialist is brought in to help the design team.
- The design of the special process work cell is farmed out to a design house that specializes in that area.
- Design team members are sent to special industry schools to get the necessary knowledge.
- Companies only hire automation designers that have a broad material processing knowledge.
- Organizations just expect the design team to learn what is necessary during the design to make the project successful.

Regardless of the company approach to the problem, your robotic and manufacturing automation career will be enhanced if you acquire some business and material processing knowledge during your education.

schedule, and on a profitable basis. Every action taken in manufacturing must be justified based on its effect on this fundamental goal. Peter Drucker, a noted educator, writer, and management expert, wrote in *The New Realities*:

> We have known for a long time that there is no one right way to analyze a proposed capital investment. To understand it we need at least six pieces of information:
>
> - Expected rate of return
> - Payout and the investment's expected life
> - Discounted present value of all returns through the productive life of the investment

- Risk in not making the investment or deferring it
- Cost and risk in case of failure
- Opportunity cost that is the return from alternative investments

Every accounting student is taught these concepts. But before the advent of [computer] data-processing capacity, the actual analyses would have taken man-years of clerical toil to complete. Now anyone with a spreadsheet [software application] should be able to do them in a few hours. The availability of information thus transforms the capital investment analysis from opinion to diagnosis, that is, into the rational weighing of alternative assumptions. Information transforms the capital-investment decision from an opportunistic, financial decision, governed by the numbers, into a business decision based on the probability of alternative strategic assumptions. As a result, the decision both presupposes a business strategy and challenges that strategy and its assumptions. What was once a budget exercise becomes an analysis of policy.

Manufacturers face enormous internal and external challenges because of the increased use of technology in the production and marketing of goods and services. As a result, manufacturing strategies such as order-winning and order-qualifying criteria, described in Chapter 1, are needed to link manufacturing and management. A focus on improving the production standards, such as setup time, inventory, quality, machine uptime, and manufacturing space ratio, makes manufacturing the strategic difference in the quest for greater market share. As Drucker noted, a business strategy is assumed, and the justification of the capital invested as a result of this strategy has changed significantly. Justification must now go beyond the tangible numbers.

9-2 CAPITAL EQUIPMENT JUSTIFICATION

The principal considerations for the justification for capital expenditures in manufacturing can be expressed in the following equations:

$$\text{Payback period} = \frac{\text{total investment}}{\text{total yearly savings}}$$

$$= \frac{\text{total investment}}{\text{savings} - (\text{savings} \times \text{TR}) + (\text{DP} \times \text{TR})}$$

The total investment includes all costs associated with the automation project. Savings includes yearly savings that result from the automation, TR is the tax rate, and DP is the yearly depreciation for the capital equipment. The following example illustrates the *payback method.*

Example 9–1
Calculate the payback period if the total investment is $250,000, savings per year is $105,000, the tax rate is 33 percent, and the depreciation on equipment is $28,000.

Solution

$$\text{Payback period} = \frac{\$250,000}{\$105,000 - (\$105,000 \times 0.33) + (\$28,000 \times 0.33)}$$

$$= \frac{\$250,000}{\$79,590} = 3.14 \text{ years}$$

Return on Investment Method

Return on investment (ROI) is the percentage return on the automation investment. The ROI is the ratio of savings to investment, in other words, the reciprocal of the payback period:

$$\text{ROI} = \frac{1}{\text{payback period}}$$

The terms in the equation are the same as those defined in the payback calculations. ROI is calculated as follows.

Example 9–2

Calculate the ROI if the payback period is 3.14 years.

Solution

$$\text{ROI} = \frac{1}{3.14}$$

$$= .318, \text{ or } 31.8 \text{ percent}$$

Cash Flow Method

Cash flow analysis is a variation of the payback technique. Stated as an equation,

$$\text{Cash flow} = \Sigma \text{ (positive and negative cash values over time)}$$

As noted in the statement of Drucker, spreadsheet software on microcomputers gives the designer of the work cell the power to make investment decisions and do what-if analysis on the alternatives for automation investments. The cash flow method is an ideal application for spreadsheets. The cash flow spreadsheet used in the West-Electric case in Chapter 12 illustrates how both tangible and intangible factors are used to make automation investment decisions.

Time Value of Money

The time value of money describes the change in the value of an investment or a savings in the future due to monetary inflation and interest. For example, as a result of inflation, $1000 invested today has a greater value than $1000 invested 2 years hence. Similarly, a savings of $1000 expected 3 years after automation is implemented is worth less than $1000 in present dollars because of the interest the money could earn in the 3 years. The benefits of automation must always be justified by future events that are affected by the time value of money. Therefore, the economic analysis methods discussed earlier must be discounted for the future events. Often the equations are rewritten as follows:

$$\text{Payback period} = \frac{\text{net present value of } \Sigma \text{ (investments over time)}}{\text{net present value of } \Sigma \text{ (savings over time)}}$$

$$\text{Discounted cash flow} = \text{net present value of } \Sigma \text{ (positive and negative cash values over time)}$$

The term *net present value* means that all future dollar investments and savings are discounted, or converted, to their *present worth*. This process requires an estimate of inflation (average over years) and market interest rates (average or yearly values) for future periods in the justification. Return on investment equations that include discounted cash flow analysis are provided in many financial analysis texts and will not be discussed in detail here. Consult the justification spreadsheet in the West-Electric case; it provides a discounted cash flow analysis that includes the impact of the discount or interest rate on money but does not include an inflation factor.

Justifying Robotics Applications

The justification principles described in this chapter for capital expenditures in manufacturing apply equally well to robot-driven automation. The cost of the automation investment must be justified with future savings and revenue. There are, however, several distinct differences when robots are involved. The introduction of robots implies the displacement of human operators; thus, a significant saving occurs in direct labor cost. Robots usually increase productivity and output because of uniform production rates and lower production cycle time. Robots enhance product quality and uniformity and support flexible and agile manufacturing. Finally, the reprogrammability of the robot gives the investment in robotics a longer lifetime, because the robot is reusable when production on the current product is stopped. These factors take on even greater significance when a breakdown of the investment dollars in robotic work cells is analyzed. The robotics application cost graph in Figure 9–1 shows the percentage of investment dollars in accessories, basic robot system, and installation costs. In each case the basic robot cost is the largest single investment. Therefore, the largest part of the automation investment is in a machine that has a production lifetime longer than other hardware in the work cell.

Justification Spreadsheet

Spreadsheet software is often used to collect all the justification data and calculate the return on investment or the break-even point. The spreadsheet also permits the designers to play what-if games where the economic consequences of design ideas and work-cell configurations are evaluated. A typical spreadsheet used for the justification of automation is included in Appendix C in three formats. The first printout in Appendix C is the spreadsheet before any work-cell

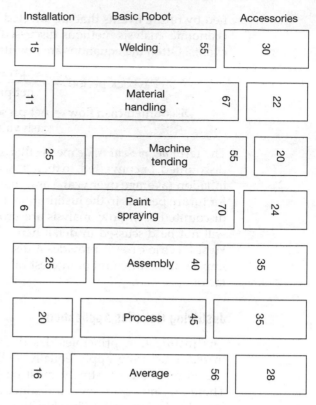

Figure 9–1 Percentage Cost Breakdowns for Robotic Cells.

Installation	Basic Robot	Accessories
15	Welding 55	30
11	Material handling 67	22
25	Machine tending 55	20
6	Paint spraying 70	24
25	Assembly 40	35
20	Process 45	35
16	Average 56	28

data have been entered. The second printout shows the formulas used in every cell, and the last item is a description of the data required for the cells. Some of the data definitions use values from the West-Electric case study in Chapter 12 to make the descriptions more meaningful. The spreadsheet template spans only 3 years, so more columns should be added if a longer payback time is needed.

9-3 AUTOMATION APPLICATIONS

Automation applications in industry vary widely; however, most robot applications would fall into one of the six categories listed in Figure 9–1. In this section, each of the applications areas are described.

Material Handling

Material handling generally involves the movement of material through manufacturing. The automation technologies used for this work include conveyers, automatic guided vehicles (AGVs), and servo and nonservo robots. The product size,

APPLICATION NOTE
Material Handling of Electrical Relays

Feme improved product quality and achieved a labor savings of two and one-half operators per shift by automating its electrical relay calibration and testing process. The system has a payback period of 1 year.

The two work cells (Figure 9–2), designed by Feme's Industrial Automation Division, consist of an AdeptOne robot and three identical calibration and test stations. The AdeptOne feeds relays in batches of 500 or more into the test stations. There are 120 different relay models differing by coil voltage, contact configuration, and number of poles. The cells run two or three shifts per day, 5 or 6 days per week, depending on market demand.

System Configuration

In each cell, the AdeptOne robot stands in the center surrounded by three test and calibration stations, and a conveyer. The conveyer delivers pallets of untested relays to the robot and removes the pallets after testing and calibration are complete. Pallets for holding rejected relays are on a pneumatic slide next to the conveyer. The robot is equipped with a two-fingered gripper.

Work-cell Operation

An operator loads the pallets containing thirty-six relays onto the conveyer. When the pallet arrives at the pickup position, the robot unloads a single

Figure 9–2   Electrical Relay Calibration Cell. (Courtesy of Adept Technology, Inc.)

relay and transfers it to a test station. The robot repeats the operation twice more to fill all three stations. The relays are picked up in a vertical position. During the transfer the relays are turned 90 degrees to a horizontal position for feeding into the test stations.

The length of the test and calibration cycle varies between 5 and 7 seconds, depending on the performance of each relay. At the end of the cycle, the test station communicates the results to the robot. If the relay is acceptable, the robot removes the relay from the test station and returns it to the same pallet on which it entered the cell. If it is rejected, the test station also communicates which of six classes of failure the relay experiences, and the robot then removes the relay from the test station and places it on the "reject" pallet classified by the type of failure.

Changeover between relay models simply requires the operator to select a different program at the terminal. The gripper is designed to accommodate all 120 models.

By automating its relay test and calibration process, Feme has improved its product quality and reduced its labor costs.

(Courtesy of Adept Technology, Inc.)

APPLICATION NOTE
Cookie Packaging

Kambly, a manufacturer of cookies located in Trubschachen, Switzerland, sought an effective solution to the rising cost of cookie assembly. Dedicated automation was ruled out because a high level of flexibility and precision was required to assemble a wide range of delicate products (Figure 9–3).

A robotic solution proposed by Adept System Integrator A.R.T. (formerly Adec Robot AG) provided Kambly with the answers they sought. The flexible automation system installed by A.R.T. provided Kambly with the desired results—improving quality, lowering costs, and increasing production line flexibility.

System Configuration

Kambly's automated packaging system is composed of four AdeptOne robots working with AdeptVision AGS-GV. The robots line each side of a conveyer that transports cookies from an oven. Each robot is equipped with a vacuum gripper that can handle delicately either seven or nine cookies, depending on the batch in production.

The robot work cell features an integrated vision system with four cameras viewing the conveyer from below. The cameras are not only responsible for guiding the robots, but also perform a variety of quality control functions.

Work-cell Operation

Cookies arrive at the robotic packaging cell randomly distributed along a 1-meter wide conveyer belt, traveling at 5 meters per second. The first two cameras locate the cookie's position and guide two upstream robots to track the cookies on the line. A seven- or nine-cup suction gripper acquires the cookies from the moving line, then packages the cookies in a blister pack on an adjacent pallet conveyer. The second pair of AdeptOne robots, also guided by two cameras, repeat the sequence to complete the package. Two cycles are necessary to fill the fourteen- or eighteen-cookie blister packs.

But the vision system does more than just instruct the robot how to pick up and orient a cookie. It also inspects the size of the product. If the cookie is not within the required limits, the cookie is left to fall off the end of the conveyer into a reject bin. In normal operation, Kambly packages four different types of cookies while operating the Adept robot system two shifts a day, 5 days a week.

Changeover of the production line takes between 15 and 30 minutes, providing Kambly with the flexibility to produce a wide variety of products. A.R.T. designed the end-of-arm tooling so that changeover is accomplished by replacing just four screws.

Figure 9–3 AdeptOne Robots with AdeptVision AGS-GV Package Cookies.
(Courtesy of Adept Technology, Inc.)

The requirements of Kambly are not unique to their industry because most companies are seeking to maximize production flexibility, improve quality, and lower production costs. If an integrated system featuring robots, vision, and flexible feeding enabled Kambly to achieve these production results, perhaps the Adept approach should be part of your "recipe" for success.

(Courtesy of Adept Technology, Inc.)

production layout and environment, and flexibility needed in the application determine if a robot is used and what type is selected. The automotive, foundry, light and heavy manufacturing, and electronics industries use large numbers of robots for material handling.

Machine Tending

In most machine tending applications, the robot replaces the human operator. The operation usually includes the loading of raw material into a production machine and the removal of the finished part when the production process is complete. In other cases, such as die casting and plastic injection molding, the robot just unloads finished parts from the machine after the molding or casting process is complete. In most applications, servo robots are used because adequate payload, a large work envelope, and dexterity are not available in pneumatic stop-to-stop machines. The robot in Figure 9–4 is unloading finished plastic parts from a plastic injection molding machine. A robot with a good horizontal reach easily handles this hot and uncomfortable task. The robot in Figure 4–22, which loads and unloads a turning center, is a good example of machine tending. Note that the robot is loading the machine from the rear, so that manual operation from the front is possible if the robot system is not operational. The West-Electric case study in Chapter 12 provides another example of machine tending with robot technology.

Figure 9–4   Machine Tending Robot.

Assembly

Assembly applications occur in all five manufacturing systems: project, job shop, repetitive, line, and continuous (Figure 3–5). The level of complexity varies from highly intricate (the assembly of high-tolerance mechanical parts) to simple (the construction of walls in residential construction). In addition, the volume varies from low to very high. As a result, the window of opportunity for robotic assembly falls in the middle. Servo robots are used primarily in assembly tasks for medium-volume repetitive and line-type production (Figure 2–16). Stop-to-stop robots are integrated into fixed automation machines (Figure 1–7) and systems and are used in high-volume line and continuous production areas. Robotic assembly is heavily integrated into the production of electronic parts and products, but used only lightly in the automotive and the light manufacturing sectors.

APPLICATION NOTE
Mechanical Assembly of Fluorescent Light Fixtures

Zumtobel, located in Dornbirn, Austria, has integrated an AdeptOne robot cell into its automated line, producing fluorescent lighting units at a rate of one every 30 seconds (Figure 9–5).

A robotic solution was chosen because of its unique ability to manage a high production rate and mix of products. Adaptable to forty variations of product, the Adept robot is responsible for assembling an average of twelve components into each lighting unit.

System Configuration
The AdeptOne robot cell, designed by A.R.T., is completely integrated with Zumtobel's automated line for manufacturing fluorescent lighting units.

A continuous conveyer passes through the center of the work cell delivering fluorescent light housings to the robot for assembly. Once the mechanical components have been assembled, the lighting units are transported to dedicated stations that add the electrical elements.

Six bowl feeders located opposite the robot orient and position the six different components that the robot has to assemble.

The AdeptOne robot is programmed to assemble forty different variations of fluorescent lighting tubes. The Adept MC Controller, linked to a VAX computer, is responsible for downloading to the Adept cell which product to assemble. At the same time, the MC Controller sends progress reports back to the VAX computer.

Work-cell Operation
As each unit arrives at the robot assembly station, it is grasped and centered by an indexing pin.

The movements of the indexing pin are controlled by the MC Controller, acting as an external linear axis. To minimize the length of each robot movement, the pin moves the lighting unit back and forth during the assembly cycle.

As assembly needs change with future lighting units, various grippers will be required. Therefore, system integrator A.R.T. provided the cell with a gripper change facility.

By installing flexible automation, Zumtobel has been able to automate completely the assembly of its lighting units and reduce required manpower. Forty different product designs can be assembled at a rate of one every 30 seconds.

(Courtesy of Adept Technology, Inc.)

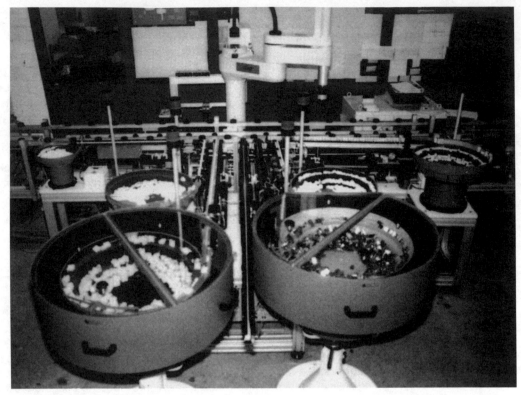

Figure 9–5 Automated Assembly of Fluorescent Light Fixtures.
(Courtesy of Adept Technology, Inc.)

APPLICATION NOTE
Mechanical Assembly of Dishwasher and Vacuum Cleaner Components

The Miele factory in Bielefeld, Germany, installed a flexible automation solution using a single AdeptThree robot (Figure 9–6) to assemble two different products: the steam exhaust unit in a dishwasher and the device in a vacuum cleaner that signals when the dust-collecting bag is full. Installation of the automated system has resulted in a labor savings of two and one-half operators.

The system, designed by Proline GmbH, is capable of running almost completely unattended, three shifts per day, and has a production rate of one finished product per 30 seconds.

System Configuration
The AdeptThree robot stands in the center of the cell, surrounded by part feeding equipment and assembly workstations. The dishwasher steam exhaust assembly consists of seven parts. Specially designed molded plastic trays, containing four of each of the seven parts, are stacked in a vertical magazine. Each tray is removed automatically from the magazine and transferred by a conveyer system to the robot pickup position. Completed assemblies exit the workstation by a conveyer system parallel to the first. Two assembly fixtures are located on either side of the robot's base.

Figure 9–6 Assembly of Dishwasher and Vacuum Cleaner Components.
(Courtesy of Adept Technology, Inc.)

The AdeptThree robot is equipped with a triangular end-effector with suction cups at two corners, an expanding three-finger gripper on the third corner, and in the center, a parallel gripper.

Work-cell Operation

The assembly process begins when the robot picks up a wheel, a mounting plate, and a motor from the plastic tray, and assembles them on one of the two fixtures.

Next, the robot picks up three more components: a rubber seal, a small wheel, and a housing. It assembles the seal and the housing on the other fixture. It then moves to a screwdriving station just behind its base and loads first the small wheel, followed by the housing assembly, and the motor, to the screwdriving fixture. The robot then loads a temperature regulating unit onto the assembly and the screwdriving operation takes place. The Adept robot removes the finished product and places it on a tray. Once the tray is full, it is transferred by conveyer to a vertical magazine.

To assemble the vacuum cleaner product, the robot changes its gripper, and follows a similar stacked-assembly procedure.

In specifying the design of the cell, Miele has attempted to achieve the greatest degree of autonomy possible. If there is a problem with one of the assembly tasks, the robot will repeat the operation three times before rejecting the problematic part. It will then take a new part from a reserve located just behind the robot base and try the operation three additional times. If it is still unsuccessful, the robot is programmed to remove all the components of a started subassembly from the different fixtures and continue with a fresh set of parts.

(Courtesy of Adept Technology, Inc.)

An automotive manufacturer using robots to insert pistons into engine blocks wanted to check for conditions such as missing piston rings, jammed rings, or collisions between pistons and crankshafts.

To perform this task, an Assurance Technologies robotic F/T sensor was mounted to a Seiko RT-5000 robot (see Figure 9–6). A three-finger gripper, mounted on the end of the F/T sensor, was designed to close around the piston rings, centering and compressing them slightly while grasping the body of the piston.

The piston has an oil ring, a bottom compression ring, and a top compression ring. As each is pushed into the cylinder, the insertion force rises until the ring is in place; it then decreases to the level required to push the piston itself into the cylinder. This change in force is detected by the F/T sensor and analyzed by an IBM AT computer to determine the presence or absence of each ring. The F/T controller provides data to the computer via an RS 232 serial port. The controller's discrete I/O port is connected to the robot to provide force threshold communication, and passes this information through the I/O port to the robot. Thresholding detects jamming, collisions between connecting rod and crankshaft, dropped pistons, and crashing.

As part of the analysis, a time derivative of the force signal is calculated and used to detect the rings as they snap into the cylinder during insertion. Using the time derivative rather than the signature itself simplifies the ring detection problem by eliminating the commonly encountered variations in absolute insertion force. Such variations are caused by slightly different insertion conditions for each piston/cylinder combination.

The piston rings appear as zero-crossing spikes in the derivative signal. The compression rings generate single spikes. The oil ring, consisting of a plastic spacer sandwiched between the steel rings, generates a double spike. The derivative signal is compared to a predetermined threshold value to detect the presence/absence of a ring.

Figure 9–8 shows the data obtained during the insertion of a piston with the bottom compression ring missing. This condition is detected by the analysis software, which produces an error condition that prevents the piston assembly from being passed along the line. The robotic force sensor, which in effect provides 100 percent inspection, dramatically improves the quality of the piston insertion procedure.

(Reprinted with permission from SENSORS—YOUR RESOURCE FOR SENSING, COMMUNICATIONS, AND CONTROL, November 1992.)

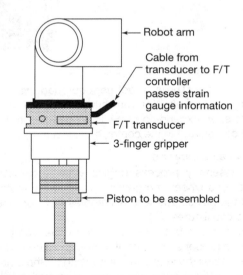

Figure 9–7 Force/Torque Sensing in Piston Assembly. (Courtesy of SENSORS—YOUR RESOURCE FOR SENSING, COMMU- NICATIONS, AND CONTROL)

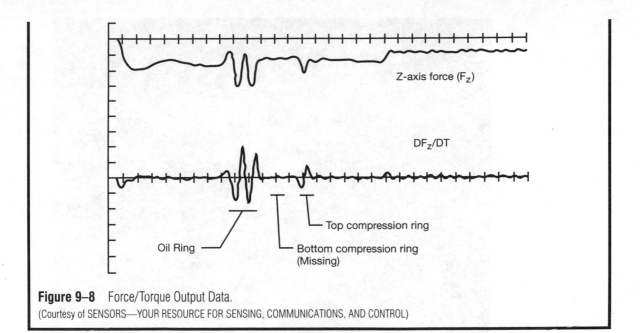

Figure 9–8 Force/Torque Output Data.
(Courtesy of SENSORS—YOUR RESOURCE FOR SENSING, COMMUNICATIONS, AND CONTROL)

Process

All applications that use the robot as a production machine are grouped into this classification. The processes performed frequently by a robot include welding, painting, sealing, deburring, grinding, drilling, and finishing. Welding and painting are major application areas for the robotic industry. Another process area, the application of glues and sealers, is time-consuming and tedious when performed by hand; as a result, industrial robots easily outperform their human counterparts. A robot offers the advantages of accurate bead placement and a high application rate. The consistency in rate of application provides an additional savings in material. Figure 9–9 shows a robot with a three-roll wrist applying sealer to a sheet metal joint in the passenger compartment of an automobile. Note the ability of the robot to enter the car through the window of the side door. In some cases, the robot can apply the sealer while the car continues to move on the conveyer.

Welding

Welding applications are classified in three categories: resistance, arc, and gas. In each application area, only servo-type robots are used because path control is critical. Resistance or spot welding was developed by the automotive industry in the mid-1970s and continues to be a major application in that industry. The robot

Figure 9–9 Sealer Application.

APPLICATION NOTE
Cookie Sandwiching

At Pepperidge Farm's award-winning plant in Denver, Pennsylvania, high-risk, repetitive motion has been reduced in the cookie sandwiching operation. Milano and Brussels cookies are now sandwiched by twenty AdeptOne robots with AdeptVision (Figure 9–10), resulting in an annual labor savings of close to $1 million through a reduction of eight jobs per shift.

Several years ago, Pepperidge Farm determined that cookie sandwiching posed a high risk of carpal tunnel syndrome. Pepperidge Farm reviewed various automation options and eventually selected robotics, because it was the only solution capable of handling the wide range of products required.

The time required from the start of baking to finished packaging is 40 minutes.

System Configuration

The system, designed by system integrator Technistar of Longmont, Colorado, consists of four AdeptOne robots with HyperDrive and sixteen standard AdeptOnes, each equipped with AdeptVision AGS-GV. Ten robots stand on each side of a moving conveyer.

Work-cell Operation

Cookies are mixed, deposited, and baked twenty across on a 1-meter wide moving conveyer. After exiting the oven, the outside lanes of cookies travel over a chocolate bottomer while the center ten lanes travel over a bypass conveyer. The chocolate-coated cookies turn upside down by falling out of the chocolate bottomer onto the main conveyer. Cookies from the bypass conveyer rejoin the

Figure 9–10   Cookie Assembly.
(Courtesy of Adept Technology, Inc.)

chocolated cookies on the main conveyer before entering the robot work cell.

The robot picks up the non-coated half from the center of the conveyer and places it on the chocolate-covered half. AdeptVision AGS-GV finds the randomly oriented cookies and guides the robot to pick up an uncoated half and wait for an unsandwiched coated half to come by on the conveyer. The robots at the beginning of the line work faster than those at the end because more unsandwiched cookies are available. After the sandwiching station, the cookies move into a cooling tunnel, then on to hand-packing in paper cups. The cups of cookies are then automatically bagged and weighed, and then manually packed into shipping boxes.

In the future, Pepperidge Farm plans to continue automating high-risk operations. In the near term, it plans to evaluate automating its cookies-cupping operation where cookies are manually placed in paper cups in preparation for bagging.

(Courtesy of Adept Technology, Inc.)

became a standard item on the assembly line due to its ability to move the 150-pound welding gun effortlessly into position on a moving car body and produce accurate resistance welds at a rate of one every 3 seconds. Figure 9–11 shows a series of Fanuc robots performing spot welding on car bodies. Robotic arc welding

Figure 9–11 Fanuc Robots Performing Spot Welding on an Assembly Line of Cars.
(Courtesy of Fanuc Robotics North American, Inc.)

is frequently used in the heavy-equipment manufacturing industries to produce off-road hauling and earth-moving equipment. Lighter applications occur in the automotive, electronics, and aerospace industries.

Metal inert gas (MIG) and tungsten inert gas (TIG) welding with real-time interactive control permit the robot to track the seam between two metal plates. The seam to be welded is located and tracked by a photo sensor, such as the vision camera in Figure 9–12, or by welding parameters. The location of the seam or welding path is passed to the robot controller, which moves the arm holding the welding tooling. The location and control are dynamic, so that corrections in the direction of the welding gun are made as the weld is in progress. Robotic arc welding is cost-effective when compared with manual and semiautomatic systems. Weld efficiency is a percentage of time that a seam is being produced with respect to total possible welding time. For example, a comparison of cycle times indicates that manual and semiautomatic welding are 30 and 45 percent efficient, respectively, whereas robotic arc welding is close to 70 percent efficient. In robot welding systems the robot controller has the capability to control all the welding variables including *arc voltage*, *wire feed rate*, and *torch speed*. In addition,

Figure 9–12 MIG Weld Tooling (Right) and Weld Vision System (Left) on Common Tool Plate Attached to Three-Roll-Wrist Robot.

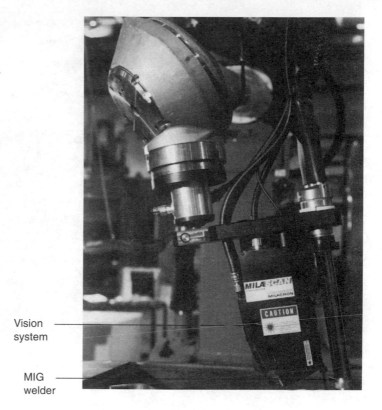

Vision system

MIG welder

the robot system can control the position of a multi-axes part positioner, like the one in Figure 9–13.

Gas welding is used to cut complex curves and shapes accurately in pipes and flat metal plates. Although it is not used as frequently as spot and arc welding, it is a major robot application.

Paint Spraying

In general, paint-spraying applications require limited automation technology. The number of sensors required is relatively low, the support equipment is primarily material handling, and robot precision is not critical. Servo paint-spraying robots have traditionally cost more than the standard servo robot, but the cost of the work cell is about the same because the support hardware is less. Several applications are supported by robotics:

- *Dip coating:* The part is lowered into the coating material reservoir and removed to allow excess material to flow off the part. In some cases parts are spun to remove excess coating material. A variety of robots are used in this application area, including servo- and nonservo-type arms.
- *Paint spraying—airless:* The coating material is pushed through a small nozzle opening by a pump in the spray head to atomize the material. The

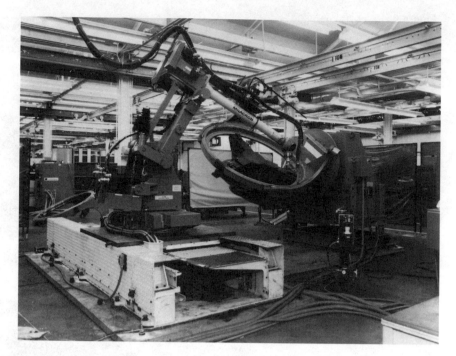

Figure 9–13   Welding Application with Part Positioner.

residual momentum from the atomizing process causes the material to coat the part. Servo arms are required when robots are used, and continuous-path control of the spray gun is necessary.

- *Paint spraying—compressed air:* The atomizing action in this system is caused by compressed air passing over a venturi. The residual momentum of the paint and air mixture carries the material to the target. Again, servo-type controllers with continuous-path control are necessary.

- *Paint spraying—electrostatic:* Paint is given an electrostatic charge as it passes through the spray head, and the part is connected to a voltage source with the opposite polarity. The paint is pulled to the part by the difference in the electrical charge between the paint droplets and the part. Continuous-path servo machines are used for robot applications in this area.

APPLICATION NOTE
Kendon Inc.'s Peacock Struts its Stuff

Peacocks and robots are not exactly birds of a feather, but at Kendon Inc., ABB Robotics' IRB 1500 is helping to turn out beautiful peacocks for the home interior products market.

Kendon Inc., a 21-year-old company based in Winnsboro, Texas, manufactures metal sculpture wall decor items and markets them through Home Interiors Inc. of Dallas and through Kendon's wholesale distribution division, "American Accents by Ditto." Its popular peacock is a 4-foot tall, brass-plated sculpture, made of $\frac{3}{16}$-inch wire, which boasts a large plume that posed a challenge for the IRB 1500 robot. In fact, according to Shane Wilson, Kendon's new products coordinator, they bought the robot with the peacock—and its unusual tail—in mind.

Last year when the new product was to be put on line, Wilson and CNC Systems Coordinator Dan Cartwright started looking at robot systems because they knew automated welding would be necessary to make production cost-effective. After looking at several different robots and visiting the 1993 AWS show in Houston, they selected ABB Robotics' IRB 1500.

The IRB 1500 is designed to ensure a minimum of maintenance and service. Its flexibility and speed make it ideal for precision arc welding, and its S3 control system can handle six internal robot axes

and six external axes. Wilson notes that the features which "cut the cake" for them were the robot's portable teach pendant and the joystick used for programming. The joystick is not only easier to use than push-buttons, it is also 25 percent faster.

Once they made their decision, Wilson and Cartwright moved quickly. They purchased the IRB 1500 straight from the AWS show floor in Houston, had it installed the following week, had their people working on building a positioning table and part-holding fixture, and headed for ABB Robotics' Robot University training course at the Welding Systems Division headquarters in Fort Collins, Colorado. When they returned, everything was ready to go. "I'd say we were in production within about 5 weeks," recalls Wilson.

The plume, or tail section, welded by the robot (Figure 9–14) is about 1 foot wide by 2 feet long. It consists of thirteen pieces, each about 6 inches

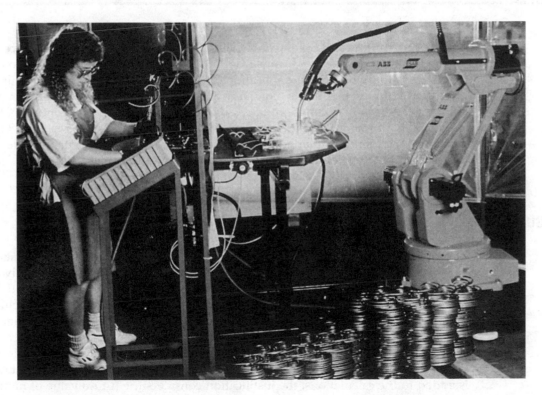

Figure 9–14   Robot Welding of Peacock Tails.
(Courtesy of ABB Flexible Automation)

long, that look like fish hooks. These pieces are all placed on a fixture, and the robot makes the peacocks' joints by connecting them with about fifty small tack welds. The positioning table has two stations so that, as the robot welds on one side, the operator can load and unload parts on the other side. When welding is complete, the table indexes and presents the next workpiece to the robot. According to Cartwright, the robot has reduced the welding time of approximately 6 minutes to 1 minute, 48 seconds.

Once the plume is completed, it is manually welded to the head and breast subassemblies, and the entire peacock is then brass plated. The entire process takes about 11 minutes from start to finish.

The robot operates for one 10-hour shift, 4 days per week and the plant, which employs about 200 people, currently produces 1100 to 1500 peacocks per week. Cartwright handles most of the programming, and he and Wilson have trained three other employees to operate the robot. Those employees are also able to handle occasional editing of the welding program.

The company plans to phase out the peacock sculpture, so it is currently in the process of designing a new product that will take advantage of the robot's capabilities. While there are no plans for another robot at this time, according to Wilson: "If we get a new product on line for the robot and it does well, you never know. If we had the need, we'd definitely get one, and it would more than likely be another ABB robot."

Cartwright agrees, adding: "Our purchase of the ABB robotic welder is indicative of Kendon's philosophy; we believe in manufacturing a quality product at an affordable price utilizing the most effective production means and equipment available. Since we purchased ABB's IRB 1500 in 1993, it has proven its capabilities and continues to present an impressive demonstration of technology."

(Courtesy of ABB Flexible Automation)

Spray painting and coating are well-established robot applications. Concerns for the health of workers and gains from productivity force manufacturers to use robots for all large painting applications. Automotive and truck manufacturers rarely use a human in a production paint area. Figure 9–15 shows a coating robot painting the back of a car. A primary advantage of the robot in painting is the repeatability of the programmed motion. If a master painter is used to record the motion of the paint program, then every robot becomes a master painter. Heavy use of painting robots occurs in the automotive and aerospace industries, and moderate use occurs in heavy-equipment industries.

9-4 SUMMARY

The goal of manufacturing is to make a profit, so justification of capital investments is necessary. In the past, justification of capital expenditures focused primarily on tangible savings and revenue changes. To meet the challenges provided by increased local and global competition, the justification process must include intangible gains that result from the automation process. Tracking these intangible gains is difficult, but transaction-based cost accounting techniques help to uncover cost savings and revenue gains from nonmeasurable sources. Three methods are used most often: payback, return on investment, and cash flow. If the discount rate is added to the calculations, the justification considers the future value of money and the analysis provides the net present value for the justification. Justification of robot applications is enhanced by the long lifetime of the machine.

Figure 9–15 Coating Robot Painting a Car.
(Courtesy of Fanuc Robotics North American, Inc.)

The applications usually associated with robots include material handling, machine tending, assembly, process, welding, and painting. The type of robot used is dictated by the application environment, size and shape of the part, and production rate that must be supported. Robots are used in all major industries; however, the greatest concentrations occur in the automotive, foundry, light manufacturing, electrical/electronic, heavy-equipment manufacturing, and aerospace groups.

QUESTIONS

1. Describe the type of industry environment that permitted justifications to be based only on tangible savings and revenue gains.
2. What forced a broader view of justification of capital investments and the introduction of intangible benefits?
3. Describe the nonmeasurable factors most frequently included in current justifications.
4. Describe the three payback methods.
5. Explain the term *future value of money.*

6. How does discounting affect the three payback methods?

7. What intangibles do robots offer in the justification process that are not found in other types of automation hardware?

PROBLEMS

1. A manufacturer was not satisfied with the irregular cycle time and production levels in the manual production of plastic garbage cans with an injection molding machine. The efficiency of the machine operators varied during the 8-hour shift. The operator ran a part every 43 seconds for the first 2 hours of a shift and every 48 seconds for the middle 4 hours. During the last 2 hours of the shift, the cycle time averaged 52 seconds. This efficiency level was essentially the same during all three shifts. The manufacturer used a robot and achieved a consistent 41-second cycle time for all three shifts. Determine the following:

   (a) How many garbage cans are produced by the robot cell over three shifts?

   (b) What was the percentage increase in garbage can production with the robot work cell?

2. A manual welding operation with two multi-axes positioning stations assembles the base for a computer mainframe in a two-shift operation. The job requires 44 seams 2 inches long from a variety of angles. In the manual mode, the welding process takes 45 minutes, with 10 additional minutes for setup and inspection. The application was moved to a single robot MIG welding work cell that included multi-axes fixtures for two bases and seam-tracking capability. After the cell was automated with a robot, a base assembly was welded in 11.6 minutes. One operator now supports the cell by loading a raw material package and inspecting finished products. Determine the following:

   (a) How many shifts will be required to produce the same number of bases with the automated cell?

   (b) How many workers can be reassigned to other production operations in the plant?

   (c) The cost of the robot was $65,000 and the 2 positioners were $12,000 each. Cell design costs were 28 percent, programming costs were 12 percent, and miscellaneous costs were an additional 14 percent (costs are percentages of robot and positioner cost). The hourly rate (labor and benefits) for workers displaced was $15.50 per hour, the tax rate is 33 percent, depreciation is 14.3 percent of the cost of the robot and positioners, and the discount rate is 10 percent. Calculate the payback time, ROI, and discounted cash flow using the West-Electric spreadsheet.

3. At a major industrial enclosure manufacturer, five manual paint spray stations operate two shifts painting the inside and outside of electrical cabinets. The company uses 200 gallons of $30 epoxy paint over the two shifts. It is estimated

that 20 percent of the paint is lost by the operator in overspraying and waste. The company wants to move to automated paint spraying and collected the following data: paint-spraying robot cells cost $150,000 each; waste and overspraying is reduced to 5 percent if robots are used; robots provide a 10 percent increase in productivity; one operator per shift is required for support of the painting cells; and the average cost of labor, including fringe benefits, is $13.50 per hour. Determine the following:

(a) How many robot cells are required and what is the cost saving per year in paint?

(b) If the tax rate is 37 percent and the depreciation is $21,000 per robot cell, what is the payback time and what is the ROI?

(c) Use the West-Electric spreadsheet analysis to determine payback time using the discounted cash flow analysis. The discount rate is 12 percent.

PROJECTS AND CASE STUDY PROBLEMS

1. Apply the concepts and spreadsheet to the justification of one or more of the following automation work cells. After you select an automation case for justification, review the solutions for that case in the previous chapters and extract any data relevant to your need in this project. The design component of the case studies varied in complexity and design detail, so it may be necessary to make some assumptions as you justify some of the cases. Document and support all assumptions.

(a) The die casting production cell started in Case Study Problem 3 in Chapter 2

(b) The kitchenware bowl production cell started in Case Study Problem 5 in Chapter 2

(c) The West-Electric slug production cell design started in Case Study Problem 8 in Chapter 2

(d) The West-Electric slug lubrication cell design started in Case Study Problem 9 in Chapter 2

(e) The kitchen products production cell design started in Case Study Problem 3 in Chapter 3

2. Read the West-Electric case study (Section 12–13 in Chapter 12) where the justification of the cell is discussed. Complete the questions and problems at the end of Chapter 12 that relate to justification.

Safety

CHAPTER GOALS AND OBJECTIVES

The primary goal of this chapter is safety in manufacturing automation. Manufacturing is primarily concerned with the safety of the workers. However, it is also important to use safety principles to ensure that production machines are not damaged by the actions of workers or other machines. At the completion of this chapter you should be able to:

- Identify the organizations that set the standards for manufacturing safety.
- Describe the procedure used to safeguard an automation work cell.
- Describe the operation of three types of presence sensing devices used in automated work cells.
- Describe the operation of power and control interlock devices and the concept of positive mode switch operation.
- List the procedures used to safeguard the operator, programmer, and maintenance personnel in a work cell.
- Develop a safety strategy including risk assessment, risk estimation, and hazard reduction.
- Develop a safety cost justification.
- List OSHA/ANSI guidelines for hazard control.
- Calculate the safe curtain distance and safeguard distance.
- Develop a complete safety system for a robotic work cell.

10-1 INTRODUCTION

In the past, people became acquainted with robots through science fiction writers such as Isaac Asimov. Robots from his writings of the 1940s and 1950s are especially memorable. From these fictional machines, Asimov developed the Three Laws of Robotics:

1. A robot must not harm a human being, nor through inaction allow one to come to harm.
2. A robot must always obey human beings, unless that is in conflict with the first law.
3. A robot must protect itself from harm, unless that is in conflict with the first and second laws.

These laws are still valid today, and no investigation of robotics would be complete without considering safety issues in their application.

Worker safety is the most important concern; of secondary importance, however, is the fact that damaged equipment causes downtime, which decreases production. It is, therefore, the responsibility of work-cell designers to consider measures to diminish the possibility of accidents to both hardware and humans. In addition, it is the responsibility of the plant safety director to alert workers to the potential hazards of robotic installations.

10-2 SAFETY STANDARDS

Safety standards have been developed by several different organizations for many different areas in society and are enforced by numerous governmental bodies. The safety agencies for robots and integrated manufacturing work cells are described below, and the primary enforcement agency, OSHA, is covered in the next section.

Robot Systems and Integrated Work Cells

Safety guidelines for applications using industrial robots result from the joint effort of the American National Standards Institute (ANSI) and the Robotics Industries Association (RIA). In the joint standard ANSI/RIA R15.06, an *industrial robot* is defined as *a reprogrammable multifunctional manipulator designed to move material, parts, tools, or other devices.* As a result of this more restrictive device definition, numerically controlled machine tools are excluded from this standard.

Two standards are used to define the safety issues associated with integrated manufacturing systems. The first is ISO 11161, Safety of Integrated Manufacturing Systems, an international standard covering requirements for the safe installation, programming, operation, maintenance, or repair of these systems. The second standard is ANSI B11.20, entitled Manufacturing Systems/Cells—Safety Requirements for Construction, Care, and Use. Both standards cover the safety of multiple machines under some type of common control. These standards define an *integrated manufacturing system* as *a group of two or more industrial machines working together in a coordinated manner normally interconnected with and operated by a supervisory controller or controllers capable of being reprogrammed for the manufacturing of discrete parts or assemblies.* When machines in an integrated system operate separately or individually, or when the safeguards are muted (temporarily turned off) or suspended, the safety standards for the individual machines should be used as a supplement. An example of an individual machine standard is IEC61496, which covers specific items such as the number of relays required, the need for a key-operated switch, lockable enclosures, transformer construction, and failure conditions. Test specifications require that the equipment be subjected to a battery of tests including moisture and dust intrusion, power supply transients, electrical interference, electrostatic discharges, component failure mode analysis, and object sensing capabilities.

Corporate Standards

To provide employees with a safe work environment, many corporations have authored their own standards for safety. These standards are frequently more stringent than those required by OSHA and can be met only by the most technically advanced products.

Governmental and Other Organizations

Guarding of specific machines in the United States is regulated by the Occupational Safety and Health Administration (OSHA). Some states have their own safety organizations with regulations that must be at least as strict as the federal OSHA standards. In addition to OSHA, other organizations provide information on proper machine guarding. The Canadian Standards Association (CSA) is a nonprofit Canadian corporation that tests and certifies the electrical integrity and safety of products. CSA is accredited by OSHA as a nationally recognized test laboratory (NRTL) that covers testing of all products under OSHA's jurisdiction. The NRTL/C mark is a counterpart to the Underwriter's Laboratory C-UL mark. Both marks indicate that a product is in compliance with both CSA and UL standards.

ANSI publishes the B11 standards to provide information on the construction, care, and use of machine tools. Certain standards are developed for specific types of machine tools. For example, B11.1 covers mechanical power presses.

10-3 OCCUPATIONAL SAFETY AND HEALTH ADMINISTRATION (OSHA)

The Occupational Safety and Health Act of 1970 represents the foundation for industrial safety standards throughout the United States. Federal regulations, commonly known as 29 CFR 1910 (OSHA), do not have any specific standards relating to robot safety; however, OSHA has issued a directive to field personnel instructing them to use the ANSI/RIA R1506 standard for safety in robot installations. In addition, the OSHA standards in Subpart O, Section 1910.212, of the code of Federal Regulations detail general requirements for safety precautions around machinery. The major emphasis of this section is on machine guarding, which protects the operator and others from danger at the point of operation (the area where the work is actually occurring).

For nonrobotic fixed machinery, OSHA requires a fixture that prevents operators from exposing any part of their body to the dangers of the operation. The addition of machine guards for human safety within the robot cell, however, may impose severe limitations on the robot's mobility and its access to all parts of the work envelope. One way to view machine guarding is to consider the entire work cell as the point of operation; then the designer can plan methods to protect the workers from hazards such as pinch points, shear points, and collisions within the entire robot work envelope.

A second way to view safety within the work cell is from the perspective of the actual robot production application (welding, material handling, grinding, and so on). Because most current production applications are covered in detail by OSHA regulations, the rules governing a particular application should be reviewed before the design of the robot work cell. Following is a list of pertinent sections:

Subpart N, Section 1910.176	Handling Materials—General
Subpart O, Section 1910.215	Abrasive Wheel Machinery
Subpart O, Section 1910.216	Mills and Calenders
Subpart O, Section 1910.217	Mechanical Power Presses
Subpart O, Section 1910.218	Forging
Subpart Q, Section 1910.252	Welding, Cutting, and Brazing

In addition, special OSHA requirements may be requested at the time of inspection.

10-4 AMERICAN NATIONAL STANDARDS INSTITUTE/ROBOTIC INDUSTRIES ASSOCIATION STANDARD FOR ROBOT SAFETY

This chapter is not intended to duplicate the RIA standard (ANSI/RIA R15.06 Safety Requirements) on robot safety but to provide an overview of the issues and areas addressed in the standard. A study of the American National Standards

Institute (ANSI) B11 series is also necessary for a full understanding of automation safety issues. The major areas covered in the standard are construction, reconstruction, and modification; safeguarding; maintenance; testing and startup; applications; continuous operation; and training. Safeguarding is addressed in the following section.

10-5 SAFEGUARDING A WORK CELL

The responsibility for safeguarding a robotic cell falls on the user. The standards address safeguarding by considering four points: devices to be used, operator safety, safety of the teacher, and safeguarding maintenance and repair personnel.

Safeguarding Devices

The type, degree, and redundancy of safeguarding must correspond directly to the type and level of hazard present in the robotic cell. The safety plan should include safeguarding devices, barriers, interlock barriers, perimeter guarding, awareness barriers, and awareness signals. The primary function of the devices is to provide a warning and to restrict traffic flow in the area. A strategy for limiting access to the areas is illustrated in Figure 10–1. The work cell is divided into zones 1 to 3, and the zones are defined as follows:

- Zone 1 is the area outside the work cell and has no restrictions on human traffic.
- Zone 2 is the area inside the work cell but out of reach of the robot arm. Only programmers, work-cell operators, and maintenance personnel are allowed in this area.
- Zone 3 is the area inside the robot work envelope, and it cannot be entered as long as the robot is in the automatic or run mode. Penetration is permitted during programming or maintenance but only after safety standards established for the type of robot present have been satisfied.

Figure 10–1   Safety Zones.

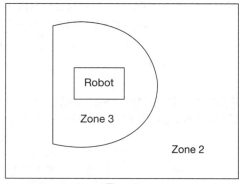

Zone 1 is typically distinguished by signs and yellow lights. A continuous barrier with electrical interlocks is usually installed to separate zones 1 and 2. Zone 3 is usually distinguished by an outline on the floor indicating the robot's maximum reach. In addition, awareness alarm electronic detectors are frequently used in zones 2 and 3. The safety systems used in the three zones are identified in the following sections.

Safety Hardware

A broad selection of safety hardware is available from several vendors who specialize in safety related solutions for industrial automation. The following list provides a brief description and an overview of the major system components available. A more detailed description of the most frequently used systems is provided in the following sections.

Physical Barriers.   Physical barriers are the first line of defense in protecting people from hazards. The barriers can include:

- Chains and guard posts
- Safety rails
- Wire mesh fencing
- Equipment within the work cell

The last item, work-cell equipment itself, when combined with guard rails and fences, provides effective safety protection at reduced cost.

Presence Sensing Devices.   Devices sense the presence of a person and cause the appropriate action to safeguard the person from any hazardous condition. The most common presence sensing devices include light curtains, pressure sensing pads, and proximity sensing devices.

E-Stop Devices.   Switches with various tripping mechanisms remove power to production machines in emergency situations. The most common E-stop devices are the mushroom pushbutton and grab wire switches.

Interlock Devices.   Interlock devices are switches that interlock a guard door with the power source of the hazard. Power interlocking and control interlocking are the two types of control most frequently used. Contact and non-contact switches are used with various tripping mechanisms that include actuator operated, hinge operated, cam operated, active actuator non-contact, unconditional guard unlocking, and conditional guard unlocking. An example of an interlock switch from Omron is pictured in Figure 10–2. Note the special key used to trip the switch.

Safety Control Units.   Safety control units are the intelligent controllers that link elements of the safety system together and interface the safety system to the rest of the automated cell hardware.

Figure 10–2 Special Limit Switches Designed for Work in Safety Systems.
(Courtesy of Omron Electronics, LLC)

The following sections provide a detailed description of the guarding hardware frequently used in robot and automated manufacturing work cells.

10-6 PRESENCE SENSING DEVICES

Three devices are frequently used in automated work cells to sense the presence of a human: sensing devices, light curtains, and pressure sensing mat. A description of each follows.

Proximity Sensing Devices

Two technologies, ultrasonic and lasers, are most frequently used to sense the presence of a human from a single sensing location. The laser provides the best results. The proximity laser sensor (PLS) in Figure 10–3 is a programmable single point safety device that can pattern a protective field to an irregularly shaped production area. The device covers a 180 degree field of view without reflectors or separate receivers. It can be used in a stationary position as a hazardous area safeguard. For mobile robots and automatic guided vehicles (AGVs), it is used to scan the forward path for problems.

The PLS creates three independent user-defined sensing fields using an infrared laser beam and time-of-flight measurements. The fields include two programmable protected zones (the *safety zone* and the *early warning zone*) and a *surveyed area zone*. A typical zone layout is illustrated in Figure 10–4. Entrance into the early warning zone initiates a warning signal, and movement into the safety zone would stop any hazardous machine motion. The surveyed area is used to support positioning and navigation for AGVs. The safeguarded zones are set using a *learn mode* in the PLS, or they are defined by the programmer using a Windows-based programming language in the *graphic input mode* or the *coordinate input mode*. After programming, the computer is removed, and the PLS maintains the zones in memory. The technical specifications for the system are illustrated in Figure 10–5.

Figure 10–3 Proximity Laser Sensor.

(Courtesy of SICK Optic-Electronic, Inc.)

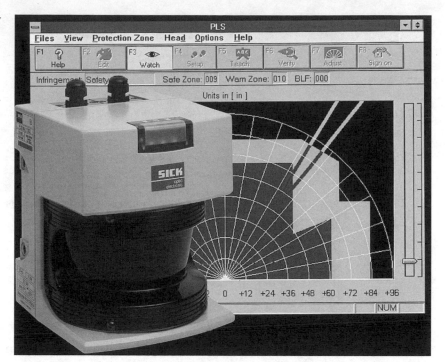

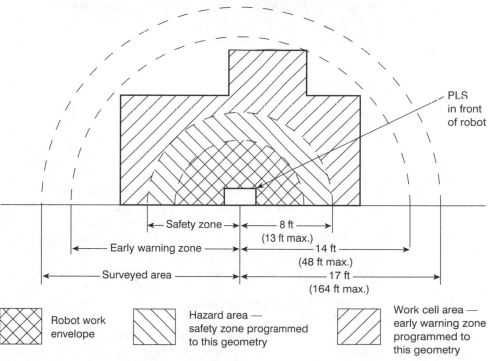

Figure 10–4 Typical Zone Layout for PLS System.

Technical Specifications

Scanning range	Safety zone radius	4 m (13 ft)
	Early warning zone radius	15 m (48 ft)
	Surveyed area radius	50 m (164 ft)
Object sensitivity (resolution)	70 mm (2.75 in.) at 4 m scanning range (variable at closer range)	
Response time	Less or equal to 80 ms safety zone, Less or equal to 40 ms early warning zone	
Safety category	Single component failure detectability; EN 954, Category 3	

Figure 10–5 PLS Specifications.

Light Curtain Introduction

Light curtains (also called light screens, optical guards, and presence sensing devices) provide guarding with a high degree of flexibility and reduced operator fatigue when compared to traditional guarding methods such as mechanical barriers, gates, and pull-back restraints. Safety light curtains simplify routine tasks such as machine setup, maintenance, and repair by replacing solid guards.

Light Curtain Operation

In a light curtain a photoelectric transmitter sends an array of synchronized, parallel infrared light beams to a receiver unit. If an opaque object interrupts one or more beams, the light curtain controller sends a stop signal to the guarded machine. The light curtain in Figure 10–6a illustrates this concept. Light emitting diodes (LEDs) in the transmitter are energized by the light curtain's timing and logic circuitry and emit pulses of invisible infrared light. The light pulses of individual diodes are energized one after another and are modulated at a specific frequency. Phototransistors and logic circuitry in the receiving unit detect only the specific pulse and designated frequency of the sending diode. This technique ensures safe operation and rejection of external light sources. A controller interfaces the curtain to guarded machine and workcell hardware, and provides user controls, status and diagnostic indicators, and power for curtain operation. A light curtain from Omron is pictured in Figure 10–6b.

The light curtain is different from a standard photoelectric sensor because of a concept called *control reliability*. Control reliability is a design standard from OSHA and ANSI (ANSI B11.19-1990, 5.5) for safety related machine control

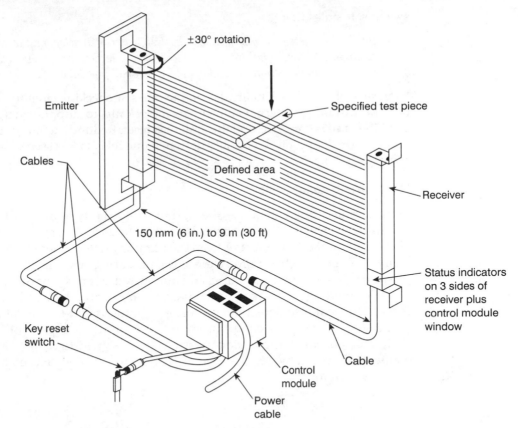

±30° rotation

Emitter

Cables

Specified test piece

Defined area

Receiver

150 mm (6 in.) to 9 m (30 ft)

Status indicators
on 3 sides of
receiver plus
control module
window

Key reset
switch

Control
module

Cable

Power
cable

Figure 10–6(a)   Light Curtain System Components.

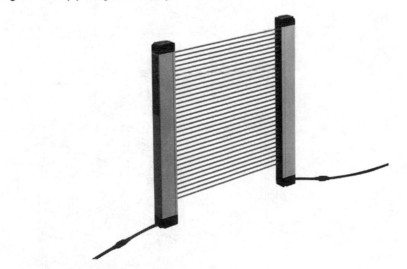

Figure 10–6(b)   Light Curtain Model from Omron.
(Courtesy of Omron Electronics, LLC)

systems. It states that

> *the device, system, or interface shall be designed, constructed, and installed such that a single component failure within the device, interface, or system shall not prevent normal stopping action from taking place but shall prevent a successive machine cycle.*

Examples of control reliability design include self-checking circuitry, which monitors the curtain for internal faults by using lockouts to stop the guarded machine until the fault is corrected and the system is reset. Redundant output relays are another example of control reliability design. If one relay fails, a second is used to stop the guarded machine.

Light Curtain Applications

Light curtain applications are often grouped by the type of guarding required. Guarding for the *point of operation,* also called the *zone of hazardous operation* or the *pinch point,* protects an operator from the hazards located where the process is performed. Finger and hand protection is the primary goal for light curtains used to guard presses, stamping, forming, and automated assembly machinery.

Perimeter guards are used to guard the boundary defined by a machine or robot. The goal is to detect the presence of humans and to control the machine to prevent hazardous conditions while personnel are present in the area. Perimeter guarding applications are usually designed to detect arm and body elements. The entrance to the robot cell in Figure 10–7 is guarded by a horizontal light curtain.

Figure 10–7 Robot Work Cell
Guarded by a Horizontal Light Curtain.

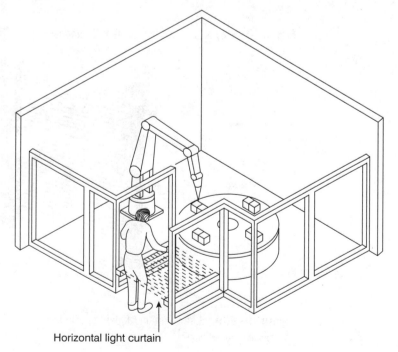

Horizontal light curtain

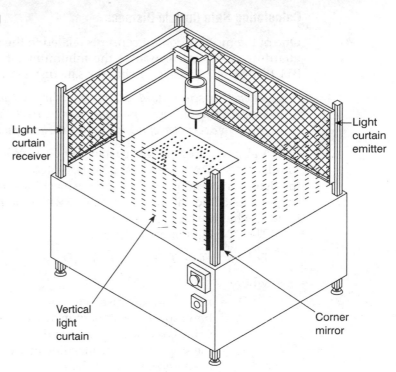

Figure 10–8 Automatic Drilling Machine Protected by a Vertical Light Curtain.

Curtains can include a single infrared beam or many parallel beams that are so close together they can detect an object as small as a finger. For example, the drilling operation in Figure 10–8 has two sides of the station guarded with a metal screen and two guarded with a light screen. Note the use of a single emitter with multiple beams, a single receiver, and a corner mirror. This is an example of point of operation guarding. When the field is interrupted, the drill machine motion is stopped. After the protective field is cleared and the safety system is reset, the drill operation resumes. The light curtain system offers two special controls, *muting* and *blanking,* to enhance production. Muting temporarily halts the operation of the curtain to permit automatic feeding of material through the curtain, and blanking allows for the passage of material through defined sections of the field.

Light curtain safety standards fall into two categories: *application standards* and *construction standards.* Application standards explain how to use a light curtain for machine guarding and how to apply a light curtain for the type of machine covered by the standard. In addition, the application standard covers how to calculate the safe mounting distance. For example, the application standard ANSI/RIA R15.06 discusses the use of presence sensing devices (light curtains) for robot guarding. Construction standards provide design, construction, and testing information on presence sensing devices. An example of the construction standard is IEC61496, *Safety of Machinery—Electrosensitive Protective Equipment.* This standard is one of the best single sources of machine tool guarding information for the American market.

Calculating Safe Curtain Distance

One of the most important concepts related to the use of light curtains in machine guarding is the calculation of the minimum safe distance. The ANSI standard B11.19-1990 (4.2.33.3.5) describes the safe distance as

> the effective sensing field [that] is located at a distance from the nearest recognized hazard such that the operator or others cannot reach the hazard with a hand or other body part before cessation of motion during the hazardous portion of the machine cycle.

OSHA also requires that the curtain not be any closer to the pinch point than the standard distance for physical guards. Stated in simple terms, determine how far the detection curtain must be from the machine pinch point so that the curtain can trigger a machine halt and the machine will be stopped by the time the operator's hand reaches the hazardous area.

The ANSI minimum formula is:

$$D_s = K * (T_s + T_c + T_r + T_{bm}) + D_{pf}$$

where

D_s = minimum safe distance.

K = hand speed constant in inches per second. The ANSI standard value is 63 inches per second.

T_s = the stop time of the machine in seconds measured from the final deenergized control element.

T_c = the response time, in seconds, of the press or machine control circuit to activate the machine's brake. *Note:* $T_s + T_c$ are usually measured together with a stopwatch.

T_r = the response, in seconds, of the safety light curtain. Use the value of 0.045 seconds in planning an installation or use the actual response time of the light curtain provided in the installation manual.

T_{bm} = the additional stopping time, in seconds, allowed by the brake performance monitor. If excessive brake wear is present, a brake monitor will halt the machine when the stop time of the machinery exceeds the brake monitor limit. Therefore, T_{bm} = brake monitor set point $- (T_s + T_c)$.

D_{pf} = the added distance for the depth penetration factor in inches. This compensates for small objects, like fingers, that can move through a curtain because the curtain's sensitivity is not sufficient to detect the presence of the object. The minimum object resolution S of the light curtain is substituted into the following equation (S has units of inches):

$$D_{pf} = \frac{S - 0.3}{0.3}$$

and the corresponding value of D_{pf} is found. Most light curtain vendors provide the D_{pf} value for each model.

The definitions of the equation variables just described satisfy a machine setup when brake monitoring is present. Other machine setups are covered in the standard ANSI B11.19-1990.

Example 10–1

A mechanical press has a stopping time for the press brake mechanism (T_s) of 0.35 seconds, and a stopping time for the control circuits (T_c) of 0.05 seconds. The brake monitor is set for 0.49 seconds. The response time of the light curtain selected is 0.045 seconds and the minimum object sensitivity (S) is 0.80 inches. Determine the safe distance D_s for the curtain to be mounted to the press.

Solution

$$K = 63 \, \frac{\text{in.}}{\text{s}} \text{ (chosen)}$$

$$
\begin{aligned}
T_{bm} &= \text{brake monitor setpoint} - (T_s + T_c) \\
&= 0.49 \text{ s} - (0.35 \text{ s} + 0.05 \text{ s}) \\
&= 0.09 \text{ s}
\end{aligned}
$$

$$D_{pf} = \frac{S - 0.3}{0.3} = \frac{0.80 \text{ in.} - 0.3}{0.3} = 1.67 \text{ in.}$$

$$
\begin{aligned}
D_s &= K \times (T_s + T_c + T_r + T_{bm}) + D_{pf} \\
&= 63 \, \frac{\text{in.}}{\text{s}} \times (0.35 \text{ s} + 0.05 \text{ s} + 0.045 \text{ s} + 0.09 \text{ s}) + 1.67 \text{ in.} \\
&= 33.705 \text{ in.} + 1.67 \text{ in.} = 35.375 \text{ in.}
\end{aligned}
$$

Presence Sensing Mats

A pressure sensing mat is a simple solution to many safety issues associated with sensing the presence of humans in an area. The full visibility and access to the work area offered by mats is their primary advantage. Another advantage is their simple mode of operation. A sensing mat is illustrated in Figure 10–9.

Sensing Mat Operation

The electrical model for a mat is a normally open switch. The switch closes when a specified minimum weight is applied to the mat. The guarded machine is halted by the mat control system when the system senses that the mat switch is closed. The mat switch is formed by two conductors separated by small insulators. Each conductor is connected to a pair of wires that are used to interface to the control circuits. The mat components are molded into a PVC structure to create a strong assembly. The resistance of the mat switch circuits falls to zero (a closed switch) when pressure is applied to the active mat area. This switch closing in the mat causes a relay in the mat controller to deenergize. When the mat controller relay turns off, contacts open and the guarded machine has power removed.

Figure 10-9 Presence Sensing Mat.
(Courtesy of Scientific Technologies Inc.)

Mats are available in various sizes and can be linked together by joining strips to provide presence sensing for a large area. Mat controllers permit multimat layouts to be combined electrically for a single machine stop signal when any mat in the group receives the minimum force. Mat controllers offer the following safety features:

- Control reliable circuitry
- Corroded-mat electrode monitor
- Mat-wiring integrity monitors
- Remote access to status, test, and reset functions
- Built-in diagnostic indicators

10-7 INTERLOCK DEVICES

The safety interlock switch is one of the most important types of protective devices because it interlocks the machine guard door with the power source of the machine. When the guard door is opened, the power to the machine is removed, which ensures the safety of the operator. There are many different types of interlock switches, and it is important to select the device type that matches the application. The general features and requirements of switches used for interlocking duties are described in the following sections.

Interlock Switch Operation

The European Standard EN 1088, *interlocking devices associated with guards,* and EN 60947-5-1 for electromechanical switches are used to design interlocks in robot and machine safety situations. The first requirement of interlock switches is reliable operation under extreme conditions and rough treatment.

A second requirement relates to security. The interlock switch must be able to overcome all attempts to *cheat* or *defeat* the mechanism. Personnel often attempt to override an interlock switch to expedite an order or to save production time. Machine usage information gathered at the risk assessment stage of the design process will help to determine if cheating is likely. Switch security levels range from resistance to impulsive tampering to being almost impossible to defeat. Interlock devices that are not accessible when the guard door is open provide the highest degree of security.

Safety standards also address the concept of *oriented failure mode*. In this mode, safety components have a known failure mode. Therefore, the device can be designed so that the guard switch contacts (the contacts that control the machine) are off when the failure occurs. Standard EN 292-2: 3.7.5 also requires that any components used in a design that are critical to the safety function should be used in duplicate. The dual circuits ensure that, should one component fail, the second will still perform the guarding function. In dual redundancy designs, it is necessary to know that one of the dual elements has failed so that the system does not continue to operate on a single working component.

Interlock Positive Mode Operation

The standard defines positive mode as the movement of one object by another through direct contact or via rigid elements. In single mechanical type interlocking switches, the movement of the guard and the switch safety contacts exhibit positive mode operation. This ensures that the contacts are physically pulled apart or *force disconnected* by the movement of the guard, and that the contacts do not rely on spring pressure to open. This feature, required by the standard, overcomes the problem of sticking or welded contacts or a weak or broken spring in the switch. Figure 10–10 illustrates negative and positive mode operation.

Forms of Interlock Devices

There are two basic types of electrical interlocking systems: *power interlocking* and *control interlocking*. In power interlocking systems, the power source of the machine is directly interrupted when the guard is opened. In control interlocking systems,

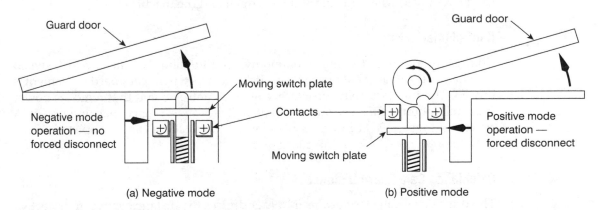

(a) Negative mode   (b) Positive mode

Figure 10–10   Negative and Positive Mode Operation of Interlock Switches.

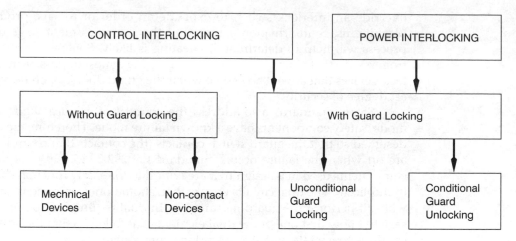

Figure 10-11 Forms of Interlocking Devices.

the power source of the hazard is interrupted by a power control device activated by the interlock device. All the current forms of interlocking are illustrated in Figure 10–11.

Power Interlocking

In this technique, guard movement is interlocked directly with the switching of the power to the machine. This means that the interlock device must be capable of switching the voltage and current required by the production machine. This works only with a small percentage of industrial machines because most machines use a three-phase voltage supply with relatively high current. When high power switching is required, one approach is to use a trapped key system, in which a key is used to open the guarded area and also to switch power to the machine. The key is trapped in position while the switch is in the on position (guard door is closed) but can be removed when the guard door is open and the machine is off.

Control Interlocking

Control interlocking, the most commonly used method, has an interlock switch attached to the guard that opens the switch contacts whenever guard movement is sensed. The interlock switch contacts trigger a control circuit that removes power from the motor starter or contactor that switches the primary machine power. A review of Figure 10–11 indicates that control interlocking can be applied with or without a physical lock on the guard door.

Calculating Safe Guard Distance

The draft European standard prEN 999 addresses the distance between protective guards and the hazard used to overcome the maximum speed an operator would use to approach a machine. The formula is

$$S = (K \times T) + C$$

where

$S =$ The minimum distance in mm from the danger zone to the opening edge of the guard.

$K =$ 1600 millimeters per second. Research data indicates that 1600 is a reasonable assumption for an approach speed by an operator. The actual application might yield a different value, but as a general guideline, a value for K from 1600 to 2500 millimeters per second is commonly used.

$T =$ The total time in (seconds) from the opening of the interlock switch contacts to the cessation of the hazard.

$C =$ An additional distance in millimeters based on how far an operator could reach over, around, or through the guard before the stop action of the safety device is initiated.

Standards EN 294 and prEN 811 provide more information on calculation of reach distances.

Example 10-2

A robot system has a protective barrier enclosing the work-cell area. The total time for the system to come to a full halt after power is removed is 1 second, and an additional distance of 4 feet is required because of the height of the guard gate. Determine the minimum distance from the guard gate to the closed part of the robot work envelope.

Solution

$$K = 1600 \text{ mm/s (chosen)}$$
$$T = 1 \text{ s}$$
$$C = 4 \text{ ft}$$
$$C = 4 \text{ ft} \times 304.8 \text{ mm/ft} = 1219 \text{ mm}$$
$$S = (K \times T) + C$$
$$= (1600 \text{ mm/s} \times 1\text{s}) + 1219 \text{ mm} = 2819 \text{ mm}$$

Mechanical Interlock Devices

Vendors supply another group of mechanical interlock devices that link movement in the guard door to positive mode switch operation. There are three main types of mechanical actuation devices: actuator operated (Figure 10–12), hinge operated (Figure 10–13), and cam operated (Figure 10–14).

The *actuator operated* device, the most common mechanical safety switch, has an actuator *tongue* (Figure 10–12) mounted to the guard gate. When the gate is closed, the tongue slides into the switch and closes the switch contacts. The contacts open when the tongue is pulled out of the switch by opening the gate. The switch pictured in Figure 10–2 is an example of this type of mechanical safety

Figure 10–12 Tongue Operated Actuation.

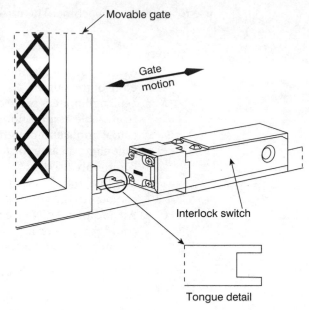

Movable gate

Gate motion

Interlock switch

Tongue detail

switch. Characteristics of this switch include the following:

- Designed to make cheating of the switch difficult.
- Easy to install.
- Reliable operation.
- Can be used on sliding, hinged, and lift-off guards.
- Alignment between the guard and switch must be maintained.
- Difficult to clean thoroughly.

Figure 10–13 Hinged Mechanical Interlock Switch.

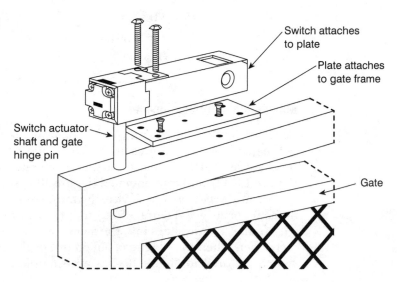

Switch attaches to plate

Plate attaches to gate frame

Switch actuator shaft and gate hinge pin

Gate

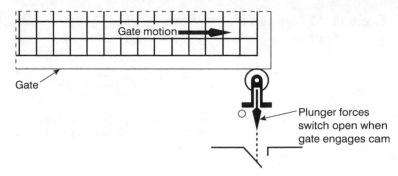

Figure 10–14 Cam Operated Mechanical Interlock.

The *hinge operated* actuation device in Figure 10–13 is mounted over the hinge pin of a hinged guard door. When the door is opened, the rotation of the hinge pin inside the switch causes a positive mode operation mechanism to open the contacts and remove power from the guarded machine. Characteristics of this switch include the following:

- Ideal for hinged guard doors where access to the hinge centerline is possible.
- Provides protection with only 3 degrees of guard door movement.
- Extremely difficult to defeat without dismantling the guard door.
- Wide guard doors have too great an opening with the 3 degree door rotation.
- Heavy doors can put too great a strain on the switch actuator shaft.

The *cam operated* actuation in Figure 10–14 is a positive mode, acting limit switch with a linear or rotary cam. As Figure 10–14 illustrates, the cam operator is most often used on sliding guards where the cam forces the plunger down to open the control circuit contacts when the guard is opened. Characteristics of this switch include the following:

- Small but reliable system.
- Not suitable for hinged or lift type guards.
- Design must ensure that cam and guard always remain in contact.
- Wear on the cam can cause failure or improper operation.

Non-contact Interlock Devices

With non-contact actuation devices, the guard door position is linked to the machine control circuit through contacts controlled by a magnetic or electric field. Figure 10–15 shows a magnetically activated reed switch used in a non-contact actuation application. These devices do not have positive mode operation, so they use other techniques to ensure that they will not compromise the safety of the system they are guarding. Two techniques discussed earlier, oriented failure mode and duplication and monitoring, are used to provide the level of performance needed. The only safety critical fault likely to occur is the welding together of the reed contacts due to excessive switch current. To prevent

Figure 10–15 Non-contact
Interlock Switch.

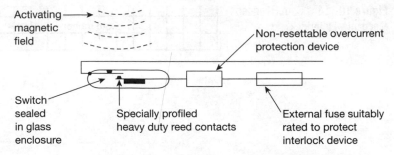

Activating magnetic field

Non-resettable overcurrent protection device

Switch sealed in glass enclosure

Specially profiled heavy duty reed contacts

External fuse suitably rated to protect interlock device

the excessive current through the contacts, an internal fuse link type overcurrent protection device, illustrated in Figure 10–15, is used. An external fuse, rated lower than the internal fuse link, is used to protect the device from inadvertent overload currents. Ordinary proximity devices activated by ferrous metal are not an appropriate replacement because they can be falsely triggered by other metal objects. An additional degree of safety is available in some non-contact switches through the use of a code that is exchanged between the actuator and the contacts. Characteristics of this switch include the following:

- Ideal for hygiene sensitive areas since the units are fully sealed and can be steam cleaned.
- Easy to install.
- Tolerate some guard wear without erratic operation.
- Offer electronic coding for a high degree of security.
- Control units (illustrated in Figure 10–16) are available to monitor up to six sensors, control the contactors, and support all system wiring.

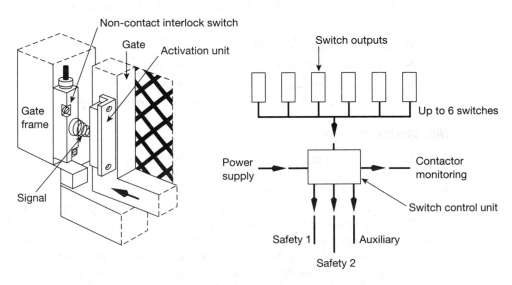

Figure 10–16 Non-contact Switch and Control Unit.

Interlock Switches with Guard Locking

A review of Figure 10–11 indicates that *guard locking* is the next level of categorizing interlocking devices. Guard locking is necessary for machines with slow rundown or stopping characteristics, but it can also provide a significant increase in protection for most types of machines. Guard locking devices are divided into two categories: unconditional guard unlocking and conditional guard unlocking.

Unconditional Guard Unlocking

A handle or knob, which is manually operated, releases the guard door lock and also opens the control circuit contacts. One example of this type of device is illustrated in Figure 10–17. The switch in the figure provides a switch contact plus a dead bolt lock for the guard gate. The bolt in Figure 10–17 is retracted by turning the knob, and the first few turns activate the switch contacts. However, the guard gate remains bolted until the operator turns the knob many more turns to extract the bolt from the gate completely. The additional turns provide about a 20-second time delay that gives the machine time to run down. Characteristics of this switch include the following:

- Easy to install.
- Rugged and reliable operation.
- Suitable mainly for sliding guards.
- Requires a predicable stopping time for the hazard being guarded.
- Stops must be used to prevent the bolt from being extended until the gate is fully closed, as indicated in Figure 10–17.

Conditional Guard Unlocking

In conditional guard unlocking, the guard can be opened if a signal shows that one of the following conditions is satisfied: the machine power contactor is off, a preset

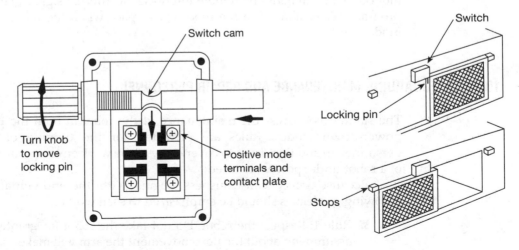

Figure 10–17 Bolt Type Unconditional Guard Unlocking Interlock.

time interval has elapsed, or dangerous motion has ceased. Signals that satisfy these conditions are derived from the auxiliary contacts on the machine power contactor, from a fail-safe timer unit, or from a stopped motion detector mounted on the machine. This technique provides interlocking of the hazard power source with guard movement but also prevents gate opening until other safe conditions are satisfied. This technique is good for use with PLC controlled machines because an externally generated signal is required for lock release.

10-8 SAFEGUARDING THE OPERATOR

The standards state that the operator safeguards should (1) prevent the operator from being in the robot work envelope when the robot is executing a program, or (2) prevent robot motion when an operator is present in the work space. In addition, operator training should prepare operators for the hazards associated with the programmed robot tasks, so they can recognize them and respond appropriately.

10-9 SAFEGUARDING THE PROGRAMMER

The standards state that the person programming robot operations or teaching translation points should have the necessary training for the programming application, check that hazards do not exist, verify that all safeguards are in place and working, and leave the work envelope before initiating the run mode. When the teach mode is used (1) the robot system and other equipment in the work envelope must be under the control of the programmer, (2) the robot system can operate at high speed under only special operational conditions, and (3) no robot motion can be initiated by remote interlocks or external signals. Finally, only the programmer is allowed in the robot work space when the robot is in the teach mode.

10-10 SAFEGUARDING MAINTENANCE AND REPAIR PERSONNEL

The standards address the maintenance of the robot in both the power-up and power-down modes. Rules are provided for the training of maintenance personnel; in addition, rules to cover entry of the work envelope and maintenance of a robot under power are listed.

Because safety consciousness begins with the individual worker, the following guidelines should be emphasized to each worker:

- Rule 1: Respect the robot. Do not take the robot for granted or make an assumption about the next movement the arm will make.
- Rule 2: Know where the closest emergency stop button is at all times.

■ Rule 3: Avoid locations in a robot work cell between the robot arm and fixed objects like metal posts or production machines. (These types of locations are called pinch points.)
■ Rule 4: Know the robot. Pay attention to unusual noises and vibrations from the machinery.

10-11 DEVELOPING A SAFETY STRATEGY

The employer has a statutory requirement to provide a safe working environment. But safety is a process that does not just happen; it occurs because of a deliberate strategy. The regulations clearly state that the safety of work equipment is viewed from three aspects: (1) its initial integrity, (2) the place where it is used, and (3) the purpose for which it is used. Therefore, a strategy to make automated work cells safe is critical. This strategy has two major elements: *risk assessment* and *risk reduction.*

Risk assessment, like analysis of all manufacturing operations, starts with a clear understanding of the work-cell limits and functions. Analyses of these cell characteristics are used to identify those that pose a potential hazard. With the hazards known, the degree of risk can be estimated. All this data is needed to determine the effectiveness of existing safety measures and the need for additional measures to reduce the risk.

The second step in the strategy is to apply risk reduction to those areas that need additional protection. The process is then repeated to verify that improved safety has been achieved by the reduction effort. This process forms the basis for the safety strategy, and a flowchart describing all the steps is illustrated in Figure 10–18. A checklist, like the one illustrated in Figure 10–19, should be used to ensure that all areas are included in the process and to make the results available for review in the future. The following sections address specific techniques used to perform risk assessment and reduction, and they can be applied to a single robot or an entire automated production system.

Risk Assessment

Risk assessment is necessary because it is the starting point for every safety strategy and, in the European Community, it is a legal requirement covered by the standard prEN 1050, *Principles for Risk Assessment.* The first step in assessment is the determination of machine limits and hazards. A complete list should be made of all single machines and systems where separate machines are linked together mechanically or with control systems. Each machine and system of machines is then analyzed to determine if it presents any hazard. This process is applied to machine installation, commissioning, maintenance, and decommissioning with consideration of correct use and operation as well as reasonably foreseeable misuse or malfunction. Types of hazards include crushing, shearing, entanglement, part ejection, fumes, radiation, toxic substances, heat, and noise. The hazards are identified and recorded for each machine. After the hazards are known, risk estimation can proceed.

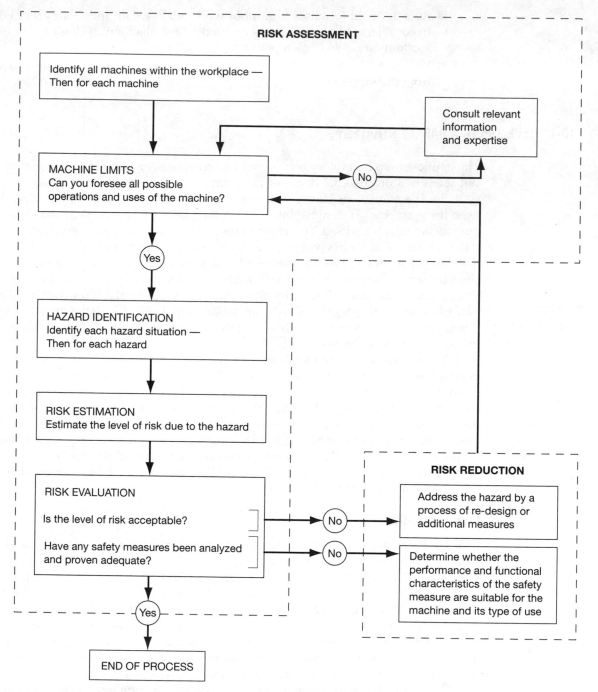

Figure 10–18 Safety Strategy Flowchart.
(Courtesy of Scientific Technologies Inc.)

Evaluation of Work Cell Safety Hazards

Evaluator: _____ Date/time: _____

Plant: _____ Building: _____ Dept: _____ Location: _____

Description of Product: _____

Check all boxes below that apply to the work cell under study.

Drive System Hazards		Current Machine Safeguarding	
Belt/pulleys	☐	Safety zone clearly marked	☐
Chains/sprockets	☐	Equipment barriers	☐
Brakes/clutches	☐	Fixed fence barriers	☐
Couplings	☐	Automatic doors	☐
Cams/cranks	☐	Sensing devices	
Gears	☐	Mats	☐
Shafts	☐	Light curtain	☐
Spindles	☐	Single beam photoelectric	☐
Pneumatic robots	☐	Presence sensors	☐
Servo robots	☐	Interlocks on gates/doors	☐
Automatic doors	☐	Safety Controls	
Other: _____	☐	Single start switch	☐
		Dual start switch	☐
Process System Hazards		Emergency stop switch	☐
Milling	☐	Push button	☐
Turning	☐	Mushroom	☐
Drilling/boring	☐	Red color	☐
Forging	☐	Within 26 inches	☐
Punching	☐	On/off	☐
Shaping	☐	Push button	☐
Forming	☐	Recessed	☐
Other: _____	☐	Selector	☐
		Emergency stop cord	☐
Support System Hazards		Slack line	☐
Pneumatic devices	☐	Tensioned	☐
Hydraulic devices	☐	Lockout/tagout procedures	☐
Belt conveyors	☐	Machine power failure restart	
Chain conveyors	☐	Manual	☐
Automatic guided vehicles	☐	Automatic	☐
Automatic inspection	☐	Power disconnects	☐
Parts feeders	☐	Warning lights	☐
Other: _____	☐	Other: _____	☐

Place a sketch of work area on the back of the form.

Comments: _____

_____

(Continue comments on the back of the form.)

J. Rehg 9/6/00

Figure 10–19 Safety Evaluation Form.

Risk Estimation

Risk estimation is the most fundamental aspect of machine safety, and it is addressed in this section using a simple but effective approach. Risk is present in all machinery; therefore, it is important to describe the level of risk on a relative scale from minimum to maximum. The following four rules should be applied:

1. A logical work pattern and full documentation of the risk estimation are required.
2. The severity of potential harm and the probability of its occurrence are used to estimate the amount of risk present.
3. Include all hazards identified in the risk assessment stage.
4. Every machine and production cell dictates a unique risk estimation process; therefore, a standard, like prEN 1050, should be the final determination tool.

With the two major elements of a safety strategy (risk assessment and risk estimation) defined, a process for performing each must be developed.

Risk Assessment—Three Steps

Risk assessment must consider two questions: What is the severity of the potential injury? and What is the probability of the injury occurring? The following three-step process is required to answer these questions:

1. Determine the severity of the potential injury.
2. Determine the frequency of exposure.
3. Determine the probability of injury.

Severity of Potential Injury.   A careful study of the hazard reveals the severity of the injury that could occur to personnel. The injuries are grouped into the following four categories:

1. *Fatal*—loss of life (10 rating).
2. *Major*—Non-reversible injury with permanent disability (6 rating).
3. *Serious*—Reversible injury but a significant loss of work time is expected (3 rating).
4. *Minor*—Injury involving cuts, bruises, or light abrasions (1 rating).

The risk assessment starts with an evaluation that places each hazard into one of the four injury categories and assigns the rating value listed.

Example 10–3
Assess the risk in the following situations and assign a potential injury rating.
   a) A skilled machine operator must unload large sheet metal stampings from a press. On occasion after a die change, she has to lean her upper body into the press to remove excess material from the back of the die.

b) Two unskilled parts handlers load plates into a fixture in a robot welding work cell and unload the finished product. The robot is in a pause mode while they work.

Solution

a) The severity is "fatal" because a cycle of the press with the upper body over the die would cause death—rating 10.

b) The severity is "serious" because a robot in the pause mode could make an inadvertent move that can cause serious injury to operators in the work envelope—rating 3.

Frequency of Exposure.   An analysis of the operation determines how frequently the personnel are exposed to the hazard. The rating for this assessment element indicates that the more frequent the exposure, the greater the danger of injury. The rating groups are listed below:

1. *Frequent*—several times per day (rating 4).
2. *Occasional*—daily (rating 2).
3. *Seldom*—weekly or less (rating 1).

Example 10-4

Assess the risk in the following situations and assign a frequency of exposure rating.

a) A skilled machine operator must unload large sheet metal stampings from a press. On occasion after a die change, she has to lean her upper body into the press to remove excess material from the back of the die.

b) Two unskilled parts handlers load plates into a fixture in a robot welding work cell and unload the finished product. The robot is in a pause mode while they work.

Solution

a) The die changes occur every few days, so the rating is between "occasional" and "seldom"—rating 1.5.

b) The cycle time for the welds is 20 minutes, so the rating is "frequent"—rating 4.

In dealing with these factors, you must use any data and expertise available, and the decisions must be based on the worst case assessment. In addition it is assumed that a protective system is not present or has failed to prevent an unprotected state. Therefore, the analysis assumes that the unexpected could happen.

Probability of Injury.   This last factor, probability of injury, considers the working relationship between the personnel and the machine and other factors, like machine speed at start-up. The ratings are determined from the following possibilities of injury:

1. *Certain* (rating 6).
2. *Probable* (rating 4).
3. *Possible* (rating 2).
4. *Unlikely* (rating 1).

Example 10-5

Assess the risk in the following situations and assign a probability of injury rating.

a) A skilled machine operator must unload large sheet metal stampings from a press. On occasion after a die change, she has to lean her upper body into the press to remove excess material from the back of the die.

b) Two unskilled parts handlers load plates into a fixture in a robot welding work cell and unload the finished product. The robot is in a pause mode while they work.

Solution

a) Based on the machine and the problem involved, the risk is possible—rating 2.

b) The machine speed and proximity of the operators to the robot make the risk probable—rating 4.

Final Assessment Rating

The final assessment rating is the sum of the three evaluations plus an adjustment for additional factors such as those shown in Table 10–1. Depending on the type and use of the machinery, other relevant factors might have to be considered at this stage.

Example 10-6

Determine the final rating for the following situations and include any special factors present.

a) A skilled machine operator must unload large sheet metal stampings from a press. On occasion after a die change, she has to lean her upper body into the press to remove excess material from the back of the die.

Table 10–1 Additional Risk Estimation Factors.

FACTOR	ACTION
More than one operator exposed to the hazard.	Multiply the severity factor by the number of people.
Protracted time in the danger zone without complete power isolation.	If time spent per access is more than 15 mins. add 1 point to the frequency factor.
Operator is unskilled or untrained.	Add 2 points to the total.

b) Two unskilled parts handlers load plates into a fixture in a robot welding work cell and unload the finished product. The robot is in a pause mode while they work.

Solution

a) The initial ratings are a *severity* of 10, a *frequency of exposure* of 1.5, and a *possibility* of 2, for a total of 13.5 out of a possible total of 20. No special factors apply to this situation, so the total of 13.5 for the safety assessment is unchanged.

b) The initial ratings are a *severity* of 3, a *frequency of exposure* of 4, and a *probability* of 4, for a total of 11 out of a possible total of 20. However, two special factors apply to this situation. First, the severity factor must be multiplied by 2 because 2 operators are present (the *severity* rating becomes $2 \times 3 = 6$), and 2 additional points must be added to the total because the operators are unskilled. So the final rating becomes $6 + 4 + 4 + 2 = 16$ out of a possible total of 20.

Analysis of the two work situations in the example problems yields results that emphasize the value of safety strategy over just a reaction to a situation. The situation described in part a, where the operator could be fatally injured, looked like a more critical safety issue when compared to the two operators working around a robot in a welding cell. However, in the final, more formal analysis, the safety rating for the press operator was 13.5 and the rating for the robot welding cell was 16. Both areas need safety emphasis and work, but the welding cell has a higher probability of the hazard producing harm to personnel. The safety strategy described in this section offers a systematic approach to analyze and rank all the hazards uncovered in the manufacturing area. What remains is the action necessary to minimize those hazards.

Hazard Reduction

Each machine and associated hazard must be addressed, and the risk present must be eliminated or reduced. The standard suggests three basic methods in priority order (the most important method is listed first):

1. Eliminate or reduce risks to the degree possible by focusing on safe machinery design.
2. Implement protection measures for every risk that cannot be eliminated in the machine design.
3. Make users aware of the residual risks due to imperfect machine design and protection measures, and emphasize the required training and the need for personal protection equipment.

The hierarchy of machine protection measures most frequently used includes the following:

1. Fixed enclosing guards.
2. Movable interlocked guards or protection devices in the form of light curtains, presence sensing mats, or other effective devices.

3. Protection appliances such as jigs, holders, and push sticks used in conjunction with guards that permit operators to feed or retrieve a work piece while keeping their bodies clear of the danger zone.

4. Easy access to information, instruction, training, and supervision.

While training in safe working methods for a machine is important, it does not permit any less diligence on the part of the employer for the first three measures listed.

Personnel Protection Equipment

Despite all the efforts of equipment manufacturers to make production machines and systems safe, the operators are often required to wear protective equipment such as special gloves, goggles, respirators, and clothing. The hierarchy of machinery protection measures listed above forms the primary safeguard against personnel injury, and the personal protective equipment used by the operators complements the machine measures. A combination of machine measures and protective equipment is frequently used to ensure a safe operating environment for the factory floor.

10-12 DESIGN GUIDELINES

The goal of every manufacturing operation is to increase productivity and profits. Safety is often compromised by operators and manufacturing administrators in an effort to meet that goal. Satisfying safety needs and meeting productivity goals is the difficult job faced by the safety engineer and system designer. The standards are the starting point for this evaluation because they provide the guidance needed to develop and implement machine safeguarding. The minimum legal requirements are covered in the federal regulations listed earlier in this chapter. The common element in the U.S. standards and European Machinery Directive is that each machine hazard should have a uniformly evaluative process for determining the safety issues present and the corrective actions required. The technique and priority for hazard control used by OSHA/ANSI is:

1. Use machine design to eliminate the hazard.
2. Use guarding to control the hazard.
3. Use warning signs and devices.
4. Use personal protective equipment.
5. Provide training to all personnel.

The safety design process must include the risk assessment and evaluation to verify that every hazard has been identified and properly evaluated. In addition, the design must have a process for addressing all five hazard control techniques list above. Safety equipment vendors offer a wide range of devices, components, and accessories described earlier in the chapter to help control hazards.

Justification of the safety management program is difficult because most changes and requirements are not value added but cost added to the product. Making the link between safety spending and product value is not easy, but the safety program protects employees, the company's most valuable asset. One tangible result of the program is a reduction in *loss costs*. Loss costs are financial burdens to the company in the form of lost revenue and cash outflows that result from safety related injury to personnel and equipment. The safety management program can be cost justified from a well-documented reduction in loss cost. The documentation of loss cost includes the immediately measurable factors like lost work hours and medical program costs. However, the *real* cost of accidents are often not factored into individual unit cost or the bottom line profit of the company.

Two theories address the real cost of accidents: (1) the *Heinrich law,* also called the *domino theory*, which identifies the relationship of unsafe acts and practices to accidents, and (2) the *Dupont accident theory,* which specifies that *all accidents are preventable through the assignment of responsibility and measured accountability.* Heinrich's studies indicated that *one lost time accident occurs for every 3000 near misses or unsafe behaviors.* The data from the Heinrich and Dupont work reveal that loss costs can be reduced significantly with a safety management program built around the safety strategy and design sequence described earlier. The real loss cost savings include many of the following hidden costs associated with unsafe machine operation:

- Lost time expense includes the time consumed by the accident or injury.
- Cost of machine downtime includes time lost on the production machine, as well as time lost on other machines in the area, due to the accident.
- Cost of repairs to a damaged machine.
- Time spent on accident investigations and completing the many statutory requirements such as filling out federal and state forms.
- Loss of production due to the loss of an experienced employee.
- The cost of federal and state penalties, which can range from $0 to $70,000 per machine for a willful violation. (Table 10–2 shows citations for some OSHA standards that cover industrial robots.)

Table 10–2 Most Frequently Cited OSHA Violations: 1990–1995.

	1995	1994	1993	1992	1991	1990
1910.212(a)(1) Machine guarding (general)	1446	1887	1728	1764	4090	4706
1910.215(b)(9) Grinding wheel guards	1252	1737	1704	1645	3405	3974
1910.219(d)(1) Mechanical power presses	1040	1375	1394	1450	—	5899
1910.212(a)(3) Point of operation guarding	984	906	1075	—	—	—

- Cost of training for the replacement worker.
- Cost of overtime to recover the lost production.

Studies at Dupont indicate that hidden costs are often 4 to 10 times the visual cost covered by the insurance carrier. Careful documentation of all loss costs associated with manufacturing accidents is the first step in justifying fully the cost of safety management.

10-14 SUMMARY

Safety is an important factor in robotic cell design—from the beginning of the design to the implementation of the process. Specific areas of safety awareness in a robotic installation are addressed in the ANSI/RIA R15.06 Safety Requirements standard. This standard is the OSHA and industry primary reference on robot safety issues governing the design and operation of robotic cells. The standard addresses the following areas: construction, reconstruction, and modification; safeguarding; maintenance; testing and start-up; applications; continuous operation; and training.

Safeguarding considers four issues: devices to be used, operator safety, safety of the teacher, and safeguarding maintenance and repair personnel. The recommended devices include barriers, interlock barriers, perimeter guarding, awareness barriers, and awareness signals. The primary functions of these devices are to provide a warning and to restrict traffic flow in the area. In many applications the robotic work cell is divided into two work zones: inside the cell but outside the work envelope and inside the cell and inside the work envelope. The barriers or markers used to identify these areas include painted lines on the floor to mark the limits of the work envelope, chains and guard posts, safety rails, wire mesh fencing, and equipment within the work cell. When detection devices are triggered, either the robot is stopped and the actuator drive power is removed, or all automatic operation of the robot and work cell is halted. Awareness signals are often provided to indicate that a security area has been entered. Electronic curtains, motion detectors, and pressure-sensitive floor pads detect unauthorized penetration.

Operators cannot be in the work envelope when a robot is executing a program, and all personnel responsible for operation of the work cell must have training in robot operation. To protect the robot programmer, the system must transfer control of all work-cell equipment to the programmer during robot programming, move the robot at reduced speed, and not be triggered by any interlocks or external signals. Also, only the programmer is allowed in the work envelope during the point teaching process.

To safeguard maintenance personnel, the following rules must be followed: respect the robot, don't take the robot for granted or make an assumption about the next movement the arm will make, know where the closest emergency stop button is at all times, avoid pinch points at all times, and pay attention to unusual noises and vibrations from the machinery.

Asimov's three rules are quite simple, but they clearly state the requirements for robot safety, namely, protect humans, other equipment, and the robots themselves from harm.

A safety strategy includes a set of techniques for making work cells safe. The two primary techniques are risk assessment and risk reduction. A major part of risk assessment is risk estimation, which determines the severity of potential injury, frequency of exposure, and probability of injury. Hazard reduction is the last step in the strategy, and it includes fixed enclosing guards, interlocks and presence sensing devices, protection equipment, and education and preparation of the workers exposed to the hazard.

The design of safe work areas uses standards data and follows a five-step process: (1) safe machine design, (2) guarding, (3) warning signs and devices, (4) personal protection equipment, and (5) training. Justification of the cost of the design must include the visible part of the loss cost due to accidents and the hidden costs that result from every injury. Some of the hidden costs include cost due to machine downtime, machine repairs, administrative time cost, overtime and time lost due to a less experienced replacement worker, and training of a new worker.

QUESTIONS

1. What are Asimov's three laws of robotics?
2. What is the greatest concern in the work cell from a safety standpoint?
3. What are OSHA recommendations on robot safety?
4. What areas of robot work-cell safety are covered by the ANSI/RIA standard on robot safety?
5. What rules should be followed to protect the operator and maintenance personnel?
6. How do corporate safety standards compare to government regulatory organizations?
7. Name two examples of physical barriers used to protect against a hazard.
8. Describe three types of presence sensing devices.
9. How are E-stops different from interlock devices?
10. Why do the standards require safety control units in safety systems instead of standard programmable logic controllers?
11. How are laser proximity-sensing devices used to guard a hazardous machine?
12. Describe the operation of light curtains.
13. Why are light curtains superior to barriers for machine guarding?
14. Describe the concept of *control reliability* and include an example.
15. Define the light curtain terms *point of operation, muting,* and *blanking.*
16. How do perimeter guards differ from machine guards?
17. What does the term *safe curtain distance* mean?

18. Describe the operation of presence sensing mats.

19. What is the basic function of interlocking devices?

20. What are the two basic requirements for interlock switch operation?

21. What does the term *oriented failure mode* mean?

22. Why is it necessary to monitor dual redundant circuits to know if one of them has failed?

23. What are the characteristics of positive mode operation switches, and what unsafe condition do they prevent?

24. Compare and contrast power interlocks with control interlocks.

25. Describe the concept of safe guarding distance.

26. Compare and contrast the three main types of mechanical interlock devices.

27. Describe how one type of non-contact interlock device operates.

28. What is guard locking and why is it used?

29. What is the difference between conditional and unconditional guard locking?

30. How is risk assessment different from risk estimation?

31. What information is risk estimation trying to predict?

32. Describe the three-step process used in hazard reduction.

33. Describe the measures used to protect machines that present a hazard.

34. Describe the five-step safety design process.

35. Define the terms *loss cost* and *real loss cost*.

PROBLEMS

1. Calculate the safe curtain distance for a machine that has a stopping time of 0.25 seconds and a control circuit response time of 0.07 seconds. The break monitor is set at 0.4 seconds, the response time of the light curtain is 0.035 seconds, and the minimum object sensitivity is 0.65 inches.

2. A robot has a stopping time and control circuit response time of 0.30 seconds, and the brake monitor factor is negligible. Due to the work-cell geometry, the guarding light curtain at the entrance to the cell must be 36 inches from the outer edge of the work envelope. Use vendor data sheets in the Thomas Register link in Appendix B or another Internet source to find a light curtain with a response time and object sensitivity that will satisfy this design. Calculate the actual minimum safe distance for the selected curtain, and attach a copy of the specifications for the selected light curtain to the back of the solution.

3. The equation for the added distance for the depth penetration factor D_{pf} uses a sensitivity value with units of inches and calculates the added distance in inches. Determine what the equivalent equation would be for the metric system. (*Hint:* The equation represents a straight line relationship between the two variables.)

4. A robot system with a turn-off time of 3.5 seconds must be guarded by an interlock gate. The five-foot-high gate requires an additional distance of 0.5 meters,

and an approach speed of 2000 mm per second is needed. Determine the minimum distance from the interlocked gate to the closest part of the robot work envelope.

5. Convert the safe guarding distance formula for use with English units of feet and seconds.

6. A robot in a work cell must be replaced. The work cell has a standard mechanical interlocked gate that is 2.5 feet from the closest edge of the robot work envelope, and the added distance for the safety system is set at 1 foot. Determine the maximum stopping time for the new robot for the current guard conditions.

7. A proximity laser scanner (PLS) is the guard device for a robot work cell. The jointed arm robot has a maximum work envelope, including the gripper, of 9.5 feet, and the response time of the robot system is 1.5 seconds. Determine the safety zone radius required for the system with a 50 percent safety factor. Use the PLS sensitivity in Figure 10–5 to determine the value for C, the additional distance. Also calculate the early warning zone radius so that a minimum warning of 2 seconds would be provided before the safety zone is entered.

8. The robot selected as the replacement in the previous problem cannot meet the maximum stopping-time specifications. Describe two other design options, including hardware specifications, that would not require relocating the guard fence and gate.

9. Complete a risk assessment for the following problems and determine the rating for the potential for injury, frequency of exposure, and probability of injury and the total rating.

   a) An unskilled operator must enter the work cell in Figure 2–9 to refill the quench water tank once every shift. The operator enters the cell when the robot is waiting for the die-cast machine to complete the cycle and adds the necessary water to the tank.

   b) An unskilled worker loads a forging press with hot metal slugs. The press ejects the finished forging automatically, but occasionally a forged part gets stuck in the die. The operator uses pliers to free the part from the die before adding a new slug.

10. Suggest a solution that would reduce the risk to the operators for the two unsafe conditions in the previous problem.

PROJECTS AND CASE STUDY PROBLEMS

1. Develop a complete safety strategy for the die casting production cell illustrated in Figure 2–9. Based on the safety strategy, develop a safety system to make the cell conform to minimum OSHA and ANSI standards. Use the Thomas Register link provided in Appendix B to select the necessary safety devices for your design. Document all assumptions made; include the checklist

in Figure 10–19; and list all the devices selected, including vendor name, device name, description, and model/part number.

2. Develop a complete safety strategy for the assembly cell illustrated in Figures 3–6 and 3–7. Based on the safety strategy, develop a safety system to make the cell conform to minimum OSHA and ANSI standards. Use the Thomas Register link in Appendix B to select the necessary safety devices for your design. Document all assumptions made and list all the devices selected, including vendor name, device name, description, and model/part number.

3. Apply the safety concepts to one or more of the following automation work cells. After you select an automation case, review the solutions for that case in the previous chapters and extract any data relevant to your needs here. The design component of the case studies varies in complexity and design detail, so it may be necessary to make some assumptions as you apply safety standards and hardware to the cases. Document and support all assumptions.

   a) The kitchenware bowl production cell started in Case Study Problem 5 in Chapter 2

   b) The West-Electric slug production cell design started in Case Study Problem 8 in Chapter 2

   c) The West-Electric slug lubrication cell design started in Case Study Problem 9 in Chapter 2

   d) The kitchen products production cell design started in Case Study Problem 3 in Chapter 3

4. Read Section 12–14 in the West-Electric case study in Chapter 12 where safety issues in the cell are addressed. Develop a complete safety strategy for the work cell design(s). Develop a design for a safety system to make the cell conform to minimum OSHA and ANSI standards. Use the Thomas Register link in Appendix B to select the necessary safety devices for your design. Document all assumptions made and list all the devices selected, including vendor name, device name, description, and model/part number. Complete the questions and problems at the end of Chapter 12 that relate to the safety system design.

Human Interface: Operator Training, Acceptance, and Problems

Chapter

11

CHAPTER GOALS AND OBJECTIVES

Industry's current focus on manufacturing automation and the enormous strides made in technology often make us forget that humans are still a vital part of manufacturing, even automated manufacturing. The goal of this chapter is to highlight the importance of the human interface present in all of automation. At the completion of this chapter you should be able to:

- Describe the general training program for all employees where robot automation is being implemented.
- Describe the operator and maintenance training program for affected employees where robot automation is being implemented.
- Describe the concept and operation of self-directed work teams.
- Discuss the area of resistance to automation normally present and some techniques used to overcome it.
- Describe the effect of 24–7 on manufacturing and on human workers.

11-1 INTRODUCTION

Automated processes are most frequently implemented because of a need to increase productivity or product quality. The impact of automation is felt by almost every segment of society, from the suppliers of automated hardware to the workers who will be operating the system, to the employees displaced by automation, to the consumers of the products of automation. The plan for implementing an automated system usually includes sufficient time and resources to solve the engineering problems but often overlooks the need for an equal emphasis on the human interface present in all such systems.

The following three distinct training activities should be implemented by every industrial plant that considers introducing robots or work-cell automation in the production process:

1. General employee awareness training with an overview of the new types of automation, especially robots
2. Operation and programming training programs for all personnel involved in the production process
3. Maintenance training programs for all personnel who will be responsible for maintaining the robot system

The need, type, and target groups for which training must be planned are included in the ANSI/RIA R15.06 Safety Requirements standard. The standard states that the user must provide the following: (1) training for employees who program, teach, operate, maintain, or repair robots or robot systems; (2) training in standard safety procedures and the robot vendor safety recommendations; and (3) integration of training in safety precautions and procedures for the specific robots, installation, and applications across all project activity.

Plans should be made to have the appropriate phases of the training occur before, during, and after the implementation process.

11-2 GENERAL TRAINING

"Robots are just another form of automation" is a statement that has often been made by those responsible for developing automated manufacturing systems. The statement can be supported from the hardware standpoint without difficulty, but because of the way robots are perceived by both management and labor, training techniques for robot projects must be especially well thought out. Although the difference between robotic automation and automation in general is more psychological than material, it is a potent force in the minds of employees and management. For workers, robots create a higher level of anxiety over possible job loss than do other automated processes, and robots also encounter greater resistance at the management level. Management resistance starts at the level of the first-line supervisor and extends up through varying levels of middle management. Therefore, a well-structured general training program is necessary to deal with personnel concerns about robot automation.

General Training Program

The general training program to prepare all levels of labor and management for the implementation of robots and integrated automation should include the following procedures:

- An overview of current automation practices used by industry in general and competitive industry in particular
- A thorough explanation of the need for broad-based manufacturing automation based on production and marketing data
- A complete description of robot automation including what robots are and what they can and cannot do
- A comprehensive statement of company policy regarding jobs eliminated or changed as a result of automation
- A detailed plan describing how all employees affected by the automation will be retrained
- A clear statement of support from the highest level of management for including automation in any productivity improvement process

Today robots are rarely used alone; they are part of a work-cell automation system. As a result, training must go beyond just that required for the robot. The training must have a broad focus and address all the computer-controlled hardware that is integrated into the manufacturing system.

The audience receiving this general training is extremely varied, both in the depth of information needed and the emphasis on information delivered. The management level, for example, would profit from a complete overview of current and future robot and system automation for the industry and from training in the specific skills required to implement the integrated robotic cells. The production worker, conversely, would be most interested in the company's plans for workers displaced as a result of automation.

Timing of the training is also important. Starting well in advance of the first robot implementation and automation project is important, but even those companies that currently use robots in automated cells would profit from a general training program before the next installation. Continued information exchange during and after installation is important to keep everyone involved in the current project or in future projects up to date.

Training techniques for general audiences are as varied as the groups of listeners they address. General information on robots and CIM may be a topic for an employee newsletter or company paper. Bulletin boards, quality circle groups, shop and department meetings, and seminars can also be effective training media.

The rationale for general training programs of this nature is simply that successful implementation requires a team effort from every worker, and informed workers are better team players.

11-3 OPERATOR TRAINING

The term *robot operator* remains undefined. In one case the human operator may simply activate the work cell at the start of the shift, in other situations the operator works side by side with the robot, and in others a team of workers may be responsible for a broad range of cell activities. The list of responsibilities for a robot operator or cell team may include programming, program loading, work-cell support loading raw material into fixtures, unloading finished parts, and routine system checkout. Currently, support engineering groups do most of the programming and system checkout, and operators perform that portion of their former job that the robot automated work cell cannot do. Operator training related to production in the work cell generally occurs in the plant facilities.

Most robot and automation system vendors recommend that training at their facilities begin 2 to 4 weeks before the machine arrives.

11-4 MAINTENANCE TRAINING

To understand the maintenance requirements for robots and automated work cells, it is necessary to review the type of hardware that will be present. Robots and integrated production hardware currently available are powered by hydraulics, pneumatics, or electric drives. The electric drives are both dc and ac servo, but ac drives are becoming the most frequently used on new equipment. The mechanism itself has the standard gear, pulley, belt and cam linkages and drives found on automated machines used in manufacturing. Every robot system has a controller, which is either a special-purpose computer or a programmable logic controller. The special-purpose computers use either microprocessor technology or minicomputer technology as the central processing unit. The robot work cell includes all the manufacturing hardware required for production, one or more robots, sensors, and material-handling equipment. The cells have some type of cell-control computer connected to an area controller by a local area network. In many installations the production cells are linked to a corporate host computer and share a single database for all production information.

Maintenance training on the robot, work-cell manufacturing hardware, and the CIM system has generally become a routine procedure. Robot vendors, like most manufacturers of production hardware, have 1- and 2-week training programs designed to train maintenance personnel. The training includes preventive maintenance and troubleshooting of hardware failures. Using the buil-in diagnostics in the system, an operator can isolate most of the problems to a printed circuit board so that troubleshooting at the board level is all that is required. If a failure cannot be located, the vendors have field service personnel who will come to a facility and do in-plant repair.

Maintenance training at the system level rests squarely on the shoulders of the end user. The work cell typically has equipment from five or more vendors, and each piece of hardware may contain an electronic control system or computer. The interfacing of the hardware to achieve coordinated operation is usually the responsibility of corporate engineering or a systems engineering company. Because of the unique nature of each package, diagnostic tests are usually not available for use by maintenance personnel, and system failures frequently require troubleshooting to the device level. Development of an effective maintenance training program for the system level in a manufacturing work cell presents the stiffest challenge in the training area.

11-5 TEAM-BASED MANUFACTURING

The dramatic changes in technology and the global market have forced the manufacturing industry to use more work teams. The worldwide market demands a rapid response to customer needs, frequent product introductions, increased productivity, lower production costs, and higher quality. Achieving these *order-winning criteria* with the work force management policies of the 1950s and 1960s is impossible. Many companies have introduced *self-directed work teams* to give them the competitive edge. The definition of a self-directed work team in manufacturing varies. The primary issue is the authority to act and the power to carry the actions to completion. In the purest definition of a self-directed work team, team members are given the broad authority to solve process and personnel problems in their work area. Most companies have not reached that level of employee enpowerment; therefore, the self-directed work team is an expansion of the concept of the *manufacturing process improvement team*.

The following sections describe the concept of the self-directed work team.

Description of a Self-directed Work Team

Several common, generic characteristics describe a work team:

- Frequently a self-directed work team is formed from a functional work group currently in a process area. The term *functional work group* means that all members of the team come from a single process area.
- The work team meets as a functional group to manage the process area and resolve problems that reduce the effectiveness and profitability of the process. In many organizations the process improvement performed by work teams is an extension of the continuous improvement process initiated for the implementation of CIM automation.
- Team members, trained to work in a self-managed mode, take responsibility of the work area in which they function. Groups are managed by natural leadership from within the group.

■ The team follows a five-step process to improve the production process in its area: (1) identification of a process problem, (2) development of a process improvement plan, (3) implementation of the process improvement, (4) tracking of plan effectiveness through data collection, and (5) evaluation of process data to determine performance.

In many companies the team members receive financial rewards from team-implemented improvements in the process.

Making Work Teams Work

The first issue is the *readiness* of the workforce. No amount of training can change a workforce that is not ready to accept the responsibility placed on a self-directed team. Readiness implies that the workers understand the operation of the process and want ownership of it. When asked, "Could you do your job better if something were changed?" workers ready for the team process know specifically what needs to be changed to improve the performance of the manufacturing process. On the other hand, workers who are not interested in being part of the improvement process make poor team members.

Successful teams generally have members who know how the production system functions, have a high level of motivation for improvement in the process, and have a sense of ownership of their jobs. If these conditions are present, the group is ready to be a self-directed work team.

The transition to a self-directed work team is smooth when four rules are followed: (1) management must understand the issues and transfer the authority to act to the functional team; (2) a plan, based on local culture and existing readiness, must be developed for the specific plant location; (3) the team must receive training in self-management skills; and (4) the transition of power to the team must be incremental and spread over a number of years.

Companies using the team concept have an advantage over manufacturers who choose to maintain the traditional supervisor-worker relationship. In organizations where the workers stop thinking when they arrive for work, only a handful of employees are working on the solution to enterprise problems. In contrast, every employee is working on the solution to enterprise problems in a company that adopts the concept of work teams successfully. Over time, the company using work teams will surpass the mindless competition.

11-6 RESISTANCE

Resistance to automation is an age-old problem. An industrial consultant who worked with General Electric Company on reducing resistance to robots suggests the following three guidelines to help overcome management and worker resistance:

1. Organizations may not install robots to the economic, social, and physical detriment of workers or management.

2. Organizations may not install robots through devious or closed strategies that reflect distrust or disregard for the workforce, because surely they will fulfill their own prophecy.

3. Organizations may install robots only on those tasks that, although currently performed by a person, are tasks that the person performs as a robot would.

Many manufacturers may find it difficult to adhere to the ethics expressed in these guidelines, but maintaining some degree of openness, as in guideline 2, will certainly aid in the implementation of manufacturing automation.

Resistance to robot and manufacturing automation at the management level stems from several causes:

- Concern by middle managers that an unsuccessful automation project will endanger their future promotions
- Fear by first-line supervisors that robot and manufacturing automation will cause additional employee problems and that they will not have management's support in solving them
- Concern by first-line supervisors that production output will drop during the transition to automation and that they will be held accountable
- Conflicts concerning responsibility for the project, along with jealousy between manufacturing units and disagreements about distribution of resources

Resistance to robot automation at the labor and management levels arises basically from a fear of the unknown. This could be fear of losing a job, fear of being transferred, or fear of the demands of retraining. Much of the resistance resulting from these fears and those of managers can, however, be neutralized by an effective training program at the three levels defined earlier.

11-7 ORGANIZED LABOR

Because organized labor believes that new union members do not want to do many of the tasks that are dirty, dangerous, or dull, it now officially welcomes robots and manufacturing automation systems. In addition, labor realizes that increased productivity is necessary. Organized labor, however, stands firm on two demands: (1) displacement of workers must be gradual, and (2) labor should get a percentage of the benefits that result from robot automation.

Organized labor acknowledges the need to remain competitive with foreign production through implementation of enterprise-wide automation, but it is reluctant to give management a free hand in the implementation. Unions still fear large-scale displacement of blue-collar workers as a result of automation. Unions are based on the principle of solidarity among workers, and as the ranks decline so does the operating revenue of unions. Job security and training continue to be the major issues in most contract negotiations. Union membership drives will continue to concentrate on the growing number of white-collar workers, especially those involved with manufacturing automation.

11-8 IMPACT OF 24–7

The expression 24–7 implies that something is present for 24 hours a day and for 7 days a week. Thirty years ago the term rarely applied to manufacturing because only a few industries operated round the clock. Many of the chemical and most of the steel manufacturers had to operate continuously because it was impossible to shut down the process at the end of the day and restart it in the morning. The production of nylon yarn or thread is a good example. Chemicals react with heat and pressure to produce liquid nylon. The liquid nylon is forced out through devices called spinnerets that produce the thin strands of nylon yarn. Once the process is started it must be continued, or the nylon will solidify in the process tanks and the many feet of pipes it passes through during the production process. In is not uncommon for a nylon process to run for many weeks before it is stopped, cleaned, and prepared for a restart. The production of steel also must run on continuously for similar reasons.

Employees working on these continuous processes usually worked 12-hour shifts that do not differentiate between the days of the week and the weekends. The continuous schedule affects the operators and many of the support staff as well. This includes maintenance and some the engineering and technician functions. They don't have the traditional 8 to 5 job most common in U.S. companies. While the continuous process is not new in manufacturing, the degree to which it is being adopted has changed.

The increased use of robot-type automation and the elimination of the operator from the work cell has led many manufacturers to consider a 24–7 production schedule with 12-hour shifts. The traditional 8-hour shift was driven more by the limits imposed by the human operator than by any limits set by the machinery. With many of the operators eliminated by robotics and automation, the shifts are extended to 12 hours for those employees needed to support the production, and the workweek is extended as well from 5 to 7 days. Another factor supporting this change is the consolidation occurring in industry. Companies are merging or they are buying up smaller competitors. The larger company closes less efficient production operations and moves all the production to the more productive automated plants. With a large investment of capital in automation technology, it is not cost effective to let the machines sit idle if a production schedule like 24–7 will keep them operating.

The impact of 24–7 on the human work force is just beginning to be understood. The time usually set aside for family, children, additional education, recreation, and leisure is now chopped into different time segments. Time off on the weekend is no longer guaranteed. Childcare for working parents is influenced as well. Some industries have established 24–hour daycare at the factory to solve some of the childcare problems. Automation and robotics will continue to change the landscape for the human worker.

11-9 SUMMARY

Training is a critical component in every CIM implementation. The training process should include general employee awareness training, operation and programming training for the technical staff, and maintenance training on robots and automation hardware for the maintenance staff. This type of training is required by the ANSI/RIA safety standard.

The general training should include an overview of current automation practices, a rationale for the introduction of automation, an explanation of what robots are and how they work, company policy regarding introduction of automation and displacement of workers, retraining planned for displaced workers, and a statement demonstrating support from top management. The timing, level, and emphasis for the training must be matched to the target audience.

Operator training should start 2 to 4 weeks before the robots are in place and include a level of detail consistent with the responsibility of the operators in the automated cell. The content of the maintenance training should be consistent with the type of robot system and automation hardware planned for the cell. Training on the individual pieces of automation in the cell, such as robots, programmable logic controllers, and computer numerical control machines, is usually provided by the vendor selling the hardware. The training for the system operation of the cell is the responsibility of the company performing the integration. In many cases the manufacturer must conduct the system training to achieve the best results.

Work teams are being developed daily in the manufacturing industry. The level of authority to make changes in the process and to manage the work area indicates the degree of self-direction in the team. Implementation of work teams requires a workforce that knows the operation of the manufacturing process and that wants to take ownership of the system. Companies implementing self-directed work teams swell their ranks with additional problem solvers and increase their competitive edge in the marketplace.

Resistance to automation is not new. To overcome worker resistance, companies should use automation for the benefit of employees; communicate about automation issues openly and frankly; and install robots to free workers of dull, dangerous, and dirty jobs. Overcoming management resistance to automation requires an understanding of the roots of resistance, for example, concerns that unsuccessful automation projects will have adverse effects on careers, too little management support will be offered when automation causes worker unrest, concern that supervisors will be held responsible for transition problems in automation implementations, and turf battles and jealousy over selection of automation projects.

In the past, organized labor has provided only lukewarm support for robot and manufacturing automation. Now the well-documented challenges of the global economy have convinced organized labor that it should be part of the solution rather than part of the problem. The issues are speed of automation implementation and the payback to the worker when automation is installed.

Robotics and automation continue to change the landscape for the human worker. The move to longer work shifts, usually 12 hours, and extending shifts though the weekend are recent examples. The concept that a production plant should run 24–7, twenty-four hours for seven days, is rooted in two factors. First, the cost of automation dictates that the machines should be producing products and profits every available minute. Second, robots and automation do not get fatigued, so limiting a shift to eight hours is not necessary. With more machines and fewer workers in automated production systems, the automation dictates how the plant will be run. As a result, the advance of automation once again alters human labor and impacts the worker.

QUESTIONS

1. What are the three training activities required in every industrial plant planning to use robots? What do the standards say about the training needs?
2. Describe the elements that should be present in a general training program.
3. What is the primary focus of a general training program for each level within an industrial organization?
4. List the major elements of an operator training program.
5. What complicates maintenance training procedures on automated systems?
6. What is the difference between maintenance at the system level and maintenance at the machine level?
7. What are the key internal and external issues affecting the management of the workforce?
8. Describe the common characteristics of a self-directed work team.
9. What are the three laws for reducing resistance to automation?
10. What causes resistance to robot automation at the management level?
11. Give organized labor's view of automation and robotics.
12. How is the training requirement affected by the type of robot selected, level and number of sensors, and system architecture?

PROJECTS AND CASE STUDY PROBLEMS

1. Analyze how the West-Electric team handles the human interface issues in the case study in Chapter 12, and write a critique of the process used in their automation project.
2. If you are working on the case study at the end of each chapter, complete any questions and problems at the end of Chapter 12 that have not been completed in previous chapters.

Work-Cell Design Case Study

CHAPTER GOALS AND OBJECTIVES

The primary goal of this chapter is to provide a comprehensive work-cell design case study that will demonstrate how production is changed from a manual operation to an automation work cell using robotics. At the completion of this chapter you should be able to:

- Completely design an automated work cell including robotics, gripper technology, sensors, safety, supporting work-cell hardware, a cell control architecture, machine and system programming, and justification.
- Describe how an automation design team interacts in the design process.
- Identify one or more members of the design team whose training and skills are similar to your own.

CAREER SPOTLIGHT

This chapter as a whole is a career insight. The chapter describes in detail how a design team composed of different engineering and technology disciplines works together to complete the design of an automated production cell. Read through the chapter with a focus on understanding the job function of each team member. Pick out one of the job functions and picture yourself performing that role. Read through the chapter again and take notes on how you would have approached each action or assignment in contrast to the person in the case study. If the responsibility, team interaction, and problem-solving challenge present in the case interest you, then a career in manufacturing automation awaits.

12-1 INTRODUCTION

The hardware, software, and design process used in the development and implementation of an automated work cell as part of a CIM plan was covered in the last eleven chapters. The case study presented in this chapter illustrates in detail how the design process is used to develop a production work cell using the techniques learned in the previous chapters.

12-2 COMPANY PROFILE

West-Electric is a $50 billion multinational corporation with more than 15 large divisions. The turbine division produces gas and steam turbines for use in the generation of electric power. This division has five manufacturing facilities, with three located in the United States and two located off-shore. The company and division are vertically integrated at every level in the production of the turbine products. Three of the manufacturing facilities produce components for the turbines and two are used for final assembly and testing. One of the assembly plants also rebuilds turbines for the electric utility industry. The production facilities produce parts for use in new products, but they also produce replacement parts to support a large international field service component of the division. The case study describes the production at one of the production plants of gas turbine blades used for assembly of new machines, in the rebuilding of used machines, and for field service needs.

The problem presented in this case is fictional, but it is based on industrial experience and integrates production techniques common to many national and international companies. Students are invited to join the West-Electric team and participate in the design process through the use of the questions and problems at the end of the chapter.

12-3 INTRODUCTION TO CIM AUTOMATION AT WEST-ELECTRIC

On the drive to work, Bill Baxter usually reviewed the work scheduled for the day, but today he had a hard time concentrating. It had been about a year since he joined West-Electric (W-E), and much of the time was spent working through the initial phase of the Customer-Driven Manufacturing (CDM) project, West-Electric's response to CIM. If implemented fully, the plan would make significant changes in the division's manufacturing operations. Bill was ready to move from the planning to the implementation stage, and if the rumors were true, the meeting at 10 this morning would begin the automation phase.

It was 9, and Bill had just enough time before the meeting to review the assessment data his team had pulled together in the past several months. They had spent about 2 days a week for 6 months working on benchmarks for several performance measures on the division's major product, steam and gas turbine blades. The process had been developed at the corporate level, but plant manager Roger Walker allowed them to take some liberties whenever it benefited the assessment process. Roger was the best plant manager with whom Bill had been associated. Rumor had it that his plant would get the first automation project because of his reputation across the company.

"Good morning, Bill . . . I need a few minutes of your time."

Bill looked up to see Roger standing in the doorway to his office. "Sure, come on in. I was just reviewing the CDM project report for the meeting this morning."

"That meeting is what I want to talk to you about. Bill, I'm naming you the project leader for the first automation project to support the CDM initiative. I will announce it in the morning meeting. You look a little surprised."

"Well, I certainly hoped to be part of the project team, but I thought the leadership would go to a more senior engineer in manufacturing."

"You're my choice for several reasons. You've had experience in installing automation in your previous job, plus you've had the most recent academic exposure to the automation technology we need to use," Roger said. "I also want a fresh look at how we make parts, and I want someone without any emotional attachment to the current processes we use. The project will focus on automation of the turbine blade manufacturing line." After a few moments, Roger asked, "Are you okay with this assignment?"

"It's a great opportunity . . . you'll get my best effort," Bill responded.

Laying a folder on Bill's desk, Roger said, "I asked Ellen Becker in Human Resources to put together a short overview of each of the team members so you know the type of support you have available. Hold any other questions until after the meeting. See you shortly."

Bill indicated he understood and picked up the overview of his team.

West-Electric Automation Team

Bill Baxter (Manufacturing Engineer). One year with W-E following the completion of a BS and MS in mechanical engineering. Two years of co-op experience with another division of W-E during graduate study. Previous experience includes 5 years installing and troubleshooting automated manufacturing systems following the completion of an associate of science (AS) degree in mechanical engineering technology.

Ellen Becker (Director of Human Resources). Ten years with W-E, with a master's degree in human resources earned while working for the company.

Marci Hatcher (Manufacturing Engineer). Seven years with W-E, with the most recent work on process improvement and standards for an MRP II project. Previous experience includes 6 years in methods and work simplification in a metal fabricating company. Educational background: BS in liberal arts, AS in industrial engineering, and an MBA.

Ted Holcome (Design Drafter).   Fifteen years with W-E in a broad range of documentation and design projects. Education includes an AS in engineering graphics and recent update training in three-dimensional and solid modeling CAD, plus training on a new work-cell design and simulation software package.

Mike Perry (Senior Designer).   Eleven years with W-E working on the design of production machines and fixtures. A BS in mechanical engineering technology and a good background in CAD and finite-element analysis.

Jerry Thompson (Senior Electrical Technician).   Eight years with W-E, with experience in installation of new equipment and the maintenance of current systems. Education includes an AS in industrial electronics technology and recent training to install and program the programmable logic controller, which was picked as the standard for W-E.

The First Meeting

Bill came to the meeting early so that he could be there as everyone else arrived. He knew Marci and Ellen from the initial CDM project and had worked through a production fixture design problem with Mike shortly after joining the company. He didn't know the other team members and wanted to meet them as they arrived. At exactly 10, Roger Walker entered the conference room, motioned for quiet, glanced at his notes, and began to speak.

> This week I will start three productivity projects as a result of the work done in the CDM initiative. The one in the product design area focuses on a defect-free design process by developing a design for a manufacturing and assembly process with a six-sigma design component [a technique that tightens allowable part variation to ensure successful part assembly]. The second project in production planning and control will work on alleviating some quality and inventory issues. Your project will add automation to the turbine blade manufacturing line. All of you participated to some degree in the CDM process. Bill, Marci, and Ellen were on the enterprise team that established the process and completed the assessment of our division in the areas of technology, human resources, and process systems. I'm not going to cover the initial CDM results in detail; however, a review of some findings will be a good introduction to the project.
>
> A study of our current manufacturing technology indicated that we were current with the technology used to achieve process-specific results. But no technology was in place to integrate and automate the processes or to make manufacturing data available to enterprise units on a real-time basis.
>
> In the human resources area the results were mixed. In general, the staff had the capability to implement and operate the technology required to meet internal and external customer expectations. The option to use consultants or to buy the automation from outside companies was considered. It was rejected, however, because I know we have the talent to do the job, and I want to increase that capability for

future automation projects. Though we had the talent, we had not done a good job of informing everyone in the division of the need for change in design, production control, and manufacturing. This shortcoming was especially apparent in the organized labor component in manufacturing. As you know, the case for a stronger customer focus has been the theme of my communications to your departments for the past year. In addition, Ellen has done a good job of getting that message to the production areas through the work-cell, continuous-improvement team meetings.

Our ISO 9000 project contributed much of the data for assessment of the process systems area. The ISO documentation gave us some insight into how the current production processes operate. When it became apparent that the turbine blade line was targeted for the first manufacturing automation project, additional assessment was performed. All of that documentation is available to you for this project.

You may be interested in the rationale for selecting the turbine blade line for the initial project. Plans are in the works to build another gas turbine assembly plant in about 30 months, including a blade manufacturing line. The design for the new line calls for tighter integration between manufacturing operations and a significant increase in automation. Automating our current blade line will serve two purposes: our productivity and capacity will be increased, and we will learn how to integrate and automate the production of blades effectively.

As you develop the plan for the automation of the line, consider the following: We are the first division in the company to get capital dollars for new automation. Second, the level of future funding is always affected by the success of current projects. Third, what we do here will be closely watched by the folks at the corporate level. What I'm saying is that we need a successful project. Do everything reasonable to guarantee that the design on paper will perform when it's built.

I have asked Bill Baxter to lead the project work. Ellen has been assigned to all three automation teams, so her support is divided accordingly. Your supervisors recommended you for the project and are aware of my interest in its success. While Bill will work full-time on the project, the rest of the team will have some other department work. I made it clear to your supervisors that this project takes priority.

One final comment before I turn the meeting over to Bill. The project must have a 2-year or less payback, and the completion date for the automation of the line must be less than 18 months. That's about it from my end. You all know that I believe in the CDM concepts and know that this project can be successful. Bill, is there anything you want to add?

Bill Baxter leaned back in his chair, looked slowly around the room, and addressed the group.

I would like to get started on the project as soon as possible. I have some background material for each of you that includes an overview of the blade line and a description of the CDM design process we will follow. Please study the material and meet back here tomorrow morning at 9.

12-4 TURBINE BLADE PRODUCTION

The turbine blade production line manufactures 70,000 blades per week in seven sizes; the production is about equally divided over the different geometries. The shape of the blades is similar (Figure 12–1), but the weight varies from 3 to 16 ounces. The blades are manufactured from seven different lengths of titanium or stainless steel bar stock that varies in diameter from $\frac{3}{4}$ to $1\frac{1}{2}$ inches. The current production system uses manual production machines operated on three shifts for 5 days a week. Saturday overtime is often necessary to meet delivery schedules when the production capacity is exceeded or when equipment problems disrupt normal production flow.

The blade production uses a six-step production process for five blade shapes and a four-step process for the other two blades. During the six-step blade manufacturing process, the raw material passes through the following five stages: *slug, extrusion, upset, block forge, final forge and trim* (Figure 12–2). In the four-step process the *extrusion* and *upset* operations are eliminated because the blades are small. The process flow for all the products, illustrated in Figure 12–3, is divided into three major operations: slug preparation, airfoil forging, and final production finishing. Detailed process flowcharts for each of the major operations are included with the following descriptions.

Figure 12–1   Forged Turbine Blade.

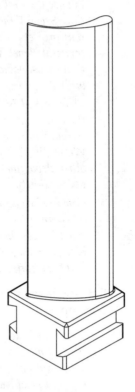

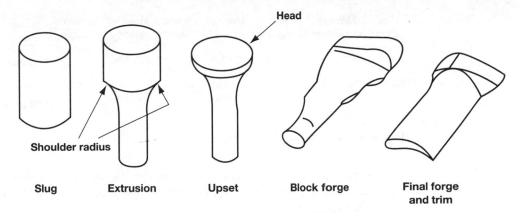

Figure 12–2 Airfoil Forging Sequence.

Step 1: Raw Material Inspection

Production begins (Figure 12–4) with the selection of a 20-foot titanium or stainless steel bar with a diameter of 0.75, 1.0, 1.125, 1.25, or 1.5 inches. An operator passes the bar through an ultrasonic inspection station that checks the rod for internal flaws in the metal, such as voids or impurities. A visual and audio output indicates a bad section of bar that cannot be used in the subsequent forging processes. The

Figure 12–3 Overall Blade Production Sequence.

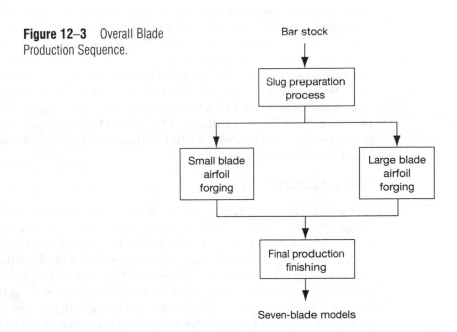

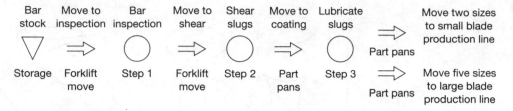

Figure 12–4 Slug Preparation Process.

bad section is located by slowly rotating the bar in the machine until the exact location is found. The spot is marked by the operator and eliminated when the bar is cut into slugs in the next operation.

Step 2: Production of Slugs

Good bar sections are fed into an automatic shearing machine (Figure 12–4) where induction heaters raise the bar temperature to 1300 degrees Fahrenheit (1300°F) in preparation for the shearing operation. The shear cuts the bar to a preset length, and the slug drops into a tote pan for transport to the next operation.

Step 3: Slug Lubrication

The slug lubrication area (Figure 12–4) coats the slugs with a ceramic silica glass lubricant that enhances the forging operation. The lubricant is applied using three manual spray booths. The operators pick up a tote of approximately 500 slugs from the shear or from inventory, place them on rectangular spray tables, and coat them in batches of 72 or 90, depending on slug size. The slugs are arranged in a nine-by-eight or a ten-by-nine array (Figure 12–5) on a rotating platform. The operator picks up the spray gun from a hook on the booth and applies the lubricant as the table is manually rotated to obtain an even coat on all visible surfaces. After a 10-minute drying time, the slugs are removed and placed in a tote for delivery to the next process. The larger size slugs go to the large blade production process (Figure 12–6), and the two smaller sizes go to the small blade production process.

Step 4: Extrusion

Three manual work cells, each equipped with an oven and a 500-ton capacity forge, produce 3000 to 3600 extrusions per shift for three shifts. The slugs are placed on a rotating circular table inside the oven and heated to 1800°F. It takes about 500 seconds for the larger diameter slugs to reach saturation temperature; the smaller parts require only 400 seconds. At the start of a shift, slugs are loaded onto the oven's indexing table at 15-second intervals until the first loaded slug has been in the oven for the required time. At that point, the operator picks up a cold-coated slug from the tote basket with 20-inch tongs and opens the automatic door of the oven with a foot switch. The cold slug is placed on the rotating shelf in an

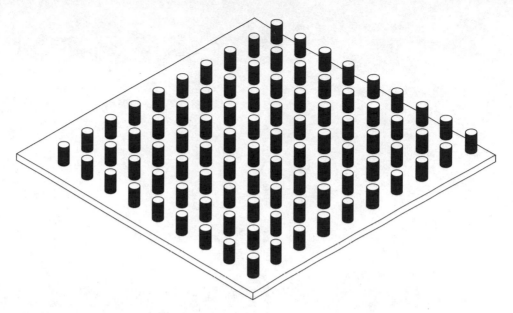

Figure 12–5 Lubrication Spray Table.

upright position, and a hot slug is removed for the extrusion process. The hot slug is dropped into the extrusion die, and the forge is cycled by the operator using two start buttons. After the 3-second extrusion process, the operator waits until a 15-second timer indicates that the extrusion has cooled sufficiently to be removed from the extrusion die. The operator uses a foot switch to activate an ejector, which pushes the extrusion out of the die so that the operator can transfer it with tongs to a finished parts tote. The operator uses a manual sprayer hanging on the forge to coat the upper and lower halves of the extrusion dies with graphite lubricant after each forge cycle. The total cycle time for the process is 30 seconds. Seventy-five to 300 tons of pressure are required, depending on the material and size of the slug, and a quick visual inspection of the surface finish is performed as the extrusion is removed from the forge. The operator works in a hot environment and must wear gloves for protection. In addition to the visual inspection, the shoulder radius (Figure 12–2) is measured with gages and with an optical comparator at 200-part intervals.

Step 5: Upset Forging

The upset forging operation (Figure 12–6) forms the head of the bar (Figure 12–2) in preparation for the block forge operation. This process does not require a ceramic silica glass lubrication coat before forging. The three manual work cells are nearly identical in layout and operation to the extrusion process with an oven and a 100-ton press present. The production rates are the same as those for the extrusion process; however, the cycle time is shorter (15 seconds) because the die does not require the graphite lubricant application between cycles and the forge cycles

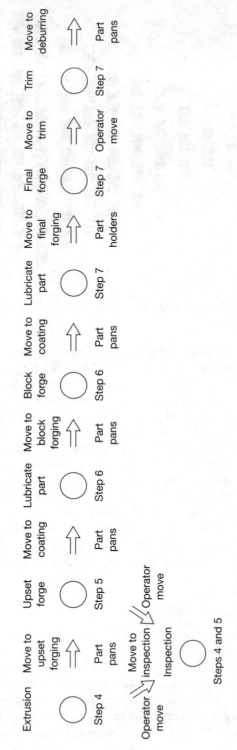

Figure 12–6 Large-Blade Production Process.

in 1.5 seconds. A cold part must be in the oven 450 seconds to reach the correct temperature; as a result, the oven must be loaded with 30 parts (a part every 15 seconds for 450 seconds) at the start of a shift. After the initial loading, an audible signal every 15 seconds indicates that the parts shelf in the oven has indexed, and a part is at 1800°F, ready for upset forging. After the upset operation, the part is removed immediately from the die and dropped randomly into a finished part basket. The production requires only 5 to 30 tons of the rated forge capacity for the five different part sizes manufactured. No visual inspection is performed on parts as they are produced, but the shoulder radius (Figure 12–2) of the upset is measured with a dial indicator at 100-part intervals.

Step 6: Block Forge

A block forge (Figure 12–2) gives the airfoil the rough geometry needed for a successful finish-forge operation. The three block-forge work cells have the same general layout and operation as the two previous steps, extrusion and upset forging. The ceramic silica glass lubrication is applied to the upset part using either a hand-dip method for larger parts or an automated electrostatic spray line for the smaller parts. In both cases, the parts are placed manually in carriers that prevent contact between coated parts. The oven-loading sequence, temperature saturation time, and process cycle time are identical to the upset operation. However, the dies have to receive a graphite lubrication spray between forge cycles. No inspection is performed on the block-forged parts.

Step 7: Final Forge and Trim

The four-cell final forging operation (Figure 12–6) has an equipment layout and process similar to the previous three processes. The primary exception is the addition of a trim press used to remove excess material after the final forge operation. The cycle time is 30 seconds (equally split between forging and trimming), and the time to reach forging temperature is 900 seconds for the largest part and 240 seconds for the smallest. As a result, the number of parts in the oven is a function of part size. The coating of the block-forged part with ceramic silica glass lubricant and the graphite spray for the upper and lower dies must be applied carefully to produce parts within design tolerances. The operator never releases the part from the tongs as the forge cycles. The round shaft at the end of the block-forged part (Figure 12–2) is used to hold the part throughout the process. After final forging, the operator cools the larger end of the airfoil by dipping it into water for 2 seconds, then places the blade in a trim press. The trim press removes the extra metal on the finished part, called flashing, that came out between the upper and lower die faces during final forging. At this point the airfoil has the desired shape.

Step 8: Alternate Process for Small Airfoils

The smaller blades skip the extrusion and upset operations and go from lubrication of slugs to block forging. The process in block forging (Figure 12–7) of the small parts has several variations. The oven is replaced with two induction heaters, and

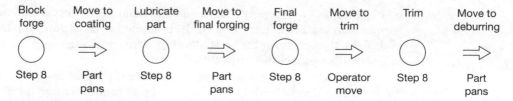

Figure 12–7 Small-Blade Production Process.

the forging operation is performed with an air hammer or drop forge. The operator has a 10-inch tongs in the right hand to remove the hot slug and holds a cold slug in the left hand. The operator removes a hot slug from one of the induction heaters using the tongs and immediately drops in the cold slug with the left hand. The block forge is performed on the slug, and the part is placed in a finished tote. The process is repeated with a slug from the other induction heater. The cycle time for the process is about two parts every 15 seconds.

The final forge on smaller parts also uses two induction heaters and an air hammer. After the final forge, however, a trimming process for removal of flash is performed. The cycle time for the final forging of one airfoil is the same as that for the block-forge operation—7.5 seconds; however, the trimming operation adds another 10 seconds to the single-part cycle time.

Step 9: Deburring

The final production finishing process (Figure 12–8) starts with deburring. Rough edges caused by the trim press operation and other irregularities on the airfoil surface are removed in this manual operation.

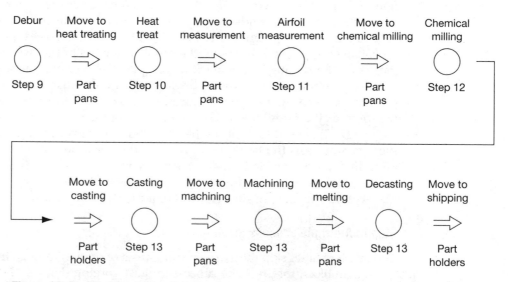

Figure 12–8 Final Production Finishing.

Step 10: Heat-treating

The blades are heat-treated in a vacuum oven to remove the stresses developed in the forging operations. The total time for this operation is 8 hours and includes manually loading 400 parts into the oven, bringing the oven up to temperature, heat-treating at specified temperature, and controlled cooling.

Step 11: Measuring Airfoil Cross-section

The complex shape of the airfoil is checked by a custom inspection machine (Figure 12–8) that simultaneously measures a minimum of three cross-sections of the blade. The operator removes a finished blade with one hand and places it in a finished part tote basket. A new part is picked up from a container with the other hand and placed in the measurement fixture. The fixture is closed manually, and the measurements are recorded in the system memory and displayed on the machine front panel. The nominal value for the cross-section at this step is 10 thousandth oversize. Oversized parts are marked and divided into five oversize ranges, from 1 thousandth to 13 thousandth, in 3-thousandth increments. Any blade over 13 requires rework, and any blade under 1 is scrapped. The measurement data are also displayed on a system monitor to support the statistical process control (SPC) analysis of the manufacturing process. The cycle time for this step is 17 seconds.

Step 12: Chemical Milling

The surface of the airfoil is milled chemically to remove the oversized stock and to produce the finish required for operation. Based on measured values, blades are placed manually in special baskets and tumbled in hydrofluoric and chromic acid for varying lengths of time. The milling time varies from 4 minutes to more than 15 minutes.

Step 13: Machining Base

The complex shape of the part (Figure 12–1) makes holding the part while machining the base difficult. To simplify the fixture problem during base machining operations, the blade is encased in a metal alloy with a melting point of 475°F. After the casting process, the entire blade has the shape of a square bar with flat ends. The casting machine has three molds arranged at equal distances around a 4-foot diameter indexing table. One mold is always in the blade load or unload position, the second is positioned under a device that directs a metered amount of the molten alloy into the mold, and the third is in a cooling position. The operator unloads a finished part, loads a new blade into the mold, presses a switch to index the table one-third of a revolution, and repeats the process. After the machining of the base, the alloy is removed by heating the finished blade to 500°F to melt off the alloy, then the finished blade is packaged for shipment.

Additional Data on the Blade Production Line

■ The average labor rate on the line is $9.00 per hour plus 42 percent for fringe benefits, with overtime paid at time and one-half.

■ The 480-minute production time per shift for the work cells is reduced to 400 minutes due to die and tooling changes, preheating dies, machine maintenance, and initial loading of ovens.

■ Extra floating operators are necessary to relieve regular cell operators for scheduled breaks.

■ Tooling in each cluster of work cells is interchangeable.

■ Each electric-heat oven has a door opening measuring 18 inches high by 15 inches wide, and the distance from the door to the parts is 16 inches.

■ The die holders are normally 24 inches square, and the distance from the front of the die to the part cavity is normally 18 inches.

■ Work in the extrusion, upset, block, and final forge areas is both physically and mentally stressful due to heat from the frequent opening of the oven door, machine noise, part weight, and the use of long tongs for part handling.

■ The dies used in the extrusion and upset operations must be removed to repair wear every 2000 operations; the block and final forge dies must be corrected for wear every 1000 cycles.

12-5 CDM IMPROVEMENT PROCESS

The customer driven manufacturing (CDM) improvement process used by W-E for manufacturing systems and work cells has three steps: assessment, simplification, and implementation with performance measures. Before starting any improvement process, W-E executives verify that the following conditions are present:

■ Management consensus favors the project and demonstrates full support.

■ Employees who could be displaced as a result of the project understand the company's plans for them in the future.

■ Management and union issues are fully resolved.

■ All levels of management in the area targeted for the improvement process are represented in the change process.

In the CDM plan, the *assessment* was performed at the corporate and division levels by teams of employees representing all areas. The results of the assessment included (1) an analysis of the technology available and the technology currently used by W-E, (2) the educational levels and technical skills in all employee classifications, (3) an analysis of the manufacturing processes and strategies used by W-E and by their competitors, and (4) a comparison of the order-qualifying and order-winning criteria of the W-E products with other brands.

In general, *simplification* in CDM focuses on the *elimination of waste*; however, the simplification procedure is dictated by the production process under study. Every process has several variables that determine the success of the final product. The variables have a target value that the process holds within some tolerance; however, waste results when the process produces parts outside that tolerance. For example, if a machined part falls below the minimum allowed variation in the cut dimension, the part is scrapped. Waste also results when the values for process variables are not chosen carefully. For example, the drying time of a sealant has a specified value, but the time will vary, depending on ambient conditions. The drying time is considered as waste because no value is added to the product while the sealant dries. Waste could be eliminated in the sealant application by using a sealant with a faster drying time or by applying heat to reduce the drying time. The CDM project emphasizes the elimination of non-value-added operations in a process; in addition, CDM places equal emphasis on process variable *elasticity*. During the improvement process, all process variables must be checked for elasticity, that is, the ability to be changed to reduce waste while not changing the final product. The elasticity of a process variable is often determined by the manufacturing process. In the sealant example, the sealer could have a much faster drying time if it were applied more swiftly by a robot rather than a human operator.

Implementation with performance measures, the last part of a CDM improvement process, is project specific. The performance measures used to determine the level of project success come from the order-winning goals for the product. Based on information gathered in assessment, simplification, and implementation, a five-step design process is initiated to determine the production and automation hardware and software, machine and process modifications, facility changes, and training required.

Five-step Design Process

To achieve the goals of CDM, changes in the design, production control, and manufacturing areas are necessary. As a result of these changes, some new production and automation equipment is usually needed, and a design process to implement the hardware and software successfully is necessary. W-E uses the following process:

1. Set up the project team. The project team typically should include representation from the following areas: *product design, quality assurance/control, manufacturing process design, production standards, economic justification, plant engineering, manufacturing, maintenance,* and *human resources.* Typically, the team is led by a representative from manufacturing engineering and could include outside consultants whose expertise is not available in the organization or team.

2. Define the improvement project objectives. The objectives support the product order-winning and order-qualifying criteria by addressing *quality, product cost, productivity, safety (product and employee), production flexibility,* and *production agility.* Product cost includes inventory and setup improvements, and production agility addresses the ability to move quickly into new market areas.

3. Select the appropriate process for improvement. The process selected for improvement should provide the necessary return on investment and have a high likelihood for successful implementation. In addition, the process targeted should be improved in one or both of the following areas:
   - ☐ Order-qualifying or order-winning criteria
   - ☐ Customer or employee safety
4. Design the work cell or system. The design process focuses on the following issues:
   - ☐ A study of the production process from three perspectives: *technical, economic,* and *human*
   - ☐ Selection of the most appropriate automation hardware and software
   - ☐ Integration of the hardware in the work cell or system
   - ☐ Integration of the work cell or system information with other enterprise systems
5. Build and operate the work cell or system. The tracking of work-cell or system performance against design goals is part of this final step.

12-6 THE AUTOMATION PLAN

After the initial meeting, Bill met with Marci, who was also a manufacturing engineer, to get another opinion on the current problems in the blade production area and to indicate that he wanted her to take a major role in the automation project. The team needed some help from the quality area, and her knowledge and experience in that area impressed Bill. It was immediately apparent that her recent experience in the MRP II system startup would be a benefit to this project. She agreed to brief the entire team on some of the critical issues at the next meeting.

The W-E project teams usually have a representative from finance to assist in the justification. Bill wondered why Roger didn't include someone to cover that area. Tom O'Brien, a golfing friend in cost accounting, might lend a hand in that area if necessary. Other than that, all the other areas were covered. It was getting late, and Bill needed to work through the design process, take a casual stroll through the blade production area, and make some to-do lists for the team and the meeting. He had to let Sandy know he would not be home for dinner.

Early the next morning Bill met with Ellen in her office because she could not attend the 9 o'clock meeting. He was glad to hear that all the production workers would be briefed on the plans for automation, starting with the first shift. Previous work on continuous improvement teams indicated to Ellen that the operators were willing to help in the process as long as the implementation would not produce layoffs. Most of the forge work-cell operators were male because of the strength required to hold the parts with the long tongs. Because of the stress and working conditions in these cells, the union was willing to agree to automation. The coating operators were primarily female, and many liked the work, so there could be a problem in this area for future automation.

Everyone arrived early for the meeting. After briefing the group on his earlier meeting with Ellen, Bill said, "Judging from my e-mail yesterday, you have some questions about this project. Before we start the discussion, I would like Marci to give us an overview of the production issues in the blade line."

Marci stood up, pushed copies of a report on blade production to the middle of the table, and paused. When everyone had a copy of the report, she glanced at her notes and said:

I did this report about 12 months ago when we were looking into production control software for mid-term and near-term scheduling of production on the blade production line. What we found out was that the investment in software could not be justified until several production issues present on the line were addressed. You can read the details later, but now I will cover just the major points. The major problems were quality and productivity. Let me discuss the quality issue first.

The present system has a quality check after extrusion and upset forging using a sampling technique; then the cross-section of each airfoil is measured after the deburring operation. The variation of airfoil cross-sections in the final test is too wide. We found that the problem results from the following: die wear, nonuniform metal flow during forging, dimensional variation in the part during the upset operation, and, to some extent, raw material variation and slug preparation. The next question was what caused these problems. Everyone working in this area knows that getting consistently good forged parts is as much an art as it is a science; however, we were able to verify several process variables that created the problems. We found that die wear increased significantly when the cycle time between cold part and forged part varied moderately. The thickness or inconsistent coverage of the lubrication coat on the block forged part had a measurable effect on final forge results.

The second issue was productivity. We are under pressure from our other divisions to pass along a lower cost to them for the blades we ship. They can get blades outside the company at more competitive prices. In addition, the divisions want more frequent shipments of smaller lot sizes on all of the parts. A major problem here is the 15 to 20 percent downtime per shift due to die change, preheating dies, maintenance, and production problems. Remember, this production process has not changed significantly in the past 20 years.

Some quality improvements have been made by the continuous improvement teams, but we can't control the cycle time variation that occurs from the start of the shift to the end. In addition, quality from shift to shift still varies more than we can accept. Work on the downtime problems has reduced it to 15 percent, but that is still too high. We need lower and more consistent cycle times at every cell. Are there any questions?

For the next 45 minutes, the team engaged in an active discussion of the production issues and the challenge of automating the line. After all the major concerns and questions were voiced, Bill brought the team back to his next agenda item, the automation design process. He suggested that everyone locate the five-step process in their handout from the last meeting. He waited as folders opened and pages turned, then said:

> As you can see, the process provides guidance but gives us the flexibility we need. Our team has representation from most of the areas listed in step 1 except quality assurance and economic justification. Marci has agreed to represent the quality issues, and I think I can get the help we need to develop the business plan and justification. Based on Marci's report and on discussions I had with Roger, our project objectives are improved quality and productivity. The productivity improvement should address lower product cost and smaller production lot sizes. As always, consideration for customer satisfaction and employee safety are a top priority.
>
> The next major decision is where to start in the blade production line. The process we select must provide the necessary return on investment, and I see no way to get the 2-year payback without eliminating direct labor cost from the process. Therefore, I suggest we consider robotics as part of the automation solution. In addition, Roger is counting on a highly successful implementation. So we will use a survey to rate the automation opportunities in all the production areas in the blade line.

Bill passed copies of the survey (see Table 12–1) to both sides of the table and gave everyone time to review the survey. Judging that everyone was ready to go on, Bill continued.

> Marci and Mike, I would like you to help me do the surveys of the work cells as a first step in this process. I suggest we each take a different shift since we may get more information by seeing as many operators as possible. After all production cells are surveyed, those that received six yes answers will be ranked by the team based on project objectives, best return on investment, and likelihood of a successful project. If robots are not the most effective path, then we should know that early in the project.
>
> After we have selected the work cell or cells for the project, the design of the cell or system will start. We have a lot of work before we reach that point. While we are doing the survey, Jerry, I would like you to write a report on the history of maintenance problems on all the major production machines used in the blade line. Ted, we will need drawings showing the layout of all the production work cells in the line. As you do these, keep in mind that three-dimensional models of the automated cell are necessary later in the design process.

Bill paused to let everyone finish writing, then said, "How long will it take each of you to get that material ready?"

After thinking for a few seconds, Jerry said, "I think a week will be necessary because some of the information is scattered across several departments."

"Some of the work cells are already in the CAD system, so a week for the rest is about right for me, also," added Ted.

Table 12–1 Plant Survey.

Robot Applications—Initial Plant Survey		

Answer the following questions for each workstation in the plant survey:

1. Can inspection by operators be eliminated from this workstation?   Yes   No
   It is difficult and expensive to include parts inspection in a robot work cell.

2. Is the shortest machine cycle 3 seconds or longer?   Yes   No
   Robot speed is limited. Human operators can work faster than robots when demanded by the process.

3. Can the robot displace one or two people for three shifts?   Yes   No
   If the average robot project costs $100,000, then it will be necessary to save the cost of one or two operators for three shifts to get a 1- or 2-year payback.

4. Can the parts be delivered in an oriented manner?   Yes   No
   Picking parts from a tote bin is easy for humans but very difficult for robots. If the parts can come oriented for easy robot pickup, then robot automation is possible.

5. Can a maximum of 6 degrees of freedom do the job?   Yes   No
   Robots have one arm that moves through a restricted work space compared with the two-armed human. A single-armed robot must be able to do the job.

6. Can a standard gripper be used or modified to lift the part or parts?   Yes   No
   The tooling is a major part of the work-cell expense. The simpler the tooling can be, the greater the likelihood of a successful project. Also, the weight of the part plus gripper must be consistent with the robot's capability.

"I have no other agenda items. Any other questions?" Bill looked at both sides of the table and then said, "We will meet again in one week . . . I will let you know the time and place."

12-7 SELECTION CRITERIA

As Bill drove to work he thought about the team meeting in the afternoon. Despite the still many unknowns, the project was falling into place. The initial work-cell surveys were complete, and he was more convinced than ever that robot automation was the best plan. There were few variations between Bill's survey of the first shift and the surveys of the second and third shifts done by Marci and Mike. From their initial discussions, Bill sensed that Marci wanted to be more aggressive in the initial integration of the cells, and Mike was taking a more conservative approach. Knowing Roger's interest in a successful implementation, Bill leaned toward the conservative approach; however, the team expressed consensus that robot automation was the correct direction. The report from Jerry on the condition of work-cell hardware supported the automation of existing production equipment. The forges were the oldest equipment in the cells, but they were in excellent condition. The ovens were only a few years old, and the internal parts shelf could be modified for automated operation. The automated coating operation was only 5 years old and was well maintained.

Marci and Mike felt the first project should be in either the extrusion or the upset areas, and Bill agreed. The only question was how much to integrate and to what degree to automate. This decision needed to be a team effort, so he gave everyone on the team a copy of Jerry's report, Ted's drawings of current work-cell layouts, and the initial work-cell surveys from all three shifts. He asked them to review the material and come to the meeting with first, second, and third choices for the initial automation project and rationales for their choices.

After briefing Roger at 9 on the status of the project, Bill started to review his handouts for the next step in the design process, the technical design checklist for the work cell and robot selection.

Work-Cell Technical Design Checklist

A good design requires a detailed study of the cell and the factors that contribute to its success. The following checklist serves as a guide to identify the technical issues that must be addressed during the design of the cell. The checklist is divided into six major areas: *performance requirements, layout requirements, product characteristics, equipment modifications, process modifications,* and *system integration.*

Performance Requirements

Cycle times	Tolerance of parts
Part handling specifications	Dwell time of tools
Feed rate of tools	Pressure on tools
Product mix	Maximum repair time
Equipment requirements	Malfunction routines
Human backup requirements	Allowable downtime
Future production requirements	

Layout Requirements

Geometry of the facility	Service availability
Environmental considerations	Floor loading
Accessibility for maintenance	Safety for machines and people
Equipment relocation requirements	

Product Characteristics

Part orientation requirements	Gripper specifications
Surface characteristics	Size, weight, and shape
Unique handling requirements	Inspection requirements

Equipment Modifications

Requirements for unattended operation	Maximum/minimum machine speed
Requirements for increased throughput	Requirements for automatic operation

Process Modifications

Lot-size changes

Process variable evaluation

Routing variations

Process data transfer

System Integration

Data interfaces and networks

Hardware integration requirements

Data integration requirements

Interface requirements

Software integration requirements

Robot Selection Checklist

A major step in the design process for a work cell that includes robotics is the selection of the robot system. After the technical issues for the robot work cell are resolved, a robot model is selected that matches application requirements. The following robot characteristics must be considered:

Positioning resolution, repeatability, accuracy

Arm geometry

Positioning flexibility

Work envelope size

Degrees of freedom

Maximum and rated payload

Maximum and payload dependent velocity

Compliance requirements

Force/torque sensing requirements

Downward force

Tool change requirements

Programming (on- and off-line)

Cost

Vision integration

Special option requirements

As everyone arrived for the meeting, Bill handed them a folder with the agenda and the two sets of checklists attached. He let everyone exchange greetings, then started the meeting by saying:

The first thing we need to do is to define the first automation project. I will record each person's top three priorities on the flip chart, and we will see if there is a consensus on the cell to automate first. Please give your rationale for the choice, as well, so any information you used to reach your decision will be shared with everyone. I will ask the three of us who completed the survey to start off. Marci, would you go first?

Marci got out her notes and said, "My first priority is an integrated system that starts with slug coating and ends with production of the upset forged part. My second priority would be just the upset cell, and the third would be just the extrusion cell. My reason for starting with those two areas is that I got yes answers for every question on the survey for both of those cells," she said. "The inspection can be handled outside the automation, and the technical issues are not that tough. If we need to be more conservative and limit the automation to a single cell, them my choice on the individual cells would be the two I mentioned." She paused to look at some notes, and then said, "If we include the direct labor cost for nine operators in each product area and the increase in production, the project should have the cost justification numbers that Roger requires. Even if we keep one operator per shift to maintain stock and handle minor automation problems, we will still get the justification in under 2 years." Marci's look at Mike said she was finished and he had the floor.

Bill nodded, and Mike said, "I think we need to be conservative on the first attempt at automation, so I just have two priorities: first, the upset cell, and second, the extrusion area. We can economically justify replacing the operators with robots as long as we get a 20 percent increase in output." Mike looked at Marci and continued, "I'm not afraid of the automation or the challenge, but I would like to stay away from dealing with coated parts for the first project. The technical issues associated with the upset cell are not a problem with current technology in robots, sensors, grippers, and work-cell controllers." Mike looked at Bill to indicate that he was finished for now.

Bill decided to wait until the others listed their priorities before indicating his preference for the first project. Everyone agreed that the project should focus first on two blade operations; most picked the upset cell for the first attempt at automation. It was time for his priorities, and so he said, "Like many of you, I picked the upset as my number one choice because I think we can get the automation working more quickly and easily. In my briefing with Roger this morning, he mentioned again his interest in a more cautious first project, but one that built in the capability of broader integration in the future. Therefore, I would suggest that we start with the upset

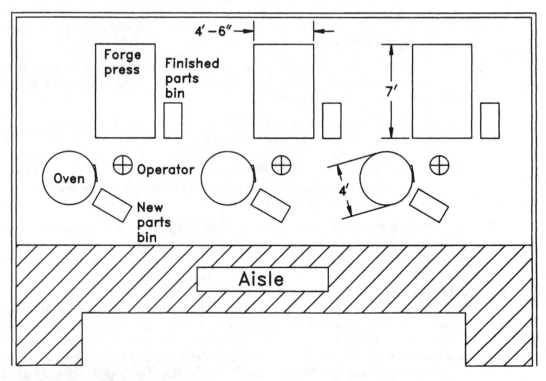

Figure 12–9 Manual Upset Forge Cells.

operation; however, I don't want to just replace operators with automation. I would like some suggestions on how we could reorganize the upset cells to reduce the downtime during die changeover and get to lower lot-size production."

After an active discussion and many trips to the flip chart to sketch alternative cell configurations, the group arrived at the initial system layout. The three manual upset cells (Figure 12–9), would be organized into a single cell (Figure 12–10) with three forges, two ovens, and two robots. One of the forges could be serviced by either robot, so that production would not stop when a die needed to be changed. Judging that most of the constructive ideas had been voiced, Bill took control of the meeting and said:

I like what I've heard today—we're on the right track. I have a list of to-do's for everyone that will carry us to the next step in the process. Before we start implementing the automation, however, we need to identify the performance measures that will be used to measure the project's success. Marci, could you do

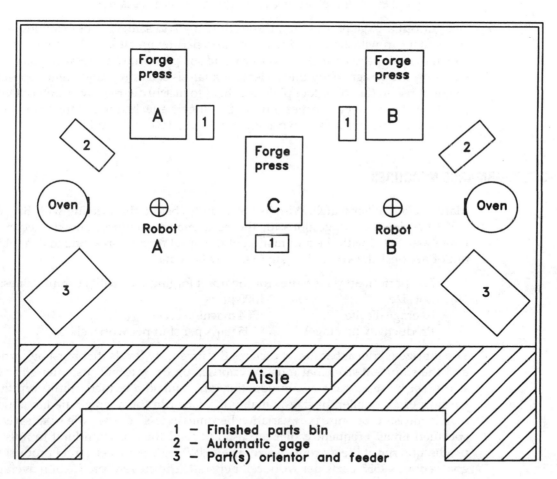

Figure 12–10 Proposed Automation Work Cell.

some research for the next meeting? In addition, I would like you to take the technical issues checklist in the folder and start identifying areas that need additional work. Mike, I would like you to work with me on the robot selection checklist to identify a robot for the cells.

Ted showed me the new Workspace 3 work-cell simulation software that we have in the design area. It should help us be more confident of our design and eliminate implementation surprises. Ted, could you give us an early look at equipment placement, cycle times, and robot work envelopes on the simulation package at next week's meeting? Jerry, please start looking into the modifications necessary for the ovens—especially, the automation of the internal parts table.

After some clarification, all members indicated that they would work on the areas listed. As the team prepared to leave, Bill said:

One last thing before we adjourn. Ted did some initial work on a gripper that could be used in either the extrusion or upset cells; the drawing [Figure 12–11] is in the folder. Ted, please give us a quick overview of the design.

Ted got out a large plot of the gripper drawing and said, "I chose stainless steel because it can withstand the high oven and part temperatures. The temperature rise in the gripper was another concern, and so I performed a stress and heat flow study on the design using finite-element analysis software. After some redesign, the heat rise in the robot tool plate was held to just 10 degrees over ambient temperature at current cycle times. I think the fingers attached to a standard parallel type pneumatic gripper will work fine on the cell layout we did today."

12-8 PERFORMANCE MEASURES

Marci looked at her watch. With just 30 minutes before the meeting with Bill, she had to review her suggestions for the performance measures and the technical issues associated with the automation of the upset forging area. She took a folder out of her desk drawer and began to review the data.

The performance measures for the upset forging area with current values:

Lot size	1500 parts
Average die life	1754 machine cycles
Productivity (average)	1105 parts per shift per work cell

The lot size is critical because it dictates the inventory levels that must be carried and the shipment sizes that must be imposed on the internal customers. The lot size is set by setup time and production cost, and so a reduction in setup time would permit smaller lot sizes. Quality is a factor in the average die life performance measure, because part variation becomes excessive when die wear exceeds specified limits. Frequently, dies are used beyond the optimum limit to finish a production run at the current lot size. The 70,000 parts per week target production goal requires 1555 parts per work cell per shift. The current production average, 1105, is 450 parts short of the goal. Overtime is required to make up the difference

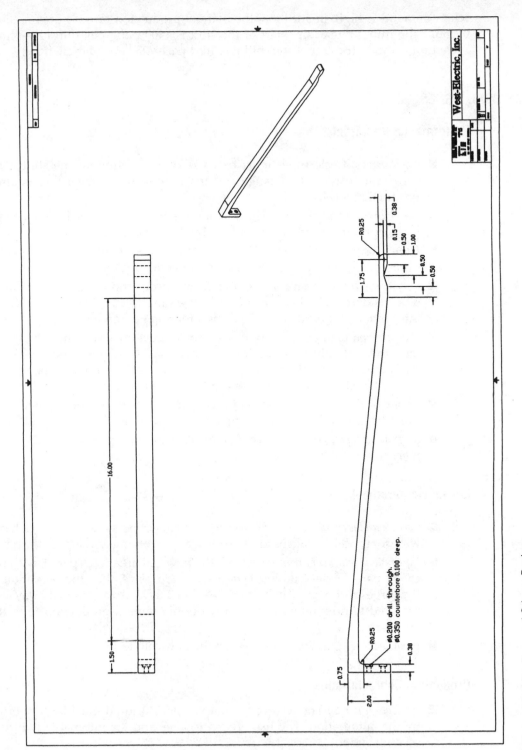

Figure 12–11 Proposed Gripper Design.

when demand is high. In addition, the manual production rate is not consistent because the initial rate at the start of the shift is 1400 parts per shift (17-second cycle time), and at the end of the shift it is 1054 parts (23-second cycle time).

12-9 TECHNICAL ISSUES

Performance Requirements

- *Cycle times:* Cycle times must be 13 seconds throughout the shift, with zero time lost for die change and machine setup to meet the required production level.
- *Part-handling specifications:* The end of the extrusion with the large diameter is the optimum location for gripping by the robot for movement in the cell.
- *Human backup requirements:* In the event of a robot failure, the cell must be capable of limited production using human operators.
- *Future production requirements:* Two future production requirements must be considered: (1) the reduction of cycle time to 10 seconds, and (2) the integration of the extrusion and upset forging operations.
- *Malfunction routines:* The control software must include a three-light system: the first light indicates low raw material stock, the second indicates an operational problem (part jammed, oven temperature low, poor quality, etc.), and the third indicates production equipment failure.
- *Allowable downtime:* A 13-second cycle permits about 6 hours per week for preventive maintenance and unplanned downtime.
- *Maximum repair time:* A goal of 2 hours to repair or replace any of the hardware in the cell is needed.

Layout Requirements

- *Service and accessibility for maintenance:* Whenever possible, the control hardware for the cell should be located outside the working envelope of the robot.
- *Equipment relocation requirements:* If the new cell area is prepared, with the robot, parts feeder, and inspection system in place, then the relocation of the forges and ovens can be completed in 2 days. An additional 2 days are required to fine-tune the system with only limited production from the automated cell.
- *Floor loading:* The floor can support the new cell design.

Production Characteristics

- *Inspection needs:* The testing of one part per hundred must be supported by the automation. The use of smart gages and an automated fixture is recommended.

Equipment Modifications

- *Requirements for unattended operation:* The system must operate three shifts with a resupply of the raw material every $2\frac{1}{2}$ hours. Changes in production models and stocking of the oven must be supported by the work-cell software.
- *Requirements for increased productivity:* The number of positions in the oven and system components must be provided for a 20 percent increase in productivity in the future.

Process Modifications

- *Lot-size changes:* A reduction of lot size from 1500 to 1000 is required on the initial design.
- *Cell operation:* The robots, ovens, parts feeders, inspection devices, and two of the three forges will operate continuously over three shifts. One forge will always be idle to permit die change and setup to minimize production downtime. The proposed operation for the cell in Figure 12–10 is described by the schedule in Figure 12–12. Note that robots *A* and *B* use forge *C* as an alternate work area after completing production on forges *A* and *B*. When the robots move to forge *C*, the dies are changed on forges *A* and *B*. At one point in the schedule, forge *C* is not in use, which allows the dies to be serviced in that forge for the next scheduled part.

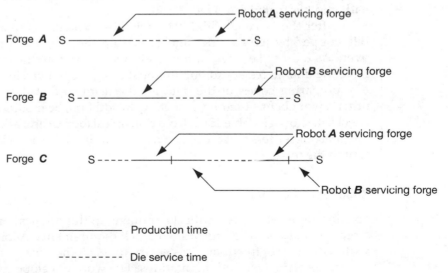

Figure 12–12 Production Sequence.

As Bill hung up the phone he looked up to see Marci standing at his office door. "Marci, come in and pull up one of those chairs. I was just talking to Mike; we are going to work on the robot specs later this afternoon. I'm anxious to see what you have on our technical requirements."

As Marci handed Bill a folder, she said, "The work is not complete, but there should be enough there for you and Mike to use. Look it over to see if you have any questions."

After studying the two documents, Bill said, "I'm surprised that you didn't recommend a quality target in the performance measures. What's your rationale for the ones you listed?"

"Bill, I want to address quality by reducing lot sizes and increasing die life. Let me explain. We measure a part every one hundred cycles to determine when the die is showing excessive wear. However, we often finish out the production lot with a marginal die, and produce one or two hundred parts out of tolerance." Marci paused to let her comment sink in, then said, "Manufacturing argues that at this stage in production, it doesn't make any difference. The truth is, we really don't know. I want to work on consistent cycle times that should improve die wear and get the lot sizes down through a reduction in setup time. Then we can be sure we run only on good dies, and see if downstream quality improves. I'm betting that it will," she said.

Bill felt that Roger was looking for a quality statement in the performance measures, but after discussing the alternatives, he agreed with Marci's analysis. He flipped to the technical issue list and said, "I don't see a path velocity for the part to complete the 15-second cycle time. Have you got a number in mind?"

"I would rather wait for Ted's simulation before we settle on a value. It should be ready tomorrow," Marci said. "My estimate is a path length of about 15 feet for the primary forge and 25 feet for the shared one. So . . . 2 feet per second should do the job." After clarifying several other values on the technical issues list, Marci gathered her material and left.

After the meeting with Mike to define a general set of robot characteristics, Bill stopped by Jerry's office and picked up the proposed modifications for the oven. As soon as he received the work-cell layout from the simulation, he would have the data necessary to put an initial design together. He spent the rest of the day preparing copies of the data for the team so that they could review each other's work before their next meeting. In addition, he modified a sensor selection checklist, shown in Table 12–2, from a technical journal and asked the team to compile a list of sensors that would be needed in the cell. The following material was sent to the team.

Robot Specifications

Position repeatability is a critical parameter in this application, and a robot with a repeatability of 0.005 satisfies work-cell requirements. A jointed spherical arm with 6 degrees of freedom is necessary to reach the numerous taught positions at various elevations and orientations. The work envelope requires a maximum reach of 5 to 6 feet from the robot's center axis and a rated payload of 20 to

Table 12–2 Sensor Selection Checklist.

Target Information

Parameter	Data Needed
Material	Is the target material ferrous, nonferrous, or nonmetallic?
Mass	How much does it weigh?
Size	How large is the target in square inches or cubic inches?
Shape	Is the target flat, cylindrical, cubic, or spherical? Is it compact, long, or narrow?
Surface	Is the surface opaque or transparent? If it is opaque, is it colored, shiny, dull, smooth, textured, porous, or nonporous?
Motion	What is the velocity in feet per second? Is the direction of movement perpendicular or oblique to the line of sight? If using photoelectric sensors, can the reflector or receiver be mounted behind the target?
Distance	How far away is the object to be detected? How close to the part or machine can the sensor be located? (This parameter alone can determine whether a proximity or photoelectric sensor should be used.)
Presentation	What is the rate of presentation to the sensor in units per minute? Are targets oriented randomly or regularly? Are parts separated or do they overlap? What is behind the target?
Precision	Is precise range needed? This is usually true in machine tool applications.

Environmental Information

Parameter	Data Needed
Enclosure	Is a NEMA qualified case required? Is the material aluminum or plastic?
Temperature	What is the typical annual range in °F? Is it very cold or very hot?
Relative humidity	What is the typical annual range in percent? Is it very dry or very wet?
Ambient lighting	What are the sources of light? Are they constant or variable? Is there any direct sunlight?
Electrical noise	What are the typical and peak intensities? A welding line would represent a worst-case example.
Mechanical vibration	What is the typical amplitude range? What is the maximum physical shock possible?
Air quality	Is the air clean or dirty? Is the contaminant dust, smoke, oil mist, or paint spray? Is the atmosphere explosive?
Mounting surface	Is the material ferrous, nonferrous, or nonmetallic? How thick is it? Is it flat, rounded, or some other shape? How much mounting space is available in square or cubic inches? Are precision clearances necessary? Is the mounting surface stationary or mobile?

Electric Circuit

Parameter	Data Needed
Power supply	120 Vac, 50/60 Hz, or 12/24/35 Vdc?
Contacts	Number and configuration?
Output form	Analog or digital?
Output load	Minimum and maximum current required?
Output voltage	Minimum and maximum voltage required?
Switching delay	Activation delay in microseconds? Release delay in microseconds?
Switching frequency	Hz?

(Reprinted with permission from *SENSORS*—Your Resource for Sensing, Communications, and Control, April 1986.

25 pounds. The maximum velocity must be 4 feet per second or greater so that a 50 percent safety factor can be applied. Both on- and off-line programming are required, along with a controller that supports easy interfacing to a programmable logic controller. The cost must be in the $40,000 to $60,000 range to achieve the desired payback. Compliance, force/torque sensing, vision, specified downward force, and tool changing are not considered necessary for the application.

Oven Modifications

The ovens used throughout the turbine blade production line have 30-inch steel tables with a center axle that rides in a bearing above and below the heated area of the oven. A motor, sprocket, and chain drive system, located in the base of the oven, indexes the table every time the door is opened and closed. The ovens would be modified to support automated operations by using a slotted table to hold the extruded parts during heating. The hanging orientation supports loading and unloading of the oven with robot automation. The suggested configuration for the table (Figure 12–13) has 30 slots; however, additional slots could be added for higher productivity. The dc motor currently used to turn the table would be replaced with a stepper motor for accurate and repeatable rotation. The oven door foot switch would be interfaced to a programmable logic controller, and the oven temperature gage would be replaced with a sensor for automatic monitoring

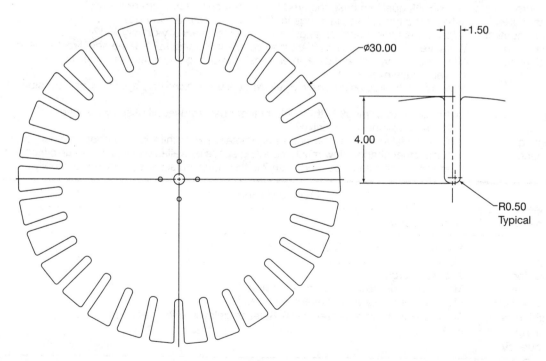

Figure 12–13   Oven Part Holder for Automated Operation.

of oven temperature. Warping of the slotted table during repeated heating and cooling of the oven is the major design problem to solve. Engineering is researching the best material to use for production of the table to minimize the warping problem.

Work-cell Simulation

The work-cell simulation illustrated in Figure 12–14 will work with an ABB Robotics model IRB 2400 robot. Other brands and models are available, but the specifications of the model 2400 closely match or exceed the initial robot specifications for the work cell. After the parts feeder and inspection station are defined, the following criteria will be established: programming of the operation, generation of minimum cycle times, optimum position of hardware, and axes movement limits during operation with each forge.

12-10 AUTOMATION CELL INTEGRATION

Bill looked at his watch. It was almost 7:30—an early start for what could be a long day. The team was meeting at 9 and he had booked the conference room for the entire day. He had just enough time to get the material organized and copies made. Bill walked into the meeting a few minutes after 9; discussion on the design was already underway. After handling everyone a folder, Bill said:

As you can see, we have a long agenda. I would like to start with some of the components of the cell, then move to cell layout. After that we need to work out the number and types of sensors needed. Before we start, I want to mention that Tom O'Brien from cost accounting and I took a rough look at the economic justification. We were conservative with the numbers, but the payback still looks good at this point.

As Bill was about to speak, Marci said, "Ted and I have been looking at the design of the automatic gage. It may be more than our design area can handle. The fixturing gets complex fast because we have to measure the radius on five different-size parts." She paused, then said, "I think we need to go outside to people who build this type of equipment. It will cost more, but I think we will get better results. What will that do to the justification?"

Before Bill could reply, Ted added, "I talked to a couple of vendors to get a ballpark figure. If we are willing to do a little setup on the fixture when we change the dies, then it looks likes $10,000 will get it built." Ted waited for Bill to finish writing, then continued, "That includes a data concentrator, which means we have only one serial interface to deal with, and we can download the quality data whenever our controller is ready."

Sensing Ted was finished, Bill said:

The cost should not be a problem. Ted, would you send them samples of our upset parts and some operational specifications and get a cost figure? Don't

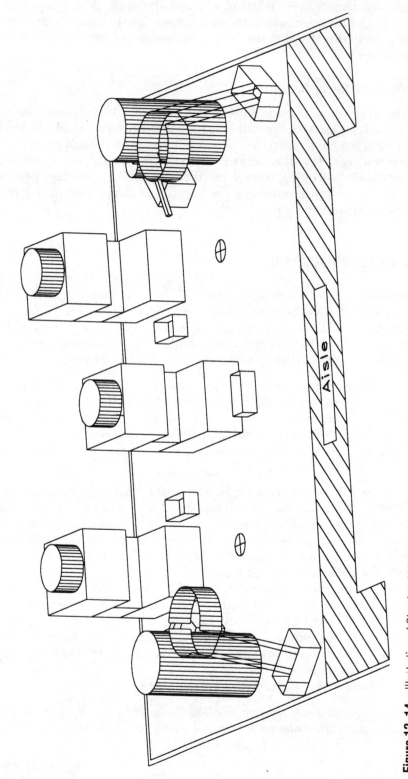

Figure 12–14 Illustration of Simulated Work Cell.

make any commitments, but I think that may be the best approach. Before we tackle the sensor selection problem, let me summarize where we are in the cell design process.

The weight of the fingers, parallel gripper, and heaviest part exceeded the payload capacity of the IRB 2400 robot that we want to use for this application. The robot is rated at 22 pounds, and I want our payload to stay under 15 pounds. Ted is working on a redesign on the fingers. The version in your folders is light enough, but we don't have the finite-element analysis data yet to know if it will hold up. We may have to use composite material for the base of the fingers if this design doesn't have the necessary rigidity.

Several of you have worked on the parts feeder and orienter problem. The parts can be oriented easily with off-the-shelf systems, so we decided to continue to use the parts bins to move the extruded parts to the automated cell. As a result, we've settled on a bowl feeder that will work for all five part sizes to get the orientation needed by the robot. The bowl feeder will be fed by an elevating hopper feeder from a storage bin that holds about 700 parts. The parts will leave the bowl feeder small end first and be held in a vertical orientation on two rails [Figure 5–26j] . I put a rough sketch of the feeder system in your folders [Figure 12–15]. Ted and Marci are working with a vendor and it looks like this part of the cell doesn't present any problems.

After Jerry and I reviewed the oven modifications, we changed the number of slots from 30 to 45. This will permit increased productivity in the future. I included a drawing of the proposed table [Figure 12–16] for everyone to review. We are still working on a table material that will not warp with repeated heating

Figure 12–15 Extrusion Feeder System.

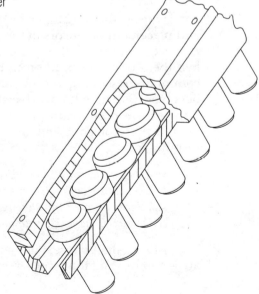

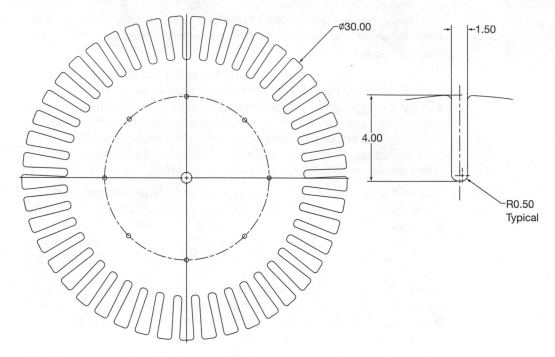

Figure 12–16   Revised Oven Parts Table with 45 Slots.

and cooling. With a robot loading and unloading the slots, we cannot afford to have more than a half-inch variation in the vertical position of the slot as the table rotates through 360 degrees.

During the remainder of the meeting the team worked on the selection and placement of sensors in the cell according to the guidelines Bill developed. A chart (Figure 12–17) was created that included a sensor identification number, the work-cell condition to be sensed, the type of sensor to be used, and comments regarding the operation of the sensor. In addition, sensor numbers were added to the work-cell layout (Figure 12–18). When the sensor selection for the production problems was complete, Marci asked about sensors to detect if someone entered the cell while the robot was operating. Bill responded:

W-E has a corporate team developing safety guidelines when robot automation is present in a work cell. Let's wait until we get more information on the safety standards we need to follow before tackling the safety sensor issues. Well, the cell is taking shape. Does anyone have any questions or concerns?

After a short pause, Bill said, "Work on the detailed designs in each of your areas and forward the results to me. We'll meet again in a few days."

Identification number	Parameter sensed	Sensor type	Comments
1,2	Oven door position	Limit switch	The limit switch lever arm rotates when the oven door is in the full open position.
3,4	Oven part table position	Absolute encoder	A digital encoder attached to the oven table shaft provides an absolute binary number indicating the position of the table.
5,6	Robot shoulder axis position	Proximity	The sensor output changes when the robot arm is fully retracted from the oven.
7,8,9	Press upper die position	Photoelectric	The sensor output changes when the upper die of the press is in the full up position.
10,11,12	Finished part ejected from the forge	Photoelectric	The sensor output changes when the air-ejected part from the press slides down the finished part chute.
13,14	Maximum desired oven temperature	Temperature switch	The sensor contacts close when the oven temperature exceeds the maximum allowed level.
15,16	Minimum desired oven temperature	Temperature switch	The sensor contacts close when the oven temperature falls below the minimum allowed level.
17,18	Oven door position	Limit switch	The limit switch lever arm rotates when the oven door is in the closed position.
19,20	Part oriented and in correct position	Photoelectric	The sensor output changes when a part is in position and ready for the robot.
21,22	Gripper finger separation distance	Proximity	The output of the proximity switch on the robot gripper indicates that the grippers are closed but no part is present.
23,24	Level of parts in bowl feeder	Proximity	The sensor output changes when the level of parts in the bowl falls below a minimum level.
25,26	Number of parts in feeder chute	Photoelectric	The sensor output changes when the number of parts remaining in the feeder chute falls below a minimum level.
27,28,29	Press cycle	Limit switch	The plunger on the limit switch has not been activated because the press failed to complete a forge cycle.
30,31,32	Hot part in lower die	Photoelectric	The sensor output changes when the radiation from the hot part triggers the sensor input.
33,34,35	Hot part is fully seated in die	Photoelectric	The sensor output changes when the part is fully seated in the die hole.

Figure 12–17 Work-cell Sensors.

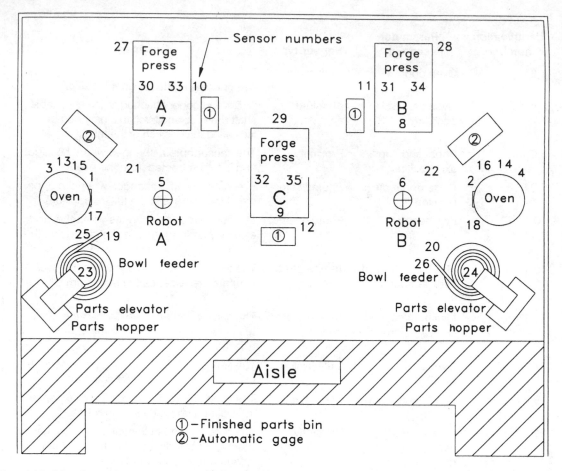

Figure 12–18   Sensor Locations in the Upset Automation Work Cell.

12-11 WORK-CELL CONTROL

The upset cell design is taking shape. Bill made copies of the upset forging cell control architecture (Figure 12–19) and the preliminary PLC interface drawing (Figure 12–20 and Figure 12–21) and forwarded copies to each team member for comment.

12-12 CELL PROGRAMMING

Components for the automated upset forging cell are in fabrication. The bowl feeder and other material-handling equipment were specified and are ready to go to the vendor. The gaging system was sent out for bids and came in under the

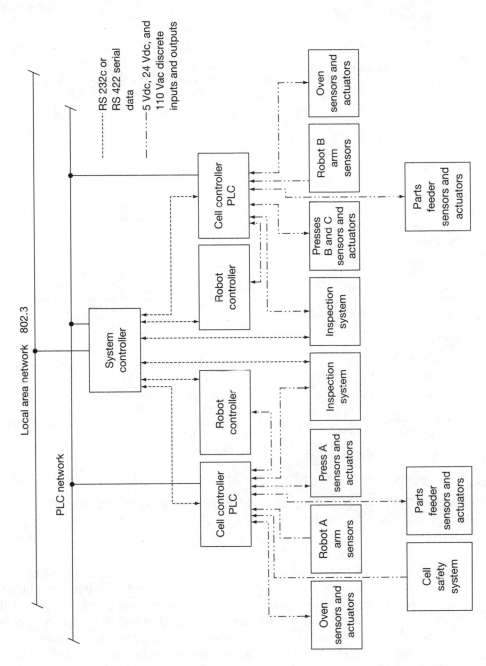

Figure 12–19 Control Architecture Upset Forging Cell.

453

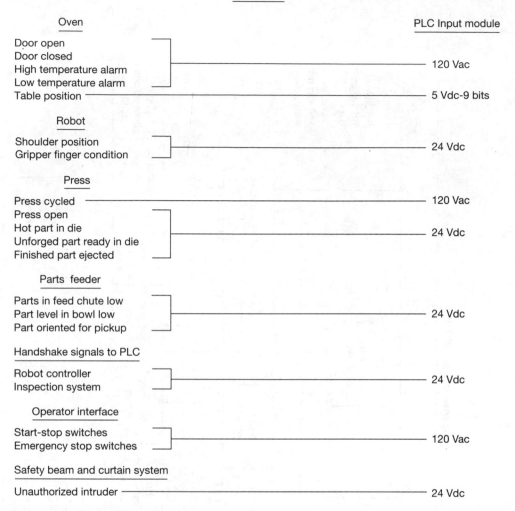

Oven

Door open	
Door closed	
High temperature alarm	120 Vac
Low temperature alarm	
Table position	5 Vdc-9 bits

Robot

Shoulder position	24 Vdc
Gripper finger condition	

Press

Press cycled	120 Vac
Press open	
Hot part in die	24 Vdc
Unforged part ready in die	
Finished part ejected	

Parts feeder

Parts in feed chute low	
Part level in bowl low	24 Vdc
Part oriented for pickup	

Handshake signals to PLC

Robot controller	24 Vdc
Inspection system	

Operator interface

Start-stop switches	120 Vac
Emergency stop switches	

Safety beam and curtain system

Unauthorized intruder	24 Vdc

PLC Input module

Figure 12–20   Programmable Logic Controller Interface Drawing for Upset Forging Cell.

budgeted amount. A spare oven was modified to test the positioning accuracy of the parts table, and the testing is planned in a week. Bill is satisfied that they will have a successful work cell. One of the last tasks is the development of the robot and PLC cell control program. He asked Jerry to develop an operational sequence (Figure 12–22) for the work cell in preparation for developing a PLC ladder logic diagram. Bill worked with Mike on the task point graph for the robot (Figure 12–23). After feedback from the team on each of these items, Ted and Mike would add the data to the WORKSPACE simulation software to test the process.

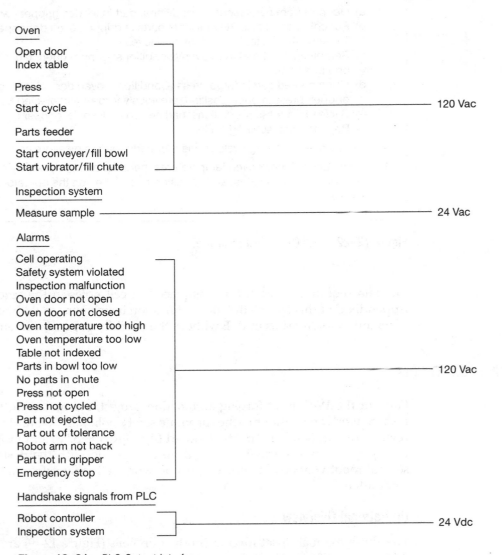

PLC output module

Oven

Open door
Index table

Press

Start cycle

Parts feeder

Start conveyer/fill bowl
Start vibrator/fill chute

120 Vac

Inspection system

Measure sample ——— 24 Vac

Alarms

Cell operating
Safety system violated
Inspection malfunction
Oven door not open
Oven door not closed
Oven temperature too high
Oven temperature too low
Table not indexed
Parts in bowl too low
No parts in chute
Press not open
Press not cycled
Part not ejected
Part out of tolerance
Robot arm not back
Part not in gripper
Emergency stop

120 Vac

Handshake signals from PLC

Robot controller
Inspection system

24 Vdc

Figure 12–21   PLC Output Interface.

12-13 COST JUSTIFICATION

Early in the blade manufacturing process, the W-E team started the development of a discounted cash flow justification spreadsheet (Figure 12–24) that would indicate payback and permit what-if analysis of automation options. A description of

Work-Cell Design Case Study   **455**

1. Load oven with parts at 15 second intervals.
2. Start upset forging operation.
   a) Get part from parts feeder (Conditions: part in feeder, gripper open).
   b) Put cold part in oven (Conditions: parts in gripper, oven door open, table in same position after last part was removed).
   c) Remove heated part from oven (Conditions: gripper open, table indexed one position).
   d) Place heated part in forge press (Conditions: oven door closed, part in gripper, press in open position, previously forged part ejected from die).
   e) Cycle forge press (Conditions: part seated in die hole, gripper clear of press).
   f) Repeat manufacturing cycle.
3. Modify manufacturing cycle at one hundredth part to measure part parameters.
4. Continuously check oven temperatures, parts feeder levels, safety screen, emergency stop buttons, and condition of switches on the operator interface panel.

Figure 12–22 Cell Operational Sequence.

the data required for each cell was placed in cell notes in the spreadsheet (see Appendix C of this text) so that the justification instrument could be distributed to other automation teams in W-E without the need for paper documentation.

12-14 SAFETY PLAN

Early in the W-E upset forging automation project, Bill met with the enterprise team to work on revisions of the corporate safety policy and plan. He was asked to review the safety requirements standard (ANSI/RIA R15.06) and propose a safety plan for the upset forging work cell. After meeting with the design team and several robot vendors, he sent a report on work-cell robot safety to the corporate committee.

Operational Overview

The three manual upset forging production cells (Figure 12–9) are modified as illustrated in Figure 12–10 to support automated production. Under the planned operation, one forge is always available for die change and setup while the other two are in production. After a change to a new forge, the previous forge die is cooled for 1 hour, then the die is changed, tested, and preheated. This die change cycle occurs about every 4 hours. The parts hopper is resupplied every $2\frac{1}{2}$ hours, and the inspected parts are collected every 8 hours. Checking and reprogramming of transition points in the work cell occurs about twice a month. As a result

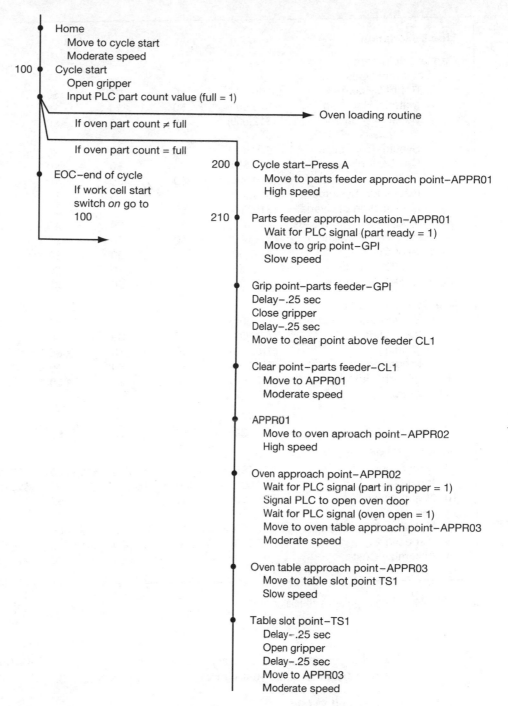

Home
    Move to cycle start
    Moderate speed
100 Cycle start
    Open gripper
    Input PLC part count value (full = 1)

                                                → Oven loading routine
    If oven part count ≠ full

    If oven part count = full

                                  200 Cycle start–Press A
                                                    Move to parts feeder approach point–APPR01
EOC–end of cycle   High speed
    If work cell start
    switch *on* go to
    100   210 Parts feeder approach location–APPR01
                                                  Wait for PLC signal (part ready = 1)
                                                Move to grip point–GPI
                                                  Slow speed

                                          Grip point–parts feeder–GPI
                                          Delay–.25 sec
                                          Close gripper
                                          Delay–.25 sec
                                          Move to clear point above feeder CL1

                                          Clear point–parts feeder–CL1
                                                  Move to APPR01
                                                    Moderate speed

                                          APPR01
                                          Move to oven approach point–APPR02
                                          High speed

                                        Oven approach point–APPR02
                                          Wait for PLC signal (part in gripper = 1)
                                          Signal PLC to open oven door
                                          Wait for PLC signal (oven open = 1)
                                          Move to oven table approach point–APPR03
                                            Moderate speed

                                          Oven table approach point–APPR03
                                        Move to table slot point TS1
                                        Slow speed

                                          Table slot point–TS1
                                        Delay–.25 sec
                                        Open gripper
                                        Delay–.25 sec
                                        Move to APPR03
                                        Moderate speed

Figure 12–23   Initial Points in the Upset Forging Task Point Graph.

Line Description	0	1	2	3	4	5
Capital Equipment						
1 Equipment Cost						
Robot Systems	($100,000)*	$0	$0	$0	$0	$0
Material Handling	($13,000)	$0	$0	$0	$0	$0
Sensors & Safety System	($12,500)	$0	$0	$0	$0	$0
System Controller	($10,000)	$0	$0	$0	$0	$0
Cell Controls	($11,500)	$0	$0	$0	$0	$0
Software	($10,500)	$0	$0	$0	$0	$0
Inspection Systems	($15,500)	$0	$0	$0	$0	$0
Modification of Ovens	($10,500)	$0	$0	$0	$0	$0
Modification of Presses	($8,500)	$0	$0	$0	$0	$0
2 Freight and Installation	($11,000)	$0	$0	$0	$0	$0
3 Sale of Old Equip.	$0	$0	$0	$0	$0	$0
4 Tax on Old Equip.	$0	$0	$0	$0	$0	$0
5 Total Net Investment	($203,000)	$0	$0	$0	$0	$0
Non-Capitalized Cost						
Moving Equipment	($6,500)	$0	$0	$0	$0	$0
Installation	($8,000)	$0	$0	$0	$0	$0
Training	($15,000)	$0	$0	$0	$0	$0
Programming	($18,000)	$0	$0	$0	$0	$0
6 Total Non-Capitalized Cost	($47,500)	$0	$0	$0	$0	$0
Inventory Change						
7 Inventory Change	$0	$0	$0	$0	$0	$0
Operating Costs						
8 Direct Labor	$79,488	$158,760	$158,760	$158,760	$158,760	$158,760
9 Indirect Labor	($7,000)	($15,000)	($15,000)	($15,000)	($15,000)	($15,000)
10 Maintenance Costs	$0	$0	$0	$0	$0	$0
11 Tooling Costs	$0	$0	$0	$0	$0	$0
12 Materials & Supplies	$500	$1,000	$1,000	$1,000	$1,000	$1,000
13 Inspection	$2,300	$6,500	$6,500	$6,500	$6,500	$6,500
14 Assembly Cost	$0	$0	$0	$0	$0	$0

Figure 12–24 Upset Forging Justification.   (Continued)

of this maintenance, a safety system is necessary to protect the operators and programmers who must work inside the operating cell.

Work-cell Safety

The automated work cell with safety hardware in place is shown in Figure 12–25. The following safety features are included:

15 Scrap & Rework	$4,000	$9,500	$9,500	$15,000	$15,000	$15,000
16 Downtime	$0	$20,000	$20,000	$20,000	$20,000	$20,000
17 Utilities	($200)	($400)	($400)	($400)	($400)	($400)
18 Taxes & Insurance	$0	$0	$0	$0	$0	$0
19 Subcontracting	$0	$0	$0	$0	$0	$0
20 Safety	$0	$0	$0	$0	$0	$0
21 Programming	($8,000)	($2,000)	($2,000)	($2,000)	($2,000)	($2,000)
22 Process Improvements	($4,500)	$2,000	$2,000	$2,000	$2,000	$2,000
23 Other	$0	$0	$0	$0	$0	$0
24 Total Operating Cost	$66,588	$180,360	$180,360	$185,860	$185,860	$185,860
Other Impact on Revenue						
25 Change in Volume	$0	$5,000	$10,000	$25,000	$25,000	$25,000
26 Reduced Lead Times	$0	$0	$0	$0	$0	$0
27 Increased Quality	$0	$0	$0	$0	$0	$0
28 New Product Introduction	$0	$0	$0	$0	$0	$0
29 Manufacturing Flexibility	$0	$6,500	$9,500	$15,000	$15,000	$15,000
30 Other Revenue	$0	$0	$0	$0	$0	$0
31 Total Revenue	$0	$11,500	$19,500	$40,000	$40,000	$40,000
Discounted Cash Flow Analysis						
32 Total Operating Cost	$66,588	$180,360	$180,360	$185,860	$185,860	$185,860
33 Total Revenue	$0	$11,500	$19,500	$40,000	$40,000	$40,000
34 Non-Capitalized Cost	($47,500)	$0	$0	$0	$0	$0
35 Total Pretax Cash	$19,088	$191,860	$199,860	$225,860	$225,860	$225,860
36 Tax Rate	0.36					
37 After-tax Cash Flow	$12,216	$122,790	$127,910	$144,550	$144,550	$144,550
38 Depreciation Rate	0.000	0.071	0.143	0.143	0.143	0.143
39 Depreciation	$0	($14,413)	($29,029)	($29,029)	($29,029)	($29,029)
40 Depreciation Cash	$0	$5,189	$10,450	$10,450	$10,450	$10,450
41 Inventory Cash	$0	$0	$0	$0	$0	$0
42 Total Net Investment	($203,000)	$0	$0	$0	$0	$0
43 Total After-tax Cash	($190,784)	$127,979	$138,361	$155,001	$155,001	$155,001
44 Discount Rate	0.15					
45 Discount Factor	1.00	0.87	0.76	0.66	0.57	0.50
46 Discounted Cash Flow	($190,784)	$111,286	$104,621	$101,916	$88,622	$77,063
47 Cum. Cash Flow	($190,784)	($79,498)	$25,123	$127,039	$215,661	$292,724

*Numbers in parenthesis indicate expenditures and capital investments. All other dollars values are dollar savings.

Figure 12–24 *Continued*

- A flashing amber light above each robot indicates that the robot is in the run mode.
- Two security zones provide for safety. Zone 1 is the area inside the work cell but outside the robot work area, and zone 2 is inside the work area assigned to the robot. Zone 1 has a 40-inch-high, three-bar rail fence as an access

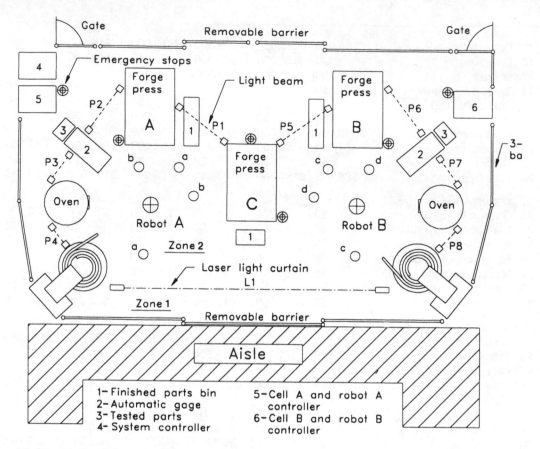

Figure 12–25   Work-cell Safety System.

barrier, and zone 2 uses a combination of a laser light curtain and photo-electric sensors to detect unauthorized traffic into the robot work area.

■ The zone 1 barrier has breaks on each side of the parts hoppers so that restocking of the hoppers does not require access to the cell. The controllers are also located in breaks in the barrier so that interlock switches used to disable parts of the safety warning equipment are accessible from outside the cell. The zone 2 barrier is composed of eight photoelectric sensor sections and a laser light curtain. Sections can be disabled to allow access to the cell for maintenance and die change while the robot is operating. Lockout switches in the cell controller allow individual sections to be disabled during die change and cell troubleshooting.

■ Two gates with position sensors provide access to the rear of the cell for maintenance personnel. The gate sensors can be locked out from the cell controllers to disable the gate-close sensor.

■ The zone 1 barrier has removable sections at the front and rear of the cell to allow access for the die change equipment. Each section has a lockout switch on the cell controller that disables the warning system in preparation for removal of the barrier.

- Detection of unauthorized entrance into the robot work area activates an audible alarm and strobe warning light. In addition, the robot is stopped and put in a wait state, and automatic operation of the cell is halted.

- Depressing any of the emergency stop switches around the cell removes power from the robot arm and all cell hardware. Use of the emergency stop switches at the robot and cell controller cabinets removes primary power from the cell. Activation of any emergency stop switch triggers an audible alarm and strobe warning light.

- Work-cell operators do not have to enter the robot work area during the resupply of raw material or the removal of finished products. Only trained maintenance personnel and work-cell programmers have keys to the lock-out switches that disable the safety warning and detection system.

- All work-cell operators and maintenance personnel receive training on robot and automated work-cell operation appropriate for their work-cell responsibility.

- Changes in the status of the safety system, such as locking out a safety gate or photoelectric detector for a die change, are collected by the work-cell controller and saved in a log file on the cell controller. The log is uploaded periodically to the area controller by the production control department.

- Maintenance staff changing dies or performing forge setup are protected from injury by the robot with software axes limits and by removable barriers (Figure 12–25). The barriers and soft limits prevent the robot arm from leaving the current work area and entering the area occupied by maintenance personnel changing a forge die. The soft axes limit prevent the robot from moving an axis beyond the limit set in software; the barriers are metal posts inserted into holes in the floor (marked by letters a, b, c, d in Figure 12–25) that physically prevent the robot from entering the maintenance area.

12-15 FINAL PERFORMANCE MEASURES

The performance of the upset forging cell was measured regularly after the automation was fully functional. At the 6-month mark, the following performance measures were recorded:

- Cycle times of 13.5 seconds per part are maintained in the operation of all forges.
- Cycle times of 9 seconds were demonstrated with forges A and B.
- Setup and die change time was reduced 40 percent.
- Lot sizes of 1000 parts can be produced economically.
- Die life has increased 25 percent.
- The number of upset forged parts that exceed the tolerance is reduced by 60 percent.

Sections 12-1 to 12-4

1. Using the case study description of the West-Electric blade production line and any other resources available, complete an initial plant robot survey for all the production cells.

2. Based on the results of the robot survey of the work cells and general case data, rank the work cells as follows:

   a. Least difficult to most difficult to automate

   b. Work cell you would recommend for automation first; the second cell area you would automate

3. How did the information and general case data provided by Roger Walker and Marci Hatcher influence your decisions in question 2b?

4. What step(s) in the blade line will be the most difficult to automate?

5. Was Bill Baxter correct to assume that the project had to include robotic cells?

6. What are the major reasons that the cycle times in the forging cells are not uniform? Could the problem be eliminated without automation?

7. What advantages would robots provide if they are used in the cells?

8. How can the team integrate the production on the entire line and avoid the problem of islands of automation?

9. Where is W-E in the three-step CIM process? What are the next steps for the W-E team after the selection of the work cell(s) for automation?

Sections 12-5 to 12-8

10. Is there a need to integrate upset forging and extrusion as Marci suggested? Does the proposed upset forging work-cell automation eliminate the opportunity for integration with the extrusion process in the future? How could the integration be implemented?

11. Develop an initial layout for the automation of slug lubrication and extrusion using the existing extrusion machines.

12. What characteristics of good gripper design are satisfied by the gripper that Ted proposed for the work cell?

13. The mass properties analysis of Ted's gripper finger indicated that the volume of the material was 6.04 cubic inches. Calculate the weight of two fingers if they are made from stainless steel. What other materials are available that would provide the high-temperature operation characteristics but would contribute less weight?

14. Using the case study description of the W-E blade production line and any other resources available, complete the work-cell technical design checklist for the team's proposed upset forging automation systems.

15. Using the case study description of the W-E blade production line and any other resources available, complete the work-cell technical design checklist for the automation system developed in Case Study Problem 11.

16. Using case data, address and resolve as many of the robot checklist issues as possible for the upset forging automation system.
17. Using case data, answer as many of the robot checklist issues as possible for the work cell developed in Case Study Problem 11.
18. Using the robot checklist data developed by Bill and Mike in the case, identify two robots that could be used in the upset forging work cell.
19. Using the robot checklist data developed in Case Study Problem 17, identify two robots that could be used in the slug lubrication and extrusion work-cell.

Sections 12-8 to 12-9

20. Using case data, the sensor selection guidelines, and the cell design for upset forging developed by the W-E team, complete a table that has the following information: sensor number, work-cell condition to be sensed, and the type and style of sensor (limit switch, lever, proximity, and photoelectric) to be used. Then comment regarding the operation in the cell.
21. Using case data, the sensor selection guidelines, the six basic reasons for sensors in an automated cell, and your work cell developed in Case Study Problem 11, complete a table that has the following information: sensor number, work-cell condition to be sensed, the type and style of sensor (limit switch, lever, proximity, and photoelectric) to be used. Then comment regarding the operation in the cell.
22. Describe any differences between the robot checklist data developed by Bill and Mike in the case study and your answer from Case Study Problem 16.
23. The team is looking strongly at the ABB IRB 2400 robot. What other robot models should be considered that would satisfy the criteria in the case study?
24. How well will Marci's performance measures for the upset forging cell indicate that the automation has improved the blade manufacturing performance? Should other performance measures be included?
25. Bill accepted Marci's recommendations for the performance measures on the upset forging cell. Should he have insisted on a performance measure directly related to quality? Why or why not?
26. Using the definition for robot payload from Chapter 1, the weight of the gripper fingers calculated in Case Study Problem 13, the maximum part weight, and the value Bill and Mike specified for robot payload range, determine the allowable weight for the parallel gripper. Will Ted's finger designs work? What are some other options?

Section 12-10

27. If the gripper base in the W-E case weighs 5 pounds, what is the total weight of the gripper, fingers (original design in stainless steel), and largest part?
28. If stainless steel fingers are used and the total payload weight is 15 pounds, what is the maximum volume for each finger of the gripper?
29. Categorize the sensors selected by the W-E team in the case into the six basic reasons why sensors are used. Are there any other sensors they should have included?

30. Categorize the sensors selected in Case Study Problem 21 into the six basic reasons why sensors are used. Are any areas not covered?

31. If the part hopper holds 700 parts and the work-cell cycle time is 12 seconds per part, how long can the cell run unattended?

32. With a 45-slot carousel and a required oven time of 450 seconds, what is the maximum number of parts that can be manufactured in a week from the automated cell?

33. Analyze the quality requirements for the blade production line and identify areas where vision could be used for inspection.

34. In addition to the vibratory feeder, what other techniques for part orienting and feeding could be used in the blade production line?

35. How could bar code technology be used to track the blade production process?

Section 12-11

36. Use the W-E cell architecture as an example and develop a work-cell architecture for the extrusion work cell developed in Case Study Problem 11.

37. Use the W-E PLC interface as an example and draw the wiring interface between the PLC, robot controller, sensors, operator interface, and other equipment in the extrusion work cell developed in Case Study Problem 11.

38. What advantages did the W-E team hope to gain by using a separate PLC for each half of the production cell?

39. What impact would the use of a single PLC have on the wiring, programming, and operation of the upset forging cell.

40. What advantages did the W-E team hope to gain by using one cell controller for both sides of the process?

Section 12-12

41. Develop a task point graph for the extrusion work cell developed in Case Study Problem 11.

42. Develop a chart showing the operational sequence for the extrusion work cell developed in Case Study Problem 11.

43. Draw a ladder logic diagram for the operational sequence chart developed in Case Study Problem 11. Use input and output module notation and functions for any available PLC.

44. Draw a ladder logic diagram for the W-E upset forging operational sequence. Use input and output module notation and functions for any available PLC.

Section 12-13

45. Assume that you are on an automation team at another W-E plant and are asked to prepare a report describing how the discounted cash flow spreadsheet is used to determine payback. Use the notes attached to all data cells in the spreadsheet (see Appendix C) and prepare a report that includes the

format required for all data entered into the spreadsheet and a description of how the calculations determine the payback period.

46. Your team at the other W-E site decides to include the impact of inflation in the discount cash flow analysis spreadsheet, and you are asked to determine how the equation for present worth (including an inflation parameter) could be added to the spreadsheet calculation. Prepare a report on your investigation.

47. Use the W-E discounted cash flow spreadsheet to determine the payback time for the extrusion work cell developed in Case Study Problem 11. Use a discount rate of 10 percent, a tax rate of 35 percent, and no intangible benefits. Make good financial and engineering assumptions and estimates for input data not directly available from the case study.

48. Identify the intangible benefits that can be used for the extrusion cell justifications, describe how they will be used as revenue entries in the spreadsheet, and compare the payback period when intangible benefits are used with the analysis in Case Study Problem 46.

49. Analyze the payback analysis for the upset forging cell presented in the case study. Determine the following:
    a. What intangible benefits did the team include in the payback analysis?
    b. Using the notes attached to the cells in the intangible section, describe how the team determined the revenue impact of the intangible benefits.

50. Determine the payback for an upset forging automation design that includes a robot for each manual work cell. Would this have been a better solution?

Sections 12-14 to 12-15

51. Assume the role of a member of the W-E enterprise safety team responsible for the review and approval of Bill's proposal for the automated robot cell. Write a response to Bill that outlines the strengths and weaknesses in the proposal and list any changes you think are necessary. If a copy of the ANSI/RIA R15.06 standard is available to you, compare Bill's plan with the minimum requirements in the standard and list any deviations or omissions in Bill's plan.

52. Prepare a safety plan for the extrusion work cell developed in Case Study Problem 16.

53. Modify the W-E cell architecture for the upset forging cell in Figure 12–19 to reflect the additional safety systems added to the work cell in Figure 12–25.

54. Modify the W-E cell interface wiring diagram in Figures 12–20 and 12-21 to reflect the additional safety systems added to the work cell in Figure 12–25.

55. As a result of development problems on the automatic gaging system, the vendor building the device requests that the robot place the part into the gage in a horizontal position with the part gripped from the small end of the part. If the robot removes the part from the forge by gripping the smaller diameter of the part just below the upset end, how could the part be reoriented to the horizontal position with the robot holding the part at the end of the smaller diameter?

Hardware Specifications

LR Mate 100™

Basic Description

Five-axis, modular construction, electric servo-driven robot designed for precision, machine load/unload, welding and material handling. The FANUC LR Mate 100 robot is engineered for maximum reliability in a variety of harsh plant or educational environments, supported by user-friendly software and our extensive service and parts network.

LR Mate 100 the Solution for:

- Dispensing
- Education
- Assembly
- Machine load/unload
- Material handling
- Material removal
- Testing and sampling
- Welding

Benefits

- Multiple mounting positions and table top size increases installation flexibility
- Hollow joint construction encloses all cable routing to eliminate snagging
- Absolute encoder positioning
- No robot modification for upright or inverted mounting
- Small footprint for tight installations on machine tools
- Integral double solenoid valves (2) at wrist

Features

- 5 axes of motion
- 4 kg (8.8 lb) load capacity
- ± 0.04mm (± 0.002") repeatability
- Axes speed up to 250° per second
- End effector connector built into the wrist

Performance Features

- Very large work envelope for this robot class
- Robot can "double-back" on itself for increased access
- Axis 2 and 3 fail-safe brakes
- Sealed for laboratory and industrial use
- Standard software permits 3-dimensional palletizing/depalletizing of rows, columns, and layers by simply teaching three points
- Small 380mm (14.9") width x 470mm (18.5") height x 320mm (12.6") depth controller for convenient placement

Reliability Features

- Brushless AC servo motors minimize motor maintenance
- Harmonic drives on all axes result in higher reliability and reduced maintenance costs

- Sealed bearings and drives allow operation in harsh factory environments
- Grease fittings on all lubrication points for quick and easy maintenance
- Only two types of motors used for simpler servicing and spare parts reduction
- Easily removable service panels for quick access to robot drivetrain
- IP54 dust and liquid intrusion protection package

Options

- Other integral valve packages
- Axis 1 brake
- Higher speed CPU for improved path and cycle times

LR Mate 100 Dimensions

Footprint

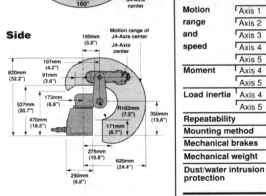

AXIS 1 ROTATING CENTER

4-DIA II THROUGH HOLE
DIA 21 COUNTERBORE DEPTH 3

160mm (6.2")
80mm (3.1")
50mm (1.9")
250mm (9.8")
60mm (2.3")
50mm (1.9")
193mm (7.5")
205mm (8.1")
165mm (6.4")
30mm (1.1")
30mm (1.1")
95mm (3.7")
190mm (7.5")

Top

Motion range of J4-Axis center

160°
R620mm (24.4")
J4-Axis center
160°

Side

Motion range of J4-Axis center
J4-Axis center
150mm (5.9")
107mm (4.2")
820mm (32.2")
91mm (3.6")
173mm (6.8")
527mm (20.7")
R182mm (7.2")
350mm (13.8")
470mm (18.5")
171mm (6.7")
275mm (10.8")
620mm (24.4")
250mm (9.8")

Front

309mm (12.2")
298mm (11.7")
147mm (5.8")
170mm (6.6")

SCALE
1/30"

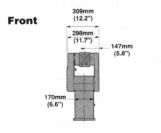

Mounting plate

Ø 5h7 DP7
45°

4-M5 DP8
PCD31 .5 EQ.SP
J5:0°

LR Mate 100 Specifications

Items		3kg (6.6 lbs) payload		4kg (8.8 lbs) payload	
		Range	Speed	Range	Speed
Motion range and speed	Axis 1	320°	150°/sec	320°	150°/sec
	Axis 2	180°	150°/sec	180°	150°/sec
	Axis 3	360°	180°/sec	360°	180°/sec
	Axis 4	240°	180°/sec	240°	100°/sec
	Axis 5	400°	250°/sec	400°	240°/sec
Moment	Axis 4	55.5kgf • cm		74.0kgf • cm	
	Axis 5	40.0kgf • cm		40.0kgf • cm	
Load inertia	Axis 4	1.1kgf • cm • s²		1.4kgf • cm • s²	
	Axis 5	0.41kgf • cm • s²		0.41kgf • cm • s²	
Repeatability		±0.04mm (±0.002") based on JISB8432			
Mounting method		Upright/inverted			
Mechanical brakes		Axis 2, axis 3 (axis 1 option)			
Mechanical weight		32kg (70.5 lbs)			
Dust/water intrusion protection		Conforms to the IP54 standard for dust and liquid intrusion protection (seals may need periodic replacement if used with chlorine or gasoline based coolants)			

FANUC
Robotics

FANUC Robotics North America
2000 South Adams Road
Auburn Hills, MI 48326-2800
Phone (810) 377-7000
Fax (810) 377-7366

Literature Request
1-800-47-ROBOT

Cincinnati, OH
Phone (513) 771-8844

Chicago, IL
Phone (708) 364-6161

Los Angeles, CA
Phone (714) 258-7075

Toronto, Canada
Phone (905) 670-5755

Mexico City, Mexico
Phone (52-5) 611-5998

Sao Paulo, Brazil
Phone (55) (11) 897-8070

LR Mate 100i™

Basic Description

Five-axis, modular construction, electric servo-driven robot designed for precision machine load/unload, welding and material handling. The FANUC LR Mate 100i robot is engineered for maximum reliability in a variety of harsh plant or educational environments, supported by user-friendly software and our extensive service and parts network.

LR Mate 100i the Solution for:

- Machine load/unload
- Material handling
- Welding
- Dispensing
- Assembly
- Material removal
- Testing and sampling

Benefits

- Multiple mounting positions and tabletop size increases installation flexibility
- Hollow joint construction encloses all cable routing to eliminate snagging
- Absolute encoder positioning
- No robot modification for upright or inverted mounting
- Small footprint for tight installations on machine tools
- Integral double solenoid valves (2) at wrist

Features

- 5 axes of motion
- 5 kg (11 lbs) load capacity
- ± 0.04mm (± 0.002") repeatability
- Axes speed up to 270° per second
- End effector connector built into the wrist
- Low voltage I/O: (20) inputs (8 dedicated), (16) outputs (4 dedicated) and (6) inputs at end-of-arm

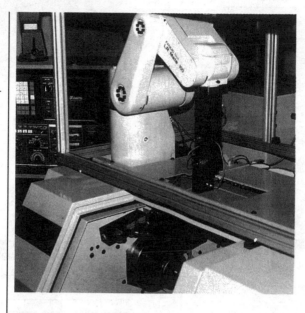

R-J2 Mate i-Controller Features

- Provides easy installation, startup, troubleshooting and maintenance
- Requires reduced installation space
- Approximately 50 kg (110 lbs)
- Small 380mm (14.9") width x 470mm (18.5") height x 320mm (12.6") depth for convenient placement

Performance Features

- Very large work envelope for this robot class
- Robot can "double-back" on itself for increased access
- Axis 2 and 3 fail-safe brakes
- Sealed for laboratory and industrial use
- Standard software permits 3-dimensional palletizing/depalletizing of rows, columns, and layers by simply teaching three points

Reliability Features

- Brushless AC servo motors minimize motor maintenance
- Harmonic drives on all axes result in higher reliability and reduced maintenance costs
- Sealed bearings and drives allow operation in harsh factory environments
- Grease fittings on all lubrication points for quick and easy maintenance
- Only two types of motors used for simpler servicing and spare parts reduction
- Easily removable service panels for quick access to robot drivetrain
- IP54 dust and liquid intrusion protection package

Options

- Other integral valve packages
- Axis 1 brake
- Higher speed CPU for improved path and cycle times

LR Mate 100*i* Dimensions

Footprint

160mm (6.3")
80mm (3.1")
50mm (1.9")
Axis 1 rotating center
4-dia. 11 through hole
dia. 21 counterbore depth 3
250mm (9.8")
193mm (7.5")
60mm (2.3")
50mm (1.9")
205mm (8.1")
165mm (6.4")
30mm (1.1")
30mm (1.1")
95mm (3.7")
190mm (7.5")

Top

Motion range of J4-axis center 3,4kg
Motion range of J4-axis center 5kg
160°
R620mm (24.4")
0°
R600mm (23.6")
160°
J4-axis center

Mounting plate

45°
Ø 5h7 DP7
4-M5 DP8
PCD31 .5 eq. sp.
J5:0°

Side

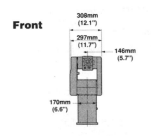

Motion range of J4-axis rotation center (5kg payload)
Motion range of J4-axis rotation center (3,4kg payload)
J4-axis rotation center
206mm (8.1") 150mm (5.9")
820mm (32.2")
250mm (9.8")
560mm (22")
R470mm (18.5")
461mm (18.1")
R170mm (6.7")
350mm (13.8")
267mm (10.5")
16mm (0.6")
620mm (24.4")
250mm (9.8")

Front

308mm (12.1")
297mm (11.7")
146mm (5.7")
170mm (6.6")

SCALE
1/30"

LR Mate 100*i* Specifications

Items		3kg (6.6 lbs) payload		4kg (8.8 lbs) payload		5kg (11 lbs) payload	
		Range	Speed	Range	Speed	Range	Speed
Motion range and speed	Axis 1	320°	180°/sec	320°	150°/sec	Same as 4kg (note 1)	
	Axis 2	185°	180°/sec	185°	150°/sec		
	Axis 3	365°	225°/sec	365°	180°/sec		
	Axis 4	240°	216°/sec	240°	100°/sec		
	Axis 5	400°	272°/sec	400°	250°/sec		
Moment	Axis 4	55.5 kgf • cm		74.0 kgf • cm		50.0 kgf • cm	
	Axis 5	40.0 kgf • cm		40.0 kgf • cm		4.4 kgf • cm (note 2)	
Load inertia	Axis 4	1.1 kgf • cm • s²		1.4 kgf • cm • s²		2.25 kgf • cm • s²	
	Axis 5	0.41 kgf • cm • s²		0.41 kgf • cm • s²		0.51 kgf • cm • s²	
Repeatability		±0.04mm (±0.002") based on JISB8432					
Mounting method		Upright/inverted					
Mechanical brakes		Axis 2, axis 3 (axis 1 option)					
Mechanical weight		32kg (70.5 lbs)					
Dust/water intrusion protection		Conforms to the IP54 standard for dust and liquid intrusion protection (seals may need periodic replacement if used with chlorine or gasoline based coolants)					
Notes		(1) In the case of 5kg wrist payload, the motion area and the wrist direction are limited (2) In the case of 5kg wrist payload, the wrist direction is limited to downward					

FANUC
Robotics

FANUC Robotics North America
2000 South Adams Road
Auburn Hills, MI 48326-2800
Phone (810) 377-7000
Fax (810) 377-7366

Literature Request
1-800-47-ROBOT

Cincinnati, OH
Phone (513) 771-8844

Chicago, IL
Phone (847) 364-6161

Los Angeles, CA
Phone (714) 258-7075

Charlotte, NC
Phone (704) 596-5121

Toronto, Canada
Phone (905) 670-5755

Montréal, Québec
Phone (514) 492-9001

Mexico City, Mexico
Phone (52-5) 611-5918

Sao Paulo, Brazil
Phone (55)(11) 3067-8070

e-mail address: fanuc.robotics@industry.net
web page: http://www.industry.net/fanuc.robotics

IRB 2400L, IRB 2400/10
IRB 2400/16

ABB Flexible Automation

ABB

Technical specifications

Design: MMC, Västerås. Printed in Sweden by Västra Aros Tryckeri, Västerås 1996.

SPECIFICATIONS

	IRB 2400L	IRB 2400/10	IRB 2400/16
Handling capacity	5-7 kg (11-15 lb.)	10 kg (22 lb.)	16 kg (35 lb.)
Reach	1.8 m (70")	1.5 m (59")	1.5 m (59")
Load offset	92/85 mm	100/100 mm	100/100 mm
	(3.6"/3.3")	(3.9"/3.9")	(3.9"/3.9")

All three robots are available in sealed "F" foundry version.

Supplementary load			
Upper arm,			
-wrist end	1 kg (2 lb.)	2 kg (4 lb.)	2 kg (4 lb.)
-rear end	10 kg (22 lb.)	10 kg (22 lb.)	10 kg (22 lb.)
Base unit	35 kg (77 lb.)	35 kg (77 lb.)	35 kg (77 lb.)
Number of axes			
Robot manipulator	6	6	6
External devices	6	6	6

Integrated signal supply	23 poles, 50 V DC
	10 poles, 250 V AC
Integrated air supply	Max. 8 bar
Memory/Instructions	
Program memory	1-5 Mb, 3,000-12,000 instructions
Storage memory	0.5-5 Mb, 3,000-35,000 instructions
3.5" diskette drive	1.44 Mb, 15.000 instructions

PERFORMANCE

Positional repeatability	±0.1 mm (0.004")

Axis movements	IRB 2400L	IRB 2400/10	IRB 2400/16
Working range			
Positioning			
Axis 1, Rotation	360°	360°	360°
Axis 2, Arm	200°	200°	200°
Axis 3, Arm	125°	125°	125°
Re-orientation			
Axis 4, Wrist	370°	400°	400°
Axis 5, Bend	240°	240°	240°
Axis 6, Rotation	800°	800°	800°
Axis 6, Option	Unlimited	Unlimited	Unlimited
Max. speed			
Positioning			
Axis 1, Rotation	135°/s	135°/s	135°/s
Axis 2, Arm	135°/s	135°/s	135°/s
Axis 3, Arm	135°/s	135°/s	135°/s
Re-orientation			
Axis 4, Wrist	310°/s	330°/s	330°/s
Axis 5, Bend	310°/s	330°/s	330°/s
Axis 6, Rotation	310°/s	400°/s	400°/s

ELECTRICAL CONNECTIONS

Supply voltage	200–600 V, 50/60 Hz
Rated power, supply transformer	4 kVA

PHYSICAL

Dimensions			
Total height (to axis 4)	1.6 m (64")	1.5 m (57")	1.5 m (57")
Manipulator bases	723x560 mm	723x560 mm	723x560 mm
	(29"x22")	(29"x22")	(29"x22")
Cabinet, H x W x D	1 300 x 915 x 530 (51 x 36 x 21")		
Weight			
Robot manipulator	380 kg	380 kg	380 kg
	(836 lb.)	(836 lb.)	(836 lb.)
Cabinet	300 kg	300 kg	300 kg
	(660 lb.)	(660 lb.)	(660 lb.)

ENVIRONMENT

Ambient temperature	
Basic manipulator in operation	5°C to 45°C (41° - 113°F)
Sealed version in operation	Surface temperature < 100°C
	(< 212°F)
Cabinet	5°C to 40°C (41° - 104°F)
Relative humidity	Max. 95%

Degree of protection

Manipulator	
Basic version	IP 54
Sealed version "F"	IP 55/56/67
Control system	IP 54
Noise level	Max. 70 dB (A)
Safety	Double circuits, emergency
	stops, safety functions
Emmission	EMC-shielded

USER INTERFACES

Operator's panel	In cabinet or external
Programming unit	Portable with joystick and keypad.
	Display 16 lines x 40 characters.
	Windows style communication.
	Enabling device, 3 positions.
	All programming and hot-edit functions.
Languages	Choice between 10 national languages
Off-line programming	RRS and ProgramMaker
PC-based training tools	QuickTeach

MACHINE INTERFACES

Digital inputs/outputs	96/96, 24 V DC, 110 V AC or 250 V AC
Analogue inputs/outputs	4/4, ±10 V and ±20 mA
Remote I/O	Allen Bradley PLC
Serial channels	Three RS 232 and one RS 485
Computer link	Robot Application Protocol, based on MMS
Process interfaces	Media and signals on upper arm
Robot vision	OptiMaster
Travel units	Integrated track motion

RobotWare SOFTWARE PRODUCTS

BaseWare™	Superior robot performance and
	communication
ProcessWare™	Includes ArcWare, GlueWare etc.
FactoryWare™	Online PC-based, user-adapted
	functionality
DeskWare™	Offline, PC-based user support including
	training, library, laboritory and
	programming

Examples of ArcWare™ functionality	
Examples of	
signal interfaces	Laser sensor interface
	Control of power source and
	wire feed from robot pendant
	Status of arc voltage, current, water,
	gas, wire-feed (DI)
	On/off of power, gas, wire-feed,
	error information (DO)
	Value of wire-feed velocity, voltage,
	current (AO)
Examples of functions	Process tuning
	Programming parameters
	Execution
	Weld retry
	Weld error report and log
	Arc start/stop
	Material preheating/cooling
	Scrape start
	Crater filling
	Wire burnback
	Weave pattern
	Monitor arc/wire/water/voltage/current/gas

6397 032-655 May 1996.

ABB Flexible Automation Inc.
2487 South Commerce Drive
New Berlin, WI 53151
414 785 3400

ABB Flexible Automation Inc.
1250 Brown Road
Auburn Hills, MI 48326
810 391 9000

ABB Flexible Automation Inc.
8401 Northwest Boulevard
Indianapolis, IN 46278
317 872 4129

ABB Flexible Automation Inc.
4600 Innovation Drive
Ft. Collins, CO 80525
970 225 7600

ABB Flexible Automation Inc.
27101 Groesbeck Highway
Warren, MI 48089
810 776 8000

Asea Brown Boveri Inc.
ABB Flexible Automation Div.
4410 Paletta Court
Burlington, Ont L7L 5R2
Canada
905 681 0565

ABB Sistemas S. A. de C. V.
Henry Ford No. 4
Esq. Dr. Gustavo Baz
54030 Tlalnepantla
Mexico
525 328 1523

Asea Brown Boveri Ltda
ABB Flexible Automation Div.
Av Dos Autonomistas, 1496
06020-902 Osasco-San Paolo
Brasil
55 11 704 9111

Knobbed Suction Cup
with sensor

SA M14x1 + NSØ
SA M16x1 + NSØ

SA M14x1   SA M16x1

Vacuum connection

NS-Suction cup

Force table (0.8 bar vacuum and smooth, dry surface)

Ø	Measure Ø:	10	15	18	24	30	40	50	60	70	85	100
	Suction power in N at 0.8 bar vacuum:	4.7	10.6	15.2	27.1	42.4	75.4	118	170	230	340	471

Air consumption with vacuum generator (see below): 13 standard liters per minute at 6 bar. For heat resistant version or food handling version, see below.

with external thread

Order Number		Ø
SA M14x1	+	NS 10
SA M14x1	+	NS 15
SA M14x1	+	NS 18

with external thread

Order Number		Ø
SA M16x1	+	NS 24
SA M16x1	+	NS 30
SA M16x1	+	NS 40
SA M16x1	+	NS 50
SA M16x1	+	NS 60
SA M16x1	+	NS 70
SA M16x1	+	NS 85
SA M16x1	+	NS 100

M14x1 For proximity switch
12
SW 17
30
M5 Vacuum connection
12

M16x1 For proximity switch
12
SW 19
30
G 1/8" Vacuum connection
22

Custom suction cup design on request

❶ Inductive sensing removes all doubt as to whether or not the suction cup has made contact. A spring loaded actuating pin is depressed when the suction cup makes contact with the part. This triggers the proximity switch (part number NJ 3-E2) while releasing the vacuum. This is valuable when using multiple suction cups with vacuum pumps. The vacuum is only applied when parts are being gripped. Air cannot leak in from the outside.

For proximity switch, order number NJ 3-E2. For further information, see Accessories starting on page 904.

Free 10 day trial offer

Knobbed Suction Cup
with sensor and vacuum generator

SAV M14x1 + NSØ
SAV M16x1 + NSØ

Vacuum Eng.

Outlet for proximity switch cable

SAV M....

Compressed air

NS - Suction cup

Standard version:	Black neoprene, oil and heat resistant from -30°C to 120°C, durometer 65 ± 3.
Heat resistant version:	Blue silicone, heat resistant to 200°C. durometer 60 ± 3. Order suffix "Si": (e.g. NS 10 Si)
Food handling version: (Special Order Only)	Tempered white silicone, heat resistant to 200°C. Order suffix "L": (e.g. NS 10 L)

Force table - see above

with external thread + vacuum generator

Order Number		Ø
SAV M14x1	+	NS 10
SAV M14x1	+	NS 15
SAV M14x1	+	NS 18

Accessories: Hose PU 4 (see page 913)

with external thread + vacuum generator

Order Number		Ø
SAV M16x1	+	NS 24
SAV M16x1	+	NS 30
SAV M16x1	+	NS 40
SAV M16x1	+	NS 50
SAV M16x1	+	NS 60
SAV M16x1	+	NS 70
SAV M16x1	+	NS 85
SAV M16x1	+	NS 100

Accessories: Hose PU 4 (see page 913)

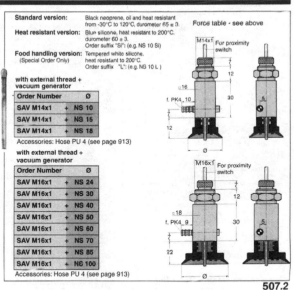

M14x1 For proximity switch
12
≥16
f. PK4_10
12
30
5
Ø

M16x1 For proximity switch
12
≥18
f. PK4_9
22
30
5
Ø

❷ Inductive sensing removes all doubt whether or not the suction cup has made contact. A spring loaded actuating pin is depressed when the suction cup makes contact with the part. This triggers the proximity switch (part number NJ 3-E2) to operate a compressed air valve. When the compressed air is applied to the vacuum generator, a vacuum is immediately applied to the suction cup. When the compressed air is removed, the vacuum immediately collapses (it is not necessary to exhaust air through the opposite direction). This is an essential condition for achieving the fastest possible cycle times. Air consumption is 13 standard liters per minute at 6 bar. For proximity switch, order number NJ 3-E2. For further information, see Accessories starting on page 904.

Free 10 day trial offer

507.2

Large Knobbed Suction Cup — SA 22 + PNSØ
for large parts-with and without vacuum generator — SAV 22 + PNSØ

1 When handling large parts, large stable suction cups are necessary. There are two versions available: 1) with an external thread for connecting to vacuum pumps; or 2) with an external thread and a built-in vacuum generator. There are separate part numbers for the threaded parts (SA 22 or SAV 22) and the oval suction pad with aluminum backing plate (PNS...A). **Note: The knobbed suction cup is also available without the aluminum plate. Just leave out the "P", e.g. order number "NS 100 A".**

with external thread

SA 22

PNS

with external thread + vacuum generator

SAV 22

PNS

Free 10 day trial offer

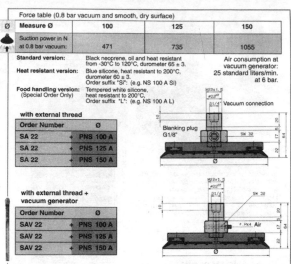

Force table (0.8 bar vacuum and smooth, dry surface)

Measure Ø	100	125	150
Suction power in N at 0.8 bar vacuum:	471	735	1055

Standard version: Black neoprene, oil and heat resistant from -30°C to 120°C, durometer 65 ± 3.

Heat resistant version: Blue silicone, heat resistant to 200°C, durometer 60 ± 3. Order suffix "Si": (e.g. NS 100 A Si)

Food handling version: Tempered white silicone, (Special Order Only) heat resistant to 200°C, Order suffix "L": (e.g. NS 100 A L)

Air consumption at vacuum generator: 25 standard liters/min. at 6 bar.

with external thread

Order Number		Ø
SA 22	+ PNS	100 A
SA 22	+ PNS	125 A
SA 22	+ PNS	150 A

Blanking plug G1/8"

Vacuum connection

with external thread + vacuum generator

Order Number		Ø
SAV 22	+ PNS	100 A
SAV 22	+ PNS	125 A
SAV 22	+ PNS	150 A

Air

Vacuum Eng.

Spring-Loaded Knobbed Suction Pads SGA22+KNSØ
with ball joint (±15°) and locking plate

2 Spring Loaded Knobbed Suction Cups (with ball joint) are characterized by their compact design and their locking system. The upper spring (2) ensures that there is no recoil when lifting the part. The return spring (4) can extend 25 mm (a 50 mm version is currently being developed), and is secured against rotation. Consequently, parts can be picked up and placed accurately. A maximum compensation angle of ± 15° is provided by a ball joint.

Note: The rubberized suction cups can be ordered separately and easily replaced. This can save time and expense during assembly and disassembly. Just leave out the "K", e.g. order number "NS 100 A".

Free 10 day trial offer
507.1

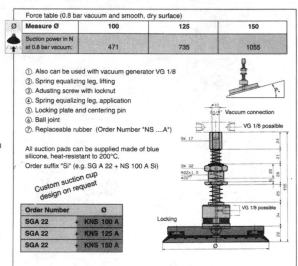

Force table (0.8 bar vacuum and smooth, dry surface)

Measure Ø	100	125	150
Suction power in N at 0.8 bar vacuum:	471	735	1055

① Also can be used with vacuum generator VG 1/8

② Spring equalizing leg, lifting

③ Adusting screw with locknut

④ Spring equalizing leg, application

⑤ Locking plate and centering pin

⑥ Ball joint

⑦ Replaceable rubber (Order Number "NS ....A")

All suction pads can be supplied made of blue silicone, heat-resistant to 200°C.
Order suffix "Si" (e.g. SG A 22 + NS 100 A Si)

Custom suction cup design on request

Vacuum connection

VG 1/8 possible

VG 1/8 possible

Order Number		Ø
SGA 22	+ KNS	100 A
SGA 22	+ KNS	125 A
SGA 22	+ KNS	150 A

Locking

OMRON

Diverse General-Purpose Limit Switch

D4A-□N

Heavy-Duty Switch Features
Plug-In Construction

- Oiltight, watertight construction meets
  NEMA 3, 4, 4X, 6P and 13
- Convenient front mounting, easy
  installation and maintenance
- Wide temperature range: -40° to
  100°C (-40° to 212°F)
- Choose operation status models with
  LED or neon lamp indicators
- Side rotary switches accept a wide
  selection of levers
- Increased mechanical durability

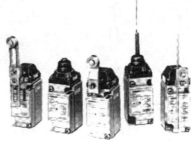

Ordering Information

■ SIDE ROTARY SWITCHES

When ordering side rotary switches, order the levers separately from the next page.

Description	SPDT double break			DPDT double break
	No indicator	Neon lamp	LED indicator	No indicator
Standard	D4A-1101N	D4A-1301N	D4A-1E01N	D4A-2501N
High-precision	D4A-1102N	D4A-1302N	D4A-1E02N	D4A-2502N
Low torque	D4A-1103N	D4A-1303N	D4A-1E03N	D4A-2503N
High precision/low torque	D4A-1104N	D4A-1304N	D4A-1E04N	D4A-2504N
Maintained contact	D4A-1105N	D4A-1305N	D4A-1E05N	D4A-2505N
Sequential operation	—	—	—	D4A-2717N
Center neutral operation	—	—	—	D4A-2918N

■ SWITCHES WITH BUILT-IN ACTUATORS

Description		SPDT double break			DPDT double break
		No indicator	Neon lamp	LED indicator	No indicator
Side plunger type	Plain	D4A-1106N	D4A-1306N	D4A-1E06N	D4A-2506N
	Vertical roller	D4A-1107-V-N	D4A-1307-V-N	D4A-1E07-V-N	D4A-2507-V-N
	Horizontal roller	D4A-1107-H-N	D4A-1307-H-N	D4A-1E07-H-N	D4A-2507-V-N
	Adjustable plunger	D4A-1108N	D4A-1308N	D4A-1E08N	D4A-2508N
Top plunger type	Plain	D4A-1109N	D4A-1309N	D4A-1E09N	D4A-2509N
	Roller	D4A-1110N	D4A-1310N	D4A-1E10N	D4A-2510N
	Adjustable plunger	D4A-1111N	D4A-1311N	D4A-1E11N	D4A-2511N
Wobble lever type	Spring wire	D4A-1112N	D4A-1312N	D4A-1E12N	D4A-2512N
	Piano wire rod	D4A-1114N	D4A-1314N	D4A-1E14N	D4A-2514N
	Stainless steel rod	D4A-1124N	D4A-1324N	D4A-1E24N	D4A-2524N
	Cat whisker	D4A-1115N	D4A-1315N	D4A-1E15N	D4A-2515N
	Coil spring	D4A-1116N	D4A-1316N	D4A-1E16N	D4A-2516N

■ OPTIONS

Add "-F" suffix for Viton® fluoroelastomer seals that withstand temperatures to 120°C (248°F). For example, **D4A-1101N-F**.

■ LEVERS FOR D4A-N SIDE ROTARY SWITCHES

Description		Lever radius	Material	Diameter	Width	Part number
Standard roller levers, front mount		38.1 mm (1.5 in)	Stainless steel	19.1 mm (0.75 in)	7.9 mm (0.31 in)	D4A-A00
		33.7 mm (1.33 in)	Stainless steel	17.5 mm (0.69 in)	15 mm (0.59 in)	D4A-B06
		38.0 mm (1.5 in)	Ball bearing	17.0 mm (0.67 in)	6 mm (0.24 in)	WL-1A400
		38.0 mm (1.5 in)	Nylon	17.5 mm (0.69 in)	30 mm (1.18 in)	WL-1A104
		50.0 mm (1.97 in)	Stainless steel	17.5 mm (0.69 in)	15 mm (0.59 in)	WL-1A200
		63.0 mm (2.48 in)	Stainless steel	17.5 mm (0.69 in)	7.0 mm (0.28 in)	WL-1A300
Standard roller lever, back mount		38.1 mm (1.5 in)	Stainless steel	19.1 mm (0.75 in)	7.9 mm (0.31 in)	D4A-A10
		38.1 mm (1.5 in)	Nylon	50 mm (1.97 in)	7.0 mm (0.28 in)	WL-1A106
Offset roller lever	Front mount	38.1 mm (1.5 in)	Stainless steel	19.1 mm (0.75 in)	7.9 mm (0.31 in)	D4A-A20
	Back mount	38.1 mm (1.5 in)	Stainless steel	19.1 mm (0.75 in)	7.9 mm (0.31 in)	D4A-A30
Adjustable roller lever, front mount		25 to 89 mm (0.98 to 3.5 in) L	Nylon	17.5 mm (0.69 in)	7.0 mm (0.28 in)	WL-2A111
		33 to 91 mm (1.3 to 3.6 in) L	Stainless steel	19.1 mm (0.75 in)	7.4 mm (0.29 in)	D4A-C00
Spring rod lever		290 mm (11.5 in) L	Stainless steel	2 mm (0.08 in)	—	WL-4A201
Adjustable rod lever		150 mm (5.91 in) L	Stainless steel	3 mm (0.12 in)	—	D4A-D00
		350 to 380 mm (13.8 to 15 in) L	Stainless steel	3.2 mm (0.13 in)	—	WL-3A100
Fork roller lever	L.H. front/ R.H. back	38.1 mm (1.5 in)	Stainless steel	19.1 mm (0.75 in)	7.9 mm (0.31 in)	D4A-E00
		38.0 mm (1.5 in)	Nylon	17.5 mm (0.69 in)	7.0 mm (0.28 in)	WL-5A100
	L.H. back/ R.H. front	38.1 mm (1.5 in)	Stainless steel	19.1 mm (0.75 in)	7.9 mm (0.31 in)	D4A-E10
		38.0 mm (1.5 in)	Nylon	17.5 mm (0.69 in)	7.0 mm (0.28 in)	WL-5A102
	Both front	38.1 mm (1.5 in)	Stainless steel	19.1 mm (0.75 in)	7.9 mm (0.35 in)	D4A-E20
Looped rod		150 mm (6 in) L	Nylon	4 mm (0.16 in)	60 mm (2.36 in)	D4A-F00

■ REPLACEMENT PARTS

Building a Part Number

The D4A-N family of switches features plug-in construction with interchangeable parts for simplified long-term maintenance. The illustration at right shows the parts that make up a whole switch, as well as how to read a part number as the sum of its parts.

Note: Compatibility of D4A-N parts with Omron's former D4A series is limited to interchangeable **receptacles** only. D4A switch bodies and operating heads cannot be used to replace similar D4A-N parts.

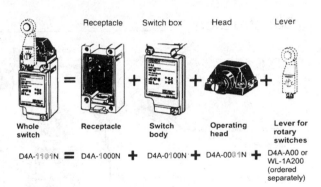

| | Receptacle | Switch box | Head | Lever |

| Whole switch | Receptacle | Switch body | Operating head | Lever for rotary switches |

D4A-1101N ═ D4A-1000N ✚ D4A-0100N ✚ D4A-0001N ✚ D4A-A00 or WL-1A200 (ordered separately)

Receptacles

Description	Part number
SPDT double break	D4A-1000N
DPDT double break, also used for sequential operation and center neutral operation switches	D4A-2000N

Switch Bodies

Omron offers two economical solutions for long-term maintenance of D4A-N limit switches. Choose a complete switch body with a built-in switch, or just replace the internal switch by removing the two screws holding it in place.

Description	Indicator type	Body with built-in switch	Internal switch only
SPDT double break	No indicator	D4A-0100N	D4A-S10N
	Neon lamp indicator	D4A-0300N	—
	LED indicator, 6 A, 24 VDC	D4A-0E00N	—
DPDT double break	No indicator	D4A-0500N	D4A-S50N
Sequential operating type, DPDT	No indicator	D4A-0700N	D4A-S70N
Center neutral operating type, DPDT	No indicator	D4A-0900N	D4A-S90N

Operating Heads

For side rotary operating heads listed below, be sure to order an actuator lever from the large selection of D4A and WL parts listed on the second page of this data sheet.

Description	Part number
Standard side rotary	D4A-0001N
High precision side rotary	D4A-0002N
Low torque side rotary	D4A-0003N
High precision/low torque side rotary	D4A-0004N
Maintained contact side rotary	D4A-0005N
Sequential operating side rotary	D4A-0017N
Center neutral operating side rotary	D4A-0018N

Description	Part number
Plain side plunger	D4A-0006N
Vertical side roller plunger	D4A-0007-VN
Horizontal side roller plunger	D4A-0007-HN
Adjustable side plunger	D4A-0008N
Plain top plunger	D4A-0009N
Top roller plunger	D4A-0010N
Adjustable top plunger	D4A-0011N

Description	Part number
Spring wire wobble lever	D4A-0012N
Plastic rod wobble lever	D4A-0014N
Stainless steel rod wobble lever	D4A-0024N
Cat whisker wobble lever	D4A-0015N
Coil spring wobble lever	D4A-0016N

Construction

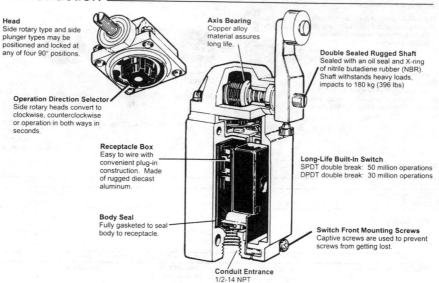

Head
Side rotary type and side plunger types may be positioned and locked at any of four 90° positions.

Operation Direction Selector
Side rotary heads convert to clockwise, counterclockwise or operation in both ways in seconds.

Receptacle Box
Easy to wire with convenient plug-in construction. Made of rugged diecast aluminum.

Body Seal
Fully gasketed to seal body to receptacle.

Axis Bearing
Copper alloy material assures long life.

Double Sealed Rugged Shaft
Sealed with an oil seal and X-ring of nitrile butadiene rubber (NBR). Shaft withstands heavy loads, impacts to 180 kg (396 lbs)

Long-Life Built-In Switch
SPDT double break: 50 million operations
DPDT double break: 30 million operations

Switch Front Mounting Screws
Captive screws are used to prevent screws from getting lost.

Conduit Entrance
1/2-14 NPT

Specifications

■ ELECTRICAL RATINGS

NEMA A600 Rating (UL/CSA)

Circuit	Rated votlage	Amperes		Continuous carrying current	Voltamperes	
		Make	Break		Make	Break
SPDT double break	120 VAC	60	6	10	7200	720
	240 VAC	30	3	10	7200	720
	480 VAC	15	1.5	10	7200	720
	600 VAC	12	1.2	10	7200	720

NEMA B600 Rating (UL/CSA)

Circuit	Rated votlage	Amperes		Continuous carrying current	Voltamperes	
		Make	Break		Make	Break
DPDT double break	120 VAC	30	3	5	3600	360
	240 VAC	15	1.5	5	3600	360
	480 VAC	7.5	0.75	5	3600	360
	600 VAC	6	0.6	5	3600	360

NEMA A300 Rating (UL/CSA) for SPDT switches with neon lamp indicators

Circuit	Rated votlage	Amperes		Continuous carrying current	Voltamperes	
		Make	Break		Make	Break
SPDT double break	120 VAC	60	6	10	7200	720
	240 VAC	30	3	10	7200	720

Note: All switches must be the same polarity.

SPDT switches with LED indicators

Circuit	Rated voltage	Load current
SPDT double break	24 VDC	6 A max. (resistive)

■ CHARACTERISTICS

Enclosure rating	UL	Types 3, 4, 4X, 6P, 13
	NEMA	Types 1, 2, 3, 4, 4X, 5, 6, 6P, 12 and 13
	IEC 144	IP67
Mechanical life		50 million operations for SPDT; 30 million operations for DPDT
Ambient operating temperature		-40° to 100°C (-40° to 212°F) for side rotary types (except low torque types)
		-20° to 100°C (-4° to 212°F) for plunger and wobble lever types
		-10° to 80°C (14° to 176°F) for indicator equipped types
		-10° to 120°C (14° to 248°F) for Viton sealed type
Vibration	Malfunction durability	10 to 55 Hz, 1.5 mm (0.06 in) double amplitude
Shock	Malfunction durability	60 G for SPDT side rotary types; 30 G for all others
	Mechanical durability	100 G

■ OPERATING CHARACTERISTICS

Legend: OF = Operating Force (max.); RF = Reset Force (min.); OT = Overtravel (min.); PT = Pretravel;
MD = Movement Differential (max.); PO = Positively Open (max.); TT = Total Travel; OP = Operating Position

Side Rotary Switches

Part number	Description	OF max.	RF min.	PT max.	OT min.	MD max. SPDT	MD max. DPDT	OP
D4A-□□01N	Standard	4 kg-cm (3.5 lb-in)	0.5 kg-cm (0.4 lb-in)	12° ±3°	70°	4° ±1°	5° ±2°	⌐
D4A-□□02N	High-precision	4 kg-cm (3.5 lb-in)	0.5 kg-cm (0.4 lb-in)	5° ±2°	75°	3° ±1°	4° ±1°	—
D4A-□□03N	Side rotary, low torque	2 kg-cm (1.7 lb-in)	—	12° ±3°	70°	4° ±1°	5° ±2°	—
D4A-□□04N	Side rotary, high-precision /low torque	2 kg-cm (1.7 lb-in)	—	5° ±2°	75°	3° ±2°	4° ±2°	—
D4A-□□05N	Side rotary, maintained contact	4 kg-cm (3.5 lb-in)	—	55° ±10°	20°	28° ±7°	28° ±7°	—
D4A-□□17N	Side rotary, sequential operation	4 kg-cm (3.5 lb-in)	0.5 kg-cm (0.4 lb-in)	1st pole 7° ±3° 2nd pole 15° ±3°	65°	Each pole 4° ±2°		—
D4A-□□18N	Side rotary, center neutral operation	4 kg-cm (3.5 lb-in)	0.2 kg-cm (0.2 lb-in)	15° ±3°	65°	4° ±1°		—

Note: The operating position depends on the lever selected.

Switches with Built-in Actuators (continued)

Part number	Description	OF max.	RF min.	OT min.	PT max.	MD max.		OP
						SPDT	DPDT	
D4A-☐☐11N	Adjustable top plunger	1.8 kg (63.5 oz.)	500 g (17.6 oz.)	5.1 mm (0.201 in)	1.6 mm (0.063 in)	0.4 mm (0.016 in)	1.0 mm (0.039 in)	55.5 to 62 mm (2.19 to 2.44 in)
D4A-☐☐12N	Spring wire wobble lever	100 g (3.5 oz.)	—	—	5° ±3°, 3.94 in radius	—	—	—
D4A-☐☐14N	Plastic rod wobble lever	150 g (5.3 oz.)	—	—	5° ±3°, 1.38 in radius	—	—	—
D4A-☐☐15N	Cat whisker wobble lever	150 g (5.3 oz.)	—	—	5° ±3°, 1.97 in radius	—	—	—
D4A-☐☐16N	Coil spring wobble lever	150 g (5.3 oz.)	—	—	5° ±3°, 1.97 in radius	—	—	—

Engineering Data

Electrical Service Life

D4A-1☐☐☐N SPDT Double Break

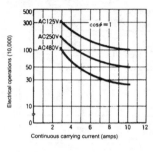 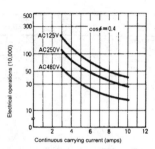

D4A-2☐☐☐N DPDT Double Break

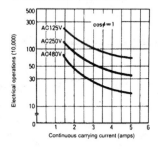 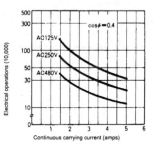

■ **CONTACT RATINGS**
NEMA A600 for SPDT models
NEMA B600 for DPDT models
NEMA A300 for SPDT models with neon lamp status indicator

■ **CONTACT RESISTANCE**
25 mΩ maximum (initial)

■ **ELECTRICAL APPROVALS**
UL Recognized, File No. E76675
CSA Certified, File No. LR45746

Operation

■ **CONTACT FORM**

SPDT Double Break with same polarity

SPDT Double Break with LED Indicator

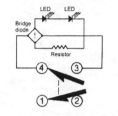

SPDT Double Break with Neon Lamp Indicator

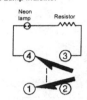

DPDT Double Break with same polarity each pole

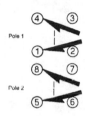

Sequential Operating Type
Pole 1 operates first
Pole 2 operates second
Either CW or CCW or both

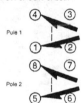

Center Neutral Operating Type
Pole 1 opertes CW
Pole 2 operates CCW

■ **TERMINAL ARRANGEMENT**

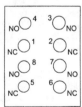

Dimensions

Unit: mm (inch)

■ SIDE ROTARY SWITCHES

D4A-☐☐01N, D4A-☐☐02N, D4A-☐☐03N, D4A-☐☐04N, D4A-☐☐17N, D4A-☐☐18N

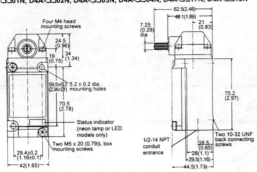

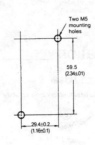

Mounting holes

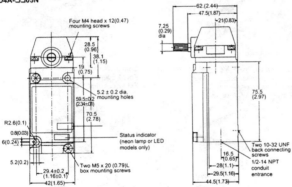

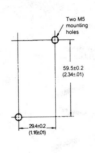

Mounting holes

D4A-☐☐05N

Internet Resources

The development of the Internet and the World Wide Web has caused an exponential increase in the information available to online users. This appendix contains universal resource locators (URLs), or links, to various vendors and automation technology resources that can be used for some of the projects and for solving the case studies provided in the text. It is important to type the URL, using uppercase letters and lowercase letters, exactly as shown.

Some of the links connect the user directly to vendor websites and others point to general sources for automation technology information. One of the best general resource sites is the online version of the Thomas Register (TR). This free online service provides an indexed search to numerous resources used in manufacturing. You must register for the service and log into the site to perform a search for vendor data. The vendor information is provided at three levels: (1) available for all listings—vendor name, address, and telephone and fax numbers; (2) available for many of the listings—online data at the TR site that is readable by a browser or the Adobe Acrobat reader; (3) available at some of the sites—a link to the company's website with online catalog data.

Another effective technique for locating information on the World Wide Web is to use general search engines. Several search engines are listed below that will locate vendors and services. The Alta Vista, Webcrawler, and Lycos search URLs frequently provide the best results for industrial vendor data.

Thomas Register

Main page   http://www4.thomasregister.com

Web Search Engines

Alta Vista   http://www.altavista.com

Excite   http://www.excite.com

Google   http://www.google.com

Infoseek   http://www.go.com

Lycos   http://www.lycos.com

Webcrawler   http://www.webcrawler.com

Standard Organizations

American National Standards Institute (ANSI)   http://web.ansi.org

American Society of Mechanical Engineers   http://www.asme.org

International Electrical and Electronic Engineers (IEEE)   http://www.ieee.org

Occupational Safety and Health Administration (OSHA)   http://www.osha.gov

Society of Manufacturing Engineers (SME)   http://www.sme.org

Robot Vendors

ABB Robotics   http://www.abb.com

CRS Robotics   http://www.crsrobotics.com

ElectroPneumatics Ltd.   http://www.electropneumatics.com

Fanuc Corp.   http://www.fanucrobotics.com

Festo   http://www.festo.com

Mitsubishi   http://www.mitsubishi.com

Reis Robotics   http://www.reisrobotics.com

Techno-Isel   http://www.techno-isel.com

Tol-O-Matic   http://www.tolomatic.com

Automation Sensors

Honeywell Inc.   http://www.honeywell.com

Omron Electronics   http://www.omron.com

Rockwell Automation/ Allen-Bradley   http://www.ab.com

Automation Support Components

Applied Robotics	http://www.arobotics.com
Compact Automation Products	http://compactair.com

Programmable Logic Controllers

Allen-Bradley	http://www.ab.com
General Electric	http://www.gefanuc.com/index.asp
Ormron	http://www.omron.com
Schneider Electric	http://www.squared.com
Siemens	http://www.aut.sea.siemens.com

Work-cell Control Software

Brooks Automation	http://www.brooks.com
IBM	http://www.ibm.com
Intellution	http://www.intellution.com
U.S. Data	http://www.usdata.com
Wonderware	http://www.wonderware.com

Material Handling Systems

Acco Systems	http://www.accosystems.com
Bosch Automation	http://www.boschautomation.com
Menziken Automation	http://www.menziken.com
Prab Conveyors	http://www.prab.com
S1 Handling Systems	http://www.sihs.com

Justification Program

The data entered into the justification program is defined in the following spreadsheet cell descriptions. For example, the description in B6 defines what the Capital Equipment section on the spreadsheet includes. The B callouts define the data to be entered, and the C and D callouts describe data calculated in the spreadsheet in Figure 12–24. The descriptions are saved as cell notes in the spreadsheet file provided on the CD-ROM that accompanies the Instructor's Manual.

B6—This section deals with the cost of buying and installing new equipment and disposing of old equipment.

B7—The total cost of equipment purchased (less any tax credits) that will be capitalized and depreciated. Enter a negative value because the expense represents a cost.

B9—The capitalized cost of shipping and installing the purchased equipment. Any component not capitalized should be entered into line 6. Enter a negative value.

B10—The cash value received for older equipment sold. It should be entered into the appropriate year as a positive number.

B11—Either the tax saving (equipment sold for less than book value) or the tax obligation (equipment sold for more than book value). Enter either a positive or negative value in the appropriate year.

B12—The total of all capitalized cost and savings. Depreciation will be calculated on this value.

B14—This is all non-capitalized cost or expensed labor necessary for the automation.

	A	B	C	D	E
1		**Justification Program**			
2					
3				YEARS	
4			0	1	2
5					
6		CAPITAL EQUIPMENT			
7	1	Equipment Cost	0	0	0
8		Add a line for all major items	0	0	0
9	2	Freight & Installation	0	0	0
10	3	Sale of Old Equip.	0	0	0
11	4	Tax on Old Equip	0	0	0
12	5	Total Net Investment	=SUM(C7:C11)	=SUM(D7:D11)	=SUM(E7:E11)
13					
14		NON-CAPITALIZED COST			
15			Moving Equipment	0	0 0
16			Installation	0	0 0
17			Training	0	0 0
18			Programming	0	0 0
19	6	Total Non-capitalized	=SUM(C15:C18)	=SUM(D15:C=D18)	=SUM(E15:E18)
20					
21		INVENTORY CHANGE			
22	7	Inventory Change	0	0	0
23					
24		OPERATING COSTS			
25	8	Direct Labor	0	0	0
26	9	Indirect Labor	0	0	0
27	10	Maintenance Costs	0	0	0
28	11	Tooling Costs	0	0	0
29	12	Materials & Supplies	0	0	0
30	13	Inspection	0	0	0
31	14	Assembly Cost	0	0	0
32	15	Scrap & Rework	0	0	0
33	16	Downtime	0	0	0
34	17	Utilities	0	0	0

B19—The total of all non-capitalized cost or expensed labor necessary for the automation. This figure can include internal labor cost for moving equipment or developing programs.

B21—This section deals with the changes in inventory levels due to the addition of the automation.

B22—The one-time change in cash due to inventory reduction based on the automation. It does not have an impact on taxes, and it is a positive value placed in the appropriate years.

B24—This section deals with the ongoing changes (increased costs are negative and reduced costs are positive values) in the operating costs as a result of the automation project.

	A	B	C	D	E
35	18	Taxes & Insurance	0	0	0
36	19	Subcontracting	0	0	0
37	20	Safety	0	0	0
38	21	Programming	0	0	0
39	22	Process Improvements	0	0	0
40	23	Other	0	0	0
41	24	Total Operating Cost	=SUM(C25:C40)	=SUM(D25:D40)	=SUM(E25:E40)
42					
43		OTHER IMPACT ON REVENUE			
44	25	Change in Volume	0	0	0
45	26	Reduced Lead Times	0	0	0
46	27	Increased Quality	0	0	0
47	28	New Product Introduction	0	0	0
48	29	Manufacturing Flexibility	0	0	0
49	30	Other Revenue	0	0	0
50	31	Total Revenue	=SUM(C44:C49)	=SUM(D44:D49)	=SUM(E44:E49)
51					
52		DISCOUNTED CASH FLOW ANALYSIS			
53	32	Total Operating Cost	=C41	=D41	=E41
54	33	Total Revenue	=C50	=D50	=E50
55	34	Non-Capitalized Cost	=C19	=D14	=E14
56	35	Total Pretax Cash	=SUM(C53:C55)	=SUM(D53:D55)	=SUM(E53:E55)
57	36	Tax Rate	0		
58	37	After-tax Cash Flow	=C56*(1-$C57)	=D56*(1-$C57)	=E56*(1-$C57)
59	38	Depreciation Rate	0	0	0
60	39	Depreciation	=C59*$C12	=D59*$C12	=E59*$C12
61	40	Depreciation Cash	=-1*C60*$C57	=-1*D60*$C57	=-1*E60*$C57
62	41	Inventory Cash	=C22	=D22	=E22
63	42	Total Net Investment	=C12	=D12	=E12
64	43	Total After-tax Cash	=C58+C61+C62+C63	=D58+D61+D62+D63	=E58+E61+E62+E63
65	44	Discount Rate	0		
66	45	Discount Factor	1	=C66*(1/(1+$C65))	=D66*(1/(1+$C65))
67	46	Discounted Cash Flow	=C64*C66	=D64*D66	=E64*E66
68	47	Cum. Cash Flow	=C67	=C68+D67	=D67+E67

B25—Enter labor cost, including fringe benefits, and consider the following: reduction in labor hours due to automation. Use increments of whole people if possible; use standard cost when possible; include production inefficiencies as necessary to reflect true direct labor costs.

C25—Six workers eliminated for 40 hours at $9.00 per hour for 25 weeks, plus 42% fringe benefits and 13 weeks of overtime for 2 workers.

D25—Six workers eliminated for 40 hours at $9.00 per hour for 50 weeks, plus 42% fringe benefits and 25 weeks of overtime for 2 workers.

B26—Changes in indirect labor with same considerations as in line 8.

C26—Elimination of manual inspection and addition of half-time technician to maintain the cell.

B27—The change in cost for maintenance materials as a result of the automation. Changes in maintenance labor costs would be included in line 9.

B28—The cost of consumable tooling. The cost of permanent tooling would be included in equipment cost.

B29—Changes in the productive supplies, like coolant and gloves.

B30—Changes in the in-process inspection component should be included here.

C30—Saving in inspection time and overhead.

B31—Changes in the assembly cost due to less lost time by the assembly area personnel.

B32—The changes in scrap and rework cost include the cost of material, cost of labor, and cost of inspection associated with the change.

B33—Lost labor time due to machine downtime from machine repair, machine preventive maintenance, or unavailability of parts to make.

B34—The change in cost of utilities.

B35—Changes in property taxes and insurance.

B36—Change in all costs associated with subcontracting parts out to other vendors.

B37—Change in cost due to reduction of lost labor time or reduced insurance as a result of automation.

B38—Changes in cost due to more efficient programming of automation over manually programmed machines.

B39—Changes that result from improved manufacturing practices.

B41—Sum of all changes in operating cost.

B43—This section deals with increases in market share as a result of the automation. In each category, the effect of increased market share on revenue dollars is added as a positive number.

B44—A greater volume of parts can be produced with the automation.

B45—Profit from a greater market share is achieved due to a reduction in the manufacturing lead time.

B46—Profit from a greater market share is achieved as a result of increased quality.

B47—Profit from a faster time to market results from a reduced time to market.

B48—Profit from a larger product mix on machines is achieved.

B50—Sum of all changes in profits due to automation.

B52—This section includes the calculations for the discounted cash flow analysis.

B53—From line 24.

B54—From line 31.

B55—From line 6.

B56—Total of lines 32 to 34.

B57—The corporate tax rate entered as a decimal.

B58—Calculated as line 35 $\times$ (1 – line 36).

B59—Enter either the straight line or MACRS as follows. Straight line is per year: 0.071; 0.143; 0.143; 0.143; 0.143; 0.143; 0.143; 0.071. MACRS is per year: 0.143; 0.245; 0.175; 0.125; 0.089; 0.089; 0.089; 0.045.

B60—The product of line 5 and line 38.

B61—The product of line 39 and line 36. It is a positive number because it represents a cash savings.

B62—The same as line 7.

B63—The same as line 5.

B64—The sum of lines 37, 40, 41 and 42.

B65—Enter either the percent discount or return rate, as a decimal, to use in the calculation.

B66—Calculated discount factor based on future values.

B67—Product of lines 43 and 45.

B68—Represents the cumulative cash flow for the investment in the automation project by year. To determine payback for the project in years, observe when the cash flow switches from negative to positive. The benefit is determined by the level of positive values of the cash flow.

Glossary

AC input module: Input module on a programmable logic controller that converts ac signals from switches, sensors, or machines to the appropriate logic levels for use within the processor.

AC output module: Output module on a programmable controller that converts the various logic level signals in the processor to ac level signals to operate external devices or peripheral equipment.

Acceleration: The change in velocity as a function of time.

Accuracy: A measurement of a robot manipulator's ability to go to a specified point in space as a result of a coordinate calculation generated inside a program.

Active accommodation: Integration of sensors with control algorithms to change a robot's preprogrammed motions in response to forces at the end effector.

Active illumination: Automatically varied illumination (intensity, patterns, or color) for vision systems to extract more visual information from a scene.

Actuator: A device that converts energy into linear or rotary motion.

A/D conversion (Analog-to-digital conversion): Conversion of information in analog form to an equivalent representation in the digital form.

Adaptive control: A method of control in which a machine's actions are continuously adjusted or corrected in response to feedback data from the machine or the environment.

Address: A name, label, or number that is given to a memory location in a computer or smart machine.

Air logic: Logic components that perform logical operations like digital integrated circuits but use compressed air in place of electrical current.

Air motor: A device that produces continuous rotary or reciprocating motion through the use of compressed air.

Algorithm: A procedure to solve a problem in a finite number of steps using a prescribed set of well-defined rules, processes, or mathematical equations.

Alphanumeric: All alphabetic, numeric, and special characters (excluding control characters) on a computer keyboard or in a computer language.

ALU (Arithmetic Logic Unit): The part of the computing system that performs logic and mathematical operations.

Ambient temperature: Steady state temperature of a material or gas.

AML (A Manufacturing Language): A robotics language developed by IBM for its SCARA robots.

Analog: An expression of values that can vary continuously over a range; e.g., the 0 to 5 volt output of a potentiometer or a transducer.

Analog control: Machine control using analog signals for information and analog devices for control.

Analog-to-digital conversion: *See* A/D conversion.

Android: A robot that has the physical appearance of a human.

ANSI (American National Standards Institute): The U.S. standards organization that sets many electrical standards.

Anthropomorphic: A machine that exhibits human-like shape or characteristics.

Anthropomorphic robot: A robot with rotary joints and motions similar to a human's body and arm. *See* jointed-arm robot.

APT (Automatically Programmed Tool): A high-level programming language for computer numerical controlled (CNC) machine tools.

Architecture: A computer or manufacturing process's physical and logical structure.

Arm: A term used to describe the links and joints on a robot that support and move the gripper.

Articulated: The ability of jointed parts to move.

AI (Artificial Intelligence): The use of computer software to perform operations normally performed by human intelligence, such as classifying, learning, problem solving, and decision making.

ASRS (Automatic Storage Retrieval System): Automated retrieval and storage of warehouse material so that human intervention is minimized.

ASCII (American Standard Code for Information Interchange): A seven- or eight-bit standard industry code used to represent 128 different alphanumeric and punctuation marks and control characters and in serial communications between computer-controlled machines.

Assembly robot: An application that uses one or more robots and programming software to assemble parts into subassemblies or complete products.

Asynchronous system: A non-synchronous system in which processes or events are not synchronized with other events or by system timing control.

AGV (Automated Guided Vehicles): Vehicles, with no on-board personnel, used to move raw material and finished products to manufacturing operations by means of automatic or computer control with the path set by a wire in the floor or by a painted line.

Axis: One of several degrees of freedom possessed by a robotic arm.

Backlash: A form of hysteresis in a power transmission system, such as the delay in motion of a gear train when the drive gear changes direction of rotation.

Back plane: A printed circuit board that provides common signal paths to interconnect other electronic cards.

Ball screw: The transformation of rotary motion to linear motion using a driven threaded shaft and a ball bearing nut to move the load.

Bang-bang control: A control unit that uses a mechanical stop to move a system to its predetermined fixed stop.

Bang-bang robot (also known as a stop-to-stop robot; pick-and-place robot; fixed stop robot): A pneumatic or hydraulic robotic with the programmed points dictated by fixed mechanical stops or by programmable fixed stops.

Barriers: Physical structures that prevent operators from entering the work envelope (work space) of a robot.

Base: The structure that supports an attached robotic arm.

Batch manufacturing: In production operations, the technique of manufacturing a group of parts in a separate run, or batch, preceded and followed by the manufacture of different part numbers or materials.

Bit: Either 1 or 0 (zero), displayed as a binary digit.

Branching: A place in a computer program where one of two different program commands is selected, based on cither a fixed-jump statement or on conditional statements, like IF-THEN.

Byte: A string of eight binary digits used to represent data or program commands in a computer or microprocessor.

Cartesian coordinate robot: A robot arm with position movements that result from changes in one of three linear axes to produce a work envelope that is a rectangular box.

Cartesian coordinate system: A coordinate system with three perpendicular and straight axes usually represented by X, Y, and Z.

CCD (Charge-Coupled Device) Camera: Solid-state camera able to transform light images into digital signals.

Chain drive: A mechanical device that uses a chain and sprocket to transmit power.

CIM (Computer-Integrated Manufacturing): The integration of the entire manufacturing enterprise through the use of integrated systems and data communications, coupled with new managerial philosophies that improve organizational and personnel efficiency.

Closed loop control: A control technique that adds a sample of the output signal back to the input signal to produce a more stable system.

CNC (Computer Numerical Control): The use of an internal computer in a machine to control the action or motions of the machine through a program or paper/magnetic tape.

Complex sensors: Sensors, such as vision, sonar, and tactile arrays, which have internal signal conditioning or information processing circuits.

Compliance: A technique that initiates or allows part movement for the purpose of alignment between mated objects.

Contact sensor: A device, such as a limit switch or tactile sensor, which must physically touch the measured object in order to make a measurement or detect a change.

Continuous-path control: A path control scheme that has programmed points closely spaced, approximately every inch or less, and that is programmed by physically moving the controlled device through the desired path.

Control hierarchy: A strategy for control that links lower level sensing devices with higher level decision-making devices.

Control system: The hardware and software used to control an output process variable using an input set point.

Controlled path: A path control scheme for robot motion in which the starting and ending points of straight line motion segments of the total program are stored in memory and executed to produce the desired robot arm movement.

Controller: The machine, robot, or cell control system, usually including a computer, network hardware, and a programming terminal.

Coordinated axis control: *See* Controlled path.

CPU (Central Processing Unit): That part of a computer, comparable to its brain, where data operations occur and commands are executed.

Cycle time: The time required to carry out one complete cycle of operations and return to the starting point.

Cycle: A sequence of operations or commands repeated regularly.

Cylindrical coordinate robot: A robot with a work envelope that has the shape of a cylinder and is usually produced by one rotary and two angular positioning axes.

Cylindrical coordinates: Spatial coordinates defined by two distances and an angle.

DC input module: Input module on a programmable logic controller that converts dc signals from switches, sensors, or machines to the appropriate logic levels for use within the processor.

DC output module: Output module on a programmable controller that converts the various logic level signals in the processor to dc level signals to operate external devices or peripheral equipment.

D/A converter (digital to analog converter): A device used to convert digital signals to analog signals.

Database: A collection of records stored on a computer system in a form that permits efficient extraction, organization, and manipulation by a database program.

Debug: To locate and fix errors in hardware or software.

Degree of freedom: Each axis on a robot arm or each movable joint on the arm.

Diagnostic routine: A test program used to locate hardware and software malfunctions in a computer or automation system.

Digital image: A representation of a TV camera picture in a computer using binary bits to represent the color and intensity of each pixel.

Digital-to-analog converter: *See* D/A converter.

Drive power: The power source used to energize actuators to produce motion.

Drum sequencer: A mechanical programming device composed of a motor-driven cylinder with adjustable tabs on the surface that can trigger external switches as the cylinder rotates.

Duty cycle: The proportion of time during which a device or system is active or at full power.

EIA (Electronics Industries Association): An electronics standards organization.

Elbow: A joint that is located between the upper arm and the forearm on a robot.

End effector: The gripper, tool, or process device attached to the tool plate on the robot wrist that performs the work.

End-of-arm tooling: *See* End effector.

Envelope: *See* work envelope.

Feature extraction: The process of locating and identifying specific objects in a digital image.

Feedback control: *See* closed loop control.

Feedback: The signal or data fed back to the input in a closed loop system.

Fixed-stop robot: *See* bang-bang robot.

Fixture: A device used to hold a work piece in the proper position for the performance of work.

Flexible automation: Automation systems that use computer-controlled machines and support equipment to produce a family or parts or products.

Flexible manufacturing system (FMS): A configuration of computer-controlled machines, like robots and CNC machines, interconnected by a material transport system. *Also see* Flexible automation.

Force sensor: A sensor designed to measure forces and torques between two objects, as in mating parts in an automated assembly.

Forearm: The robot arm link between the elbow joint and the wrist.

Gantry robot: A robot geometry that places the robot arm with two dimensions of horizontal movement on a four-legged structure.

Gray scale: A vision system's black and white scale that generates picture data with colors and intensity represented by different gray shade levels from pure white to pure black.

Gripper: A two-, three-, or four-fingered device used to grasp, hold, transport, and deposit parts and objects. *See* End effector.

Guard: A barrier, screen, or fence used in a factory environment to protect personnel and equipment.

Guard-distance: The distance from the guard to the protected machinery that is set by determining the time required for the machine to reach a safe state for human contact when the guard is crossed.

Guard-fixed: A barrier formed by permanent equipment or industrial fences that are not readily removed.

Hand: *See* Gripper.

Handshake: A communications strategy between two smart machines that gives either machine the ability to indicate that data was received or that data should be sent.

Hard automation: Production hardware and software designed and built to manufacture a single or small number of different products usually at very high speed.

Hierarchical control: *See* Control hierarchy.

Hydraulic cylinder: A piston and cylinder combination that produces linear motion by moving the piston with pressurized oil.

Hydraulic motor: An actuator powered by high pressure oil to produce rotary mechanical motion.

IEEE 488 Standard: A digital data communications standard for instrumentation data.

IEEE 802 Standard: A protocol standard for local area network communication.

IEEE (Institute of Electrical and Electronics Engineers): A non-profit professional society.

Industrial robot: A reprogrammable, multifunctional manipulator designed to move material, parts, tools, or specialized devices through variable programmed motions for the performance of a variety of tasks.

Input devices: Devices that provide environmental, operational, and system status data to system and other machine controllers.

Interface: The mechanical, electrical, optical, or pneumatic interface between two machines or systems.

Interlock: To link the control of a machine or device to the operation of a second machine or device to assure their proper coordination—often a safety feature.

Islands of automation: Stand-alone machines or devices not integrated to form a total operating system.

Joint: A rotary or linear articulation between two links on a robot arm.

Jointed arm robot: A robot with an arm configuration similar to the human body with vertically jointed axes at the shoulder and elbow and three wrist axes.

Joint space: The translational vector that specifies the displacement of each robot joint to a reference displacement for each other joint.

Kinematics: The prediction of the movement or motion of a jointed or hinged manipulator as it moves through its full range of motion.

LAN (Local Area Network): A group of computers and smart devices connected with some type of serial data highway.

Limit switch: A contact switch actuated when the motion of a part moves an operator on the switch.

Linear array camera: A solid-state digital camera that has one row of photo-sensitive elements.

Local area network (LAN): *See* LAN.

Machine vision: The process of using digital TV cameras to capture data useful for machine control.

Manipulator: A mechanism made up of links and joints with the capability to perform some manufacturing operation with either manual or automatic control.

Material handling: A system comprised of conveyors, automatic retrieval and storage systems, lift trucks, automatic guided vehicles, and robots, for moving parts from one location to another.

Maximum speed: The highest velocity that a robot or other production tool can attain.

Mean-time-between-failures (MTBF): The average length of time that a device will operate before a failure.

Mean-time-to-repair (MTTR): The average time needed to repair or service a device after failure.

NEMA (National Electric Manufacturers Association): A standards organization for electrical equipment.

NEMA standard: An industry standard set by the National Electric Manufacturers Association.

Network: The interconnection of a number of computer-controlled devices by data communications protocol.

Numerical control (NC): The automatic control of a machine tool using programmed information in symbolic form to represent the machining sequence.

Off-line programming: A method used to program robots in which the program development does not require any motion from the robot arm or interaction by the robot controller.

Open loop: A system without feedback.

Open-loop control: System operation where the location of the output actuators is not measured, and feedback to the system controller does not occur.

Operating system: The software that manages the operation of a computer or smart machine or device.

Orientation axes: The wrist joints (pitch, roll, and yaw) of a robot used to align the tooling with the work piece.

OSHA (Occupational Safety and Health Administration): The enforcing agency of the U.S. government that sets and enforces work rules and safety practices throughout business and industry.

Output devices: Devices such as valves, motor starters, and heaters that trigger a change in a process or machine.

Palletizing: The process of placing parts in rows, columns, and layers on a pallet.

Part orientation: The angular displacement of a part relative to the coordinate system of the process machine.

Pattern recognition: A process used in digital imaging to identify objects based on patterns stored in the system's memory.

Payload: The maximum weight of material or a part that can be safely handled by a machine or robot in a continuous operation.

Peripheral equipment: A device(s) attached to a robot or other industrial machine to support the production process.

Pick-and-place robot: *See* Bang-bang robot.

Pinch point: Any point between a robot link and other fixed hardware where injury can occur if part of the human body is trapped there during a robot move.

Pitch: The vertical movement of the pitch orientation axis in the robot wrist.

Pixel: The smallest region of a digital screen image that contains color and intensity information.

PLC (Programmable Logic Controller): A factory computer used to replace relay ladder logic hardware with programmed software control using input modules, process modules, and output modules.

Pneumatic cylinder: A combination of a piston and a cylinder, designed to provide linear movement as a result of pressurized air applied to either side of the piston.

Point-to-point control: A robot control strategy in which only a limited number of programmed points are used to specify the desired path of motion.

Polar coordinate system: A coordinate system using only one angular dimension and one linear dimension to specify a point in a two-dimensional plane.

Potentiometer: A resistor with a continuously variable wiper that is used to select a resistance value between the device's minimum and maximum values.

Positioning axes: The robot joints (base, shoulder, and elbow) used to move the robot arm into position to perform the desired task.

Presence sensing device: A device used to sense the presence of a human inside a specified detection volume.

Process machine: A production machine that performs a manufacturing process—like milling, forming, forging, shearing, joining, or extruding—on raw material.

Programmable controller: *See* PLC.

Programming: The process of developing a set of control instructions required for a robot or machine controller to perform the desired task.

Proximity sensor: A non-contact measuring device that senses objects without making physical contact using magnetic field changes, reflected sound, or other techniques.

Rated-load capacity: The equivalent of maximum load capacity for some robots, but often considered the amount of weight a robot is capable of lifting reduced by a factor of safety.

Real-time operation: Machine control that makes changes in the process operation at nearly the same time as the need for the process change is sensed and communicated to the machine.

Rectangular coordinate system: *See* Cartesian coordinate system.

Relative coordinate system: A coordinate system with an origin that is related to another coordinate system by some constant offset or a variable offset.

Reliability: The probability that a device will operate failure-free for a specified time period or amount of usage.

Remote center compliance (RCC): A passive compliant device that corrects for angular and displacement differences between mating parts.

Repeatability: The deviation between the location of the robot tooling's taught point and the location of the tooling when that point is reached during program execution under identical conditions of load and velocity.

Resolver: A transducer with rotary and fixed windings that converts changes in the angular displacement between the rotary winding and the fixed winding into and an analog electrical signal.

Robot systems: An industrial robot system includes the robot(s) hardware and software, consisting of the manipulator, power supply, and controller; the end effector(s); any equipment, devices, and sensors with which the robot is directly interfacing; any equipment, devices, and sensors required for the robot to perform its tasks; and any communications interface that is operating and monitoring the robot, equipment, and sensors.

Robot: A reprogrammable multifunctional manipulator designed to move materials, parts, tools, or specialized devices through variable programmed motions for the performance of various tasks.

Robotic cell: One or more robots integrated into a production cell that includes a process machine and other support hardware, such as material handlers, conveyors, parts feeders, inspection systems, vision systems, sensors, and a cell controller.

Robotics: The field of study focusing on the design and application of robot hardware and software for use in manufacturing, domestic applications, and other service roles.

Roll: Angular rotation of the robot tool plate and attached end effector about an axis along its principal direction of motion.

RS 232-C, RS 422, RS 423, RS 449: Standard electrical serial interfaces for exchanging data between two computer-controlled systems.

SCARA (Selective Compliance Assembly Robot Arm): A robot with a horizontal jointed shoulder and elbow axes attached to a base rotation axis with a linear Z axis located at the end of the elbow link.

Seam tracking: A robot welding system with feedback from vision sensors or other measuring devices that permits the robot controller to automatically cause the welding tooling to follow the seam between two metal parts that are to be welded together.

Sensors: A device that measures some physical condition, such as position, speed, temperature, or pressure, and generates an electrical response proportional to the physical input.

Serial interface: A data transmission strategy that uses a stream of binary bits, including start bits, stop bits, parity bits, and data bits, to exchange information between smart machines or devices.

Servomechanism: A closed-loop control system that continually samples the position and rate of change of the output, so that the system can drive the output to its desired location, along the required path and at the desired speed and acceleration.

Servo-controlled robot: A robot driven by a servomechanism to control the tooling.

Servomotor: An ac or dc motor designed for use in a servomechanism.

Shaft encoder: An encoder used to convert a mechanical shaft position to an electrical signal.

Shoulder: The joint that connects the upper robot arm link to the base.

Simulator: A device or program that imitates another device.

Slew rate: The maximum rate that a system output can change.

Smart machine: A term used to indicate that the machine has a microprocessor or computer integrated into the operational system.

Smart sensor: A sensor that has a built-in computer to process sensor data.

Solenoid: An electromagnet with a movable core that is used to produce linear displacement.

Solid-state camera: *See* CCD camera.

Speed: A robot parameter to indicate the maximum speed that the robot tooling can move.

Spherical coordinate system: A coordinate system with two angular dimensions and a third linear distance from the point of origin.

Stepper motor: A bidirectional permanent magnet motor that moves in small angular increments, based on the number of pulses applied.

Stiffness: The amount of force applied by the control system to the tooling when two mating parts are jointed.

Stop: The mechanical limit of a pneumatic or hydraulic piston, or a mechanical device or movable block used to stop or limit mechanical motion of an actuator.

Structured light: A projected light pattern on an object that permits the dimensions and geometry to be measured by a vision system.

Subroutine: Computer program code that is used frequently to perform a specific task or set of tasks that can be called from any point with the main program.

Synchro: Two interconnected transducers with rotary and fixed windings, which convert changes in the angular displacement between the rotary winding and the fixed winding in one transducer into a corresponding angular change in the other transducer.

Synchronous: Actions that occur at the same time.

Syntax: The rules governing the structure of a language.

Tactile: The sense of touch.

Tactile sensor: A sensor with the capability to extract information about an object by grasping, thus imitating the sense of touch in the fingers of the human hand.

TCP (Tool Center Point): The point on the robot tooling where the action required for production occurs.

Teach pendant: A hand-held control unit used to program and control the robot when programmed points are taught.

Teleoperator: A robot-like device that is remotely controlled by a human operator whose hands are in gloves equipped with sensors and feedback pressure actuators.

Template matching: A strategy used in vision software to identify an object extracted from a digital image by comparing it with similar object data stored in memory.

Tolerance: A specified deviation or error from a desired value.

Tool center point: *See* TCP.

Tool: A robotic term used to refer to the end effector attached to the robot tool plate.

Torque: The force that cause rotation of a body.

Transducer: A device that converts energy from one form to another.

Transfer line: A manufacturing system designed to produce a specific part or product using individual process machines linked with material handling systems.

Transfer machine: A robot or mechanical device used to grasp a work piece and move it automatically through a manufacturing process.

Translation: Movement of a robot between two programmed points.

VAL: The Unimate robot programming language.

Velocity: The time rate of change in position.

Videcon: An electron vacuum tube used in a TV camera that converts the optical image projected onto the tube's photosensitive window into an electrical signal.

Vision system: A system that includes a video camera and a camera controller and is used to capture two-dimensional digital images.

Work cell: A manufacturing area with one or more production stations.

Work piece coordinates: The coordinate system referenced to the work piece or part.

Work envelope: The volume that describes the working limits of a robot's motion.

Work station: A manufacturing area that includes one robot and the support equipment.

World coordinates: A coordinate system that includes or is referenced to all other coordinate systems.

Wrist: A set of rotary joints at the end of the robot's forearm link that includes at least two degrees of freedom in yaw, pitch, or roll, and to which the robot tooling is attached.

XYZ coordinates: Coordinates for the Cartesian coordinate system.

Yaw: The orientation axis that produces horizontal angular displacement of the robot tooling.

Index